Advances in Chromatography

Advances in Chromatography

Volume 55

Edited by
Nelu Grinberg

CRC Press
Taylor & Francis Group
Boca Raton London New York

CRC Press is an imprint of the
Taylor & Francis Group, an **informa** business

ISSN 0065-2415 (print)

CRC Press
Taylor & Francis Group
6000 Broken Sound Parkway NW, Suite 300
Boca Raton, FL 33487-2742

First issued in paperback 2022

© 2018 by Taylor & Francis Group, LLC
CRC Press is an imprint of Taylor & Francis Group, an Informa business

No claim to original U.S. Government works

ISBN 13: 978-1-03-240204-8 (pbk)
ISBN 13: 978-1-1380-6831-5 (hbk)
ISBN 13: 978-1-315-15807-5 (ebk)

DOI: 10.1201/9781315158075

Publisher's Note
The publisher has gone to great lengths to ensure the quality of this reprint
but points out that some imperfections in the original copies may be apparent.

Visit the Taylor & Francis Web site at
http://www.taylorandfrancis.com

and the CRC Press Web site at
http://www.crcpress.com

Contents

Preface

This volume of *Advances in Chromatography* is dedicated to the memory of Professor Eli Grushka, who served as a coeditor of the series for several decades. All the authors have graciously made their contributions as a testimony to the connections they felt with Eli, both professionally and personally. To write now about the late Eli Grushka seems somehow unreal. He was a giant in the field of chromatography, a man ahead of his time, a good friend, and a mensch.

In 1983, I was doing a postdoctoral fellowship at the Weizmann Institute of Science, Rehovot, Israel. Somebody from Weizmann advised me to continue my postdoctoral fellowship with Eli Grushka at the Hebrew University of Jerusalem, Jerusalem, Israel. I went there, and that was my first meeting with Eli. He appeared to me as a very humble man, despite his reputation in the field. I spoke with him about my research plans, and he looked at me and said they sounded very interesting. I was so proud that a person of Eli's statue was appreciating a freshly arrived immigrant in that country. For some reason, it did not work out and I pursued my second postdoc in the United States, in Professor Barry Karger's laboratories. Eli visited Karger's labs several times and every time he gave a talk. It was a pleasure listening to his lectures. A lot of science was delivered in a very professional manner. He was not so much interested in the experiment as in the underlying phenomena, trying to explain the results in depth. It was during those times that I learned, from Eli's example, how to deliver a talk to captivate an audience.

Our collaboration on the *Advances in Chromatography* book series was very fruitful. Eli and I succeeded in maintaining the high level initially created by the founder of the series, Calvin Giddings. Eli was instrumental in suggesting the best authors and topics, a contribution that made the series so successful. During our collaboration, we developed a friendship, a gift for which I will always be grateful.

Throughout the years, our friendship became stronger, and every fall we met in New York at a family party. I was very humbled that he considered me a friend. His kindness, generosity, hospitality, empathy, and sense of humor were without equal. He loved to talk science, but he also loved to talk about his family; when he did, his love for and pride in them was evident. Despite his busy schedule, he always made time to connect with people, a rare trait in someone so accomplished and dedicated to his profession.

Like so many, I lost a very good friend and a talented colleague. I miss him a lot. May his memory be a blessing.

Nelu Grinberg

Tributes for Eli Grushka

Professor Eli (Elimelech) Grushka was born in Tel Aviv, Israel on July 14, 1938, the son of Chaya (Wolfson) and Shmuel Grushka. Being born on the Bastille Day—the French national holiday—he loved to be in Paris on his birthday! He always claimed that only the French really knew how to celebrate his birthday! As a small child he lived in Magdiel and in Kfar Saba where he and his parents occupied one room in a small house that had no indoor plumbing! How times have changed!

Eli attended elementary and high school in Israel and, as he often said modestly, he was *not* the best student in the class!!! He claimed to have learned English by watching movies and by reading paperback detective novels from the United States. He served in the Israel Defense Force and credited his army service with waking him up to the value of studying. He completed all his army courses as *Outstanding Trainee* and realized that if he worked hard he could achieve whatever he wanted.

In January 1960, Eli opened a new chapter as he began studying for a BS in chemistry at Long Island University (LIU), Brooklyn, NY. He had applied to study chemistry at the Hebrew University of Jerusalem (HUJI), Jerusalem, Israel but was not admitted because his high school grades were too low. Years later, as a professor at HUJI, he claimed that it was easier to become a professor at HUJI than to be admitted as a student!!!

Excellent grades at LIU led to graduate school at the Cornell University, Ithaca, NY, where he majored in physical chemistry and analytical chemistry under the direction of Professor W. D. Cooke. Studying at the Cornell in the 1960s introduced Eli to some of the leading scholars of the time, including Benjamin Widom (who was his role model as a lecturer), Harold Scheraga, Peter Debye, Robert Plane, Roald Hoffman, and Michell Sienko. After completing his PhD in 1967, Eli went on to do postdoctoral research at the University of Utah, Salt Lake City, Utah with Professor J. Calvin Giddings, one of the giants in the field of chemical separations. During the time in Utah, Cal was not only Eli's mentor, but also became a close friend.

Eli was appointed assistant professor of chemistry at the State University of New York (SUNY), Buffalo, New York in the fall of 1969. He rose through the academic hierarchy, advancing to associate professor in 1974 and to full professor in 1978.

Eli was a Kolthoff fellow at the HUJI in 1976, at the end of which he was appointed an associate professor. He returned to Buffalo for a short time, later becoming a permanent faculty member at HUJI in 1978 and a full professor from 1979 onward. He served on many university wide committees and was head of the Institute of Chemistry for several years in the early 1990s. He officially became professor emeritus in 2006, but continued to teach both beginners and advanced students until his health failed in February 2016. He maintained a research group and even in the hospital continued giving instructions to his students to *keep working.*

Over the years, in addition to Cal Giddings, Eli collaborated with many prominent scientists. Among them were Georges Guiochon, Phyllis Brown, Ray Scott, Lloyd Snyder, John Knox, Chaim Gilon, and Jack Kirkland. He also worked very closely with people in the industry, in particular Zvulon Tomer, owner of Unipharm Ltd., Ramat Gan, Israel.

Over the years, Eli received many honors, including a Tswett medal and a Horvath medal. In 2014, the Israel Analytical Chemistry Society honored him with a Lifetime Achievement Award and in 2015 he received the Hebrew University Rector's Award for continued excellence in teaching.

Eli's years at the Cornell University contributed more than just his doctoral degree. On his first day in Ithaca in September 1963, he met Donna Gellis who was then a senior at the university and was volunteering to help new foreign students to become oriented to campus life. Even though he had been in the United States for 3½ years by that time and surely did not need *orienting*, Eli came to the orientation desk to seek out other Israelis and to learn about the Cornell. He inquired about what there was to do in a student's spare time. The young lady offered her assistance and thus began a 15-month courtship. They were married on December 27, 1964 in New York and celebrated their 51st anniversary on December 2015.

Donna and Eli were the proud parents of Elana Grushka, who is in charge of content promotion for a TV cable company in Israel; Yael Grushka-Cockayne, professor of decision sciences and project management at the University of Virginia

Darden School of Business, Charlottesville, Virginia; and Tamar Grushka-Shtern who works at the School for Overseas Students at Tel Aviv University, Tel Aviv, Israel. In his later years, Eli's greatest joy was his grandchildren—Gabriel and Abigail Cockayne, and Eitan Shalem-Grushka. A fourth grandchild, Omri Shalem-Grushka, was born in June 2016. Eli's love for his grandchildren was returned without reservation.

Eli's death is a great loss to his entire extended family, including his brother, brother- and sisters-in-law, cousins, and nieces, his many close friends and colleagues around the world, and current and former students. On the day of his funeral, hundreds of people clogged the roads around the tiny cemetery at Ramat Raziel, the village where the Grushkas lived for 25 years.

In September 2016, one of Eli's research students gave his doctoral lecture at the Hebrew University. He concluded with a lovely photo of his mentor, and perhaps the most appropriate tribute, a quote from Henry Adams, "A teacher affects eternity; he can never tell where his influence stops."

Donna Grushka

My dad, Professor Eli Grushka, or *Aba* as I call him, loved puzzles. When he came across a new puzzle, be it a math puzzle in a second-hand book that he bought, a wooden 3D object, or a video game, his eyes would light up and he would get lost in the world of trying to solve the puzzle. I always felt that, for him, his research was a huge puzzle. Research was an exciting game that caught his attention. Research was a challenge that was to be solved, not only for his personal satisfaction, but also to help others to learn how it could be solved. Whether it was sitting with his grandchildren working through a riddle or with his students in the lab, Aba was committed to going on the exploration journey with whomever he was sitting, in order to solve the puzzle.

"It is not work; it is my life." "I love what I do and how I spend my time." "I am my own boss." "I can work on what I like, the way I like to, and learn something new every day." "I get to spend my time with smart people who also love what they do." These are some statements I quite often heard Aba make about his career as a professor in academia. Growing up, I always knew he felt he was engaged in activities he loved and felt lucky to have found his calling (and the job that pays him to do what he loved).

Knowing that such a match could exist and that work could be more than just to pay the bills, I sought out my match. When time came for me to choose from where to start my academic career, I looked for a school where I could feel about my job like Aba did. Pride in what I do, excited to wake up every morning, lucky to walk into a place where I own my time, and choose what to work on, all while feeling honored to do so. For me, this meant a school where I get to work with students and colleagues I respect, whom I can help shape and develop, and a place where my research inspires actual managers in business.

True, I do not recall much from the *hardship* years Aba went through in the 1960s and 1970s. I was not born until after the Cornell doctoral days and I was just a baby while he was a junior faculty at Buffalo, working to get tenure. He used to tell me that it took him 10 years to master his way in the classroom, but to me he always seemed like the most natural and personal teacher one could have. I mainly recall the latter 40 years or so of a career that was filled with success and pride. It seemed effortless, although I know that he devoted a lot of effort to it.

Reflecting on my life, my path has been hugely impacted by my dad's choices, perhaps like most children. I was born in New Jersey, due to a leave Aba took from SUNY, Buffalo just before he got tenure. We moved to Israel when I was nearly 4, and I grew up with Jerusalem as my home, as he accepted a position at the Hebrew University. I recently found what I now realize was my first research project, a science project I completed in the 8th grade. We were living in Wilmington, Delaware at the time, since my dad was on a sabbatical from the Hebrew University, working at DuPont. My research paper was titled "Fresh Water from Salt Water." As I read it today, it is well-written, to the point, well-presented, and focused. It has his influence all over it. As I experienced it at the time, it was hours of fun time spent with my dad. He enjoyed it to no end and got me excited about it too. I think he was as proud of that work we did together as he was later when I got my PhD and eventually tenure.

Aba had so much passion for what he did. He was excited and fascinated by the world of separation and chromatography. Beyond chemistry, he loved to see the world develop and to learn new things. I realize now that it is part of what made him a good academician. Every paper, book, or poster, when done well, provides an opportunity for new learning. He was drawn to discovery and innovation. Always with the latest technology or gadget. In addition to his personal drive, Aba enjoyed getting everyone around him excited about these innovations too. He taught me about the Internet, always knew more about computers than I did and wanted to teach me, and invested hours teaching himself new skills such as MATLAB® or Photoshop.

I did not always, and still do not, understand chemistry. For a few years, it was mainly about access to a cool lab with *dry ice* or helium. But he made his world ours. We would often get to meet Aba's students for holiday parties or other celebrations. Especially those studying for advanced degrees. There was no separation between work and home; the students were part of Aba's life and therefore part of the family. Part of our life. Today, I often host students at my home and my kids get to know them. It is all part of my life, part of who I am.

A few years ago, I won a teaching award at Darden. Aba wrote to me "Receiving the Mead Endowment {award} is quite an honor and an amazing accomplishment. More important, it is also a responsibility... It will be hard work but I think that it will also be great fun... You are very lucky!" I feel so lucky. Not because of my job, my family, my health, or other successes. I feel my biggest award and luck was to have my Aba, who loved me, inspired me, put a smile on my face, and showed me that kindness and passion can take me a long way. I will always try to make him proud.

Yael Grushka-Cockayne

It was my decided pleasure to have been a colleague and trusted friend of Eli for perhaps the past several decades, ever since I was a Weizmann fellow at The Weizmann Institute of Science in Israel in the early 1970s. However, it was only after Eli had moved back to Israel and his tenured position at HUJI that we more closely interacted on both a professional and social basis. Whenever possible, during any of numerous visits to Israel, Eli would invite me to speak before his research group or the entire Department of Chemistry at the HUJI in Jerusalem, Safra campus. We also socialized on various te'uls, day trips around the country but especially in the Judean Hills near Jerusalem. Eli was also mainly responsible for my receiving a Kolthoff fellowship in 2009, so as to spend an entire semester on campus within the Institute of Chemistry, offering a graduate-level course in analytical biotechnology to students from any HUJI campus. During that time, I attended many seminars and invited lectures within the Department of Chemistry, and collaborated with several colleagues in areas of my own expertise. It was also my decidedly good fortune to be able to interact with people such as Dani Mandler, Yossi Almog, and others of the chemistry faculty and graduate students or postdocs.

During my stay in Israel, in winter 2016, where we lived in Tel Aviv, I attended another Isranalytica meeting, presented a keynote lecture, and socialized with Eli and Donna on several occasions throughout that winter stay in Israel. Eli was an organizer, founder, and active e-Board member for all Isranalytica meetings, and president of the Israel Analytical Chemistry Society for many years. We never realized that those professional and social interactions would be our last. Eli was a unique individual in many ways, and not just in his professional life in which he was truly dedicated to analytical chemistry, teaching at the undergraduate and graduate levels in these subjects, directing independent R&D among numerous students at HUJI, but also even after (forced) full retirement, he continued doing everything that any other, full time, fully employed faculty in chemistry would be pursuing. However, without any salary or compensation, he was then professor emeritus. He continued to enjoy, thrive, and encourage everyone he interacted with during his retirement, to become as successful as possible in their careers. There are dozens of his former students, postdocs, and colleagues who are still fully active in chemistry, not just in Israel but around the world, as a direct result of such interactions, and due to Eli's forceful personality as an advisor, mentor, and a friend. When I attended his talks at various Isranalytica meetings, there were always packed rooms, no matter what day or the time of day. It was almost always a standing room only, and everyone wanted to be up front to hear every word of his talks and see every slide, me too.

Being involved in the Massachusetts Separations Society in the Boston area, for decades already, I had been instrumental in inviting Eli to be our 2016 plenary speaker, this past May 4. It was not meant to happen, as Eli passed away only weeks before the event happened. He was sorely missed by all the attendees, and they missed his final invited presentation, which I also heard earlier this year at Isranalytica 2016. Analytical chemistry lost a giant with Eli's passing but there are so many former students, postdocs, and colleagues who are still very active. In addition, all of them were positively influenced and educated by Eli during his career. What a truly wonderful legacy for a giant in analytical chemistry.

Ira Krull

Eli Grushka and I were approximately contemporaries in academia with our first jobs: Eli at SUNY, Buffalo and I at Indiana University, Bloomington, Indiana. This was in the 1970s when we attended the same chromatography symposia. I still remember Eli's first presentations on the resolution optimization/time normalization in chromatography presented at the Advances in Chromatography symposia series organized for a number of years by my postdoctoral advisor, Professor Albert Zlatkis. Dr. Grushka also published then an excellent set of articles on the mechanisms of retention on the liquid crystalline stationary phases. As the method of HPLC rapidly advanced and developed its own symposia, we often met each other and became good friends.

In August 1982, I had a great opportunity to engage Eli in a small meeting organized under the joint support of the U.S. National Science Foundation and the Japan Society for Promotion of Science. The subject of this meeting was microcolumn separation techniques, a subject of much controversy at that time. While the miniaturized columns were met with a limited acceptance and even adversity from the leading HPLC authorities, notably excluding John Knox and Cal Giddings, Professor Daido Ishii from Nagoya University, Japan and I arranged this specialized meeting for the sake of idea exchanges in the field that seemingly showed much promise to younger investigators and scientists in other fields. The meeting was a great success, with the subjects as important as capillary electrophoresis (CE), use of microelectrodes as detectors, laser-based detection, and the prospects of capillary LC-MS that were discussed in the very early stages of development. The rest is history.

Not only were we financially supported from the respective Japanese and U.S. agencies, but we also invited several industrial observers whose companies were asked to provide additional support for both a good meeting and related social activities. As I was able to fully support several *observer attendees* from abroad, I thought of Eli Grushka who meanwhile had moved to Israel. After several unsuccessful

attempts to reach him, I finally got Eli on the phone: "Yes, we will pay your travel expenses to Honolulu, put you in a hotel in Waikiki Beach, you don't even have to lecture, just participate in the discussions…" Eli had some difficulties to believe this offer, so I had to repeat it again. He came and we all had a great time in Honolulu. A professional photograph of all participants are shown in the following figure, with Eli in the right corner, next to Bob Brownlee (Brownlee Laboratories) and Uwe Neue (Waters)—regrettably, all three now passed away.

Eli and I saw each other often at additional HPLC symposia for a number of years. I fondly remember my 1988 visit to Israel and meeting again Eli and his wife. After a couple of nice dinners at the beach, we discussed chromatography, religious matters, and the Middle East political situation for a long time. I still treasure those memories.

In Dr. Grushka's premature death, we have lost a great scientist and an outstanding scientific colleague and friend.

Milos V. Novotny

I first met Eli Grushka when I was quite new to publishing, I believe at an editorial board meeting for the *Journal of Liquid Chromatography*. Jack Cazes introduced us, and Eli seemed at the time like he did not want to be bothered. I asked Jack afterward if I had done anything to offend Eli, and Jack assured me that it took a while for Eli to warm up to people. Jack turned out to be right, of course, but it took longer than expected. Eli did not particularly like my ideas or suggestions for *JLC* or the *Advances in Chromatography* series. He did not like how long peer review and production often took. He did not like that I did not travel to Israel more often (although I usually visited once a year). There was a long list of things Eli did not like about me, and he was not afraid to let me know them, but I eventually learned that that was Eli's way. He did not tell you that you were doing an acceptable job or that he was pleased with recent promotional campaigns or impact factor numbers or sales—he was more about getting things done.

About 3 years ago, I was visiting Israel and contacted Eli for a hopeful visit in Jerusalem. Eli agreed that he would pick me up at the Hebrew University's Mount Scopus campus and I mentioned that after our meeting, I needed to go to Tel Aviv. We met as planned, had coffee at a cafe, and as we were finishing up Eli told me that he would drive me to Tel Aviv (about a 90 minute journey), as he needed to visit his son anyway. I told him that was not necessary, as I was planning to take the train. But Eli was decided, and we had the most cordial and personal conversation during our car ride. I did not realize at the time that that would be the last time I would see Eli, but it was certainly the nicest. That is how I will remember Eli. Telling me with his eyes fixed on the road ahead that Marcel Dekker, Inc. was okay to work with. That he had had a pretty good career. And that he had enjoyed it.

Russell Dekker

When I first began working with Eli in 2006, we fell into a very comfortable working pattern. I was blessed with a dedicated, detail-oriented, and responsive editor who patiently taught me about the aspects of the series and the importance of the authors who generously gave of their time and expertise to participate in each volume. The following year, I had the opportunity to meet Eli in person at the Csaba Horvath symposium in Connecticut. He was gracious, enthusiastic, curious, and possessed a winning smile that I was so fortunate to have experienced, given the normal distance between us. In the years that followed our initial correspondence, we worked on 9 volumes together and as my experience as an editor grew, so too did my respect and appreciation for Eli's reliable, accurate, responsive, and unfailing editorship. He was a wonderful and treasured rarity in publishing and is missed by all who had the good fortune to work with him.

Barbara (Glunn) Knott

For 10 years, it was my honor to work with Dr. Eli Grushka as the senior project coordinator at Taylor & Francis for the *Advances in Chromatography* series. His collaboration with Dr. Nelu Grinberg resulted in robust contributions from scientists and academicians. Dr. Grushka was kind, thoughtful, and gracious. He will be remembered as a creative scientist, a mentor, a teacher, and one whose work will continue because of the countless students he inspired.

Jill J. Jurgensen

Contributors

Jared R. Auclair
Biopharmaceutical Analysis Training
  Laboratory
Department of Chemistry and Chemical
  Biology
Northeastern University
Boston, Massachusetts

Ivett Bacskay
MTA–PTE Molecular Interactions in
  Separation Science Research Group
Pécs, Hungary

Amaris C. Borges-Muñoz
Department of Chemistry
University at Buffalo
State University of New York
Buffalo, New York

Peter W. Carr
Department of Chemistry
University of Minnesota
Minneapolis, Minnesota

Luis A. Colón
Department of Chemistry
University at Buffalo
State University of New York
Buffalo, New York

Joe M. Davis
Department of Chemistry and
  Biochemistry
Southern Illinois University
  at Carbondale
Carbondale, Illinois

Ziad El Rassi
Department of Chemistry
Oklahoma State University–Stillwater
Stillwater, Oklahoma

Attila Felinger
Department of Analytical and
  Environmental Chemistry
University of Pécs
Pécs, Hungary

Ira S. Krull
Department of Chemistry and Chemical
  Biology
Northeastern University
Boston, Massachusetts

Anneli Kruve
Institute of Chemistry
Faculty of Science and Technology
University of Tartu
Tartu, Estonia

Shulamit Levin
Waters (TC) Israel Ltd.
Petach Tikva, Israel

Lisandra Santiago-Capeles
Merck Sharp & Dohme, Corp.
  International GmbH
Las Piedras, Puerto Rico

Israel Schechter
Schulich Faculty of Chemistry
Technion–Israel Institute of Technology
Haifa, Israel

Mark R. Schure
Theoretical Separation Science
  Laboratory
Kroungold Analytical, Inc.
Blue Bell, Pennsylvania

Annamária Sepsey
MTA–PTE Molecular Interactions in
  Separation Science Research Group
Pécs, Hungary

Joseph Sherma
Department of Chemistry
Lafayette College
Easton, Pennsylvania

Dwight R. Stoll
Department of Chemistry
Gustavus Adolphus College
St. Peter, Minnesota

Karina M. Tirado-González
Department of Chemistry
University at Buffalo
State University of New York
Buffalo, New York

Stefan Vujcic
Saint-Gobain Ceramic Materials
Malvern, Pennsylvania

1 A Study of Peak Capacity Optimization in One-Dimensional Gradient Elution Reversed-Phase Chromatography
A Memorial to Eli Grushka

Peter W. Carr and Dwight R. Stoll

CONTENTS

1.1 INTRODUCTION

We are going to use one of Eli Grushka's first papers written at the University of Buffalo, Buffalo, New York, immediately after leaving Cal Giddings lab at the University of Utah, Salt Lake City, Utah [1] as the springboard for our contribution to this memorial issue of *Advances in Chromatography*. The paper titled "Chromatographic Peak Capacity and the Factors Influencing It" was written solely by Eli and appears to be the fifth paper ever written on the topic of peak capacity, a concept that was developed and introduced by Eli's mentor Cal Giddings [2,3]. On looking at all of Eli's papers found by SciFinder it is clear that he has contributed to a wide variety of areas of interest to chromatographers, including capillary electrophoresis, linear solvation energy relationships, structure retention relationships, retention mechanisms especially in reversed-phase liquid chromatography (RPLC), liquid crystal stationary phases, silica bonding chemistry, the use of metal ions as mobile phase components in RPLC, the measurement of diffusion coefficients by chromatography, the theory of peak shapes by peak moments, and time normalization chromatography. It was thus difficult to focus on one contribution, so we picked a topic that is very current and interesting to us in our work on multidimensional liquid chromatography (LC).

The Giddings paper [2] treated peak capacity under conditions of constant elution, that is, isothermal and isocratic elution. It was followed closely by a communication from Horvath and Lipsky [4] on peak capacity under conditions of varying elution, that is, temperature-programmed gas chromatography (GC) and eluent gradient elution in LC. The important distinction between elution under constant and time varying conditions is that the band width under constant elution conditions becomes progressively wider as elution time or volume increase, whereas with a properly designed gradient the band width is approximately the same for all peaks [4]. Grushka's paper was by far the most thorough treatment of peak capacity in that he considered the impact of the column temperature, eluent velocity, column length, and the amount of stationary phase as well as elution under time normalized conditions and gradient conditions [1].

The subject of peak capacity in chromatography is of great importance. Peak capacity is a metric of the separating power of a chromatographic method that is most relevant when dealing with samples of high complexity, that is, samples that contain many species. Its importance in fields such as polymer analysis, proteomics, and metabolomics can hardly be exaggerated. When a chromatogram becomes so crowded with peaks that an improvement in band spacing in one part of the chromatogram must result in a diminution in the room available for separation in another part of the chromatogram, then the only possible approach to improve the separation is to increase the peak capacity. Simply varying the chromatographic selectivity will not be helpful. One must then increase the total separation space relative to the zone width; this ratio is directly related to the peak capacity [5].

1.2   WHAT IS PEAK CAPACITY

Peak capacity is the maximum number of well-resolved peaks that can be fit into a separation space defined as the earliest time (t_o) at which a peak can elute from a column and the time at which the nth peak elutes (t_n). Clearly, all peaks should have the same resolution (R) otherwise some peaks will be too well separated thereby wasting space. Generally, the requisite resolution is taken as unity. Here, B and A denote the later and earlier of a pair of adjacent peaks, t_R and σ are the retention times and standard deviations of these peaks

$$R = \frac{t_{R,B} - t_{R,A}}{2(\sigma_A + \sigma_B)} \tag{1.1}$$

If R is set to unity, and when the two peaks have essentially the same widths, then the two retention times will be separated by 4σ. It is clear that in a given chromatographic space the total number of 4σ intervals is finite and controls the number of peaks that can fit in this space.

Giddings argued that under conditions of constant elution we can take the 4σ width as

$$4\sigma_i = 4\frac{t_{R,i}}{\sqrt{N}} \tag{1.2}$$

The $i + 1$th peak must elute at a time relative to the ith peak equal to the mean of the two peak widths thus:

$$t_{R,i+1} - t_{R,i} = \frac{2(t_{R,i+1} + t_{R,i})}{\sqrt{N}} \tag{1.3}$$

Here, it is assumed that all peaks have the same plate number (N). This seems to be an assumption in all models of the peak capacity. The basic theories of column dynamics [6] tell us that this assumption is not true as both the longitudinal dispersion and interphase resistance to mass transfer terms in the equations that describe the plate height (HETP) are dependent on solute characteristics, including diffusion coefficients and retention factors (k). Although some authors suggest using N for an unretained species, others suggest using the average of several representative peaks. If one wishes to use a resolution other than unity, then Equation 1.3 should be written as

$$t_{R,i+1} - t_{R,i} = \frac{2R(t_{R,i+1} + t_{R,i})}{\sqrt{N}} \tag{1.4}$$

This is easily transformed to

$$\frac{t_{R,i+1}}{t_{R,i}} = \frac{1 + 2R/\sqrt{N}}{1 - 2R/\sqrt{N}} \tag{1.5}$$

Now if the first peak elutes at $t_{R,1}$ equal to t_o, then the elution of the nth (final) peak relative to the first peak is the $n - 1$ products of the ratio on the right-hand side of Equation 1.5 thus:

$$\frac{t_{R,n}}{t_o} = \left[\frac{1 + 2R/\sqrt{N}}{1 - 2R/\sqrt{N}}\right]^{n-1} \tag{1.6}$$

Taking the logarithms gives

$$n = 1 + \frac{\ln(t_{R,n}/t_o)}{\ln\left[\left(1 + 2R/\sqrt{N}\right)\Big/\left(1 - 2R/\sqrt{N}\right)\right]} \tag{1.7}$$

where n is the peak capacity, that is, the number of peaks that fit between $t_{R,n}$ and t_o with a resolution equal to R. By and large R is sufficiently small (≈ 1) and N sufficiently large (>100), so that the argument of the logarithm in the denominator can be replaced with $1 + 4R/\sqrt{N}$ with high accuracy (ca. 1%), thus

$$n \approx 1 + \frac{\ln(t_{R,n}/t_o)}{\ln(1 + 4R/\sqrt{N})} \tag{1.8}$$

Now we see that under almost all circumstances of interest in chromatography $4R/\sqrt{N} \ll 1$; thus $\ln(1 + 4R/\sqrt{N}) \approx 4R/\sqrt{N}$, consequently,

$$n \approx 1 + \frac{\sqrt{N}}{4R} \ln(t_{R,n}/t_o) \tag{1.9}$$

In Grushka's paper [1], he introduced an alternative approach to the computation of the peak capacity, which is quite straightforward, based on his observation that the number of peaks in a very small unit of time is given by

$$dn = \frac{dt}{4R\sigma} \tag{1.10}$$

On integrating and then substituting 4σ from Equation 1.2:

$$\int_1^n dn = \int_{t_o}^{t_{R,n}} \frac{dt}{4R\sigma} = \frac{\sqrt{N}}{4R} \int_{t_o}^{t_{R,n}} \frac{dt}{t} \tag{1.11}$$

Consequently, for isocratic chromatography:

$$n = 1 + \frac{\sqrt{N}}{4R} \ln \frac{t_{R,n}}{t_o} \tag{1.12}$$

As Horvath and Lipsky pointed out [4] under appropriate time-varying elution conditions the widths of all peaks are approximately the same, thus Grushka's Equation 1.10 assuming a constant σ equal to that of the peak eluting at t_o becomes

$$dn = \frac{\sqrt{N}}{4Rt_o} dt \tag{1.13}$$

Finally, integrating with the same limits as in Equation 1.11 gives

$$n = 1 + \frac{\sqrt{N}}{4Rt_o} (t_{R,n} - t_o) \tag{1.14}$$

Comparison of Equations 1.12 and 1.14 makes it quite evident that the amount of time available for the last peak to come out of the column with elution under gradient or programmed conditions produces many more peaks than elution under constant conditions. In essence, programmed elution produces more peaks per unit time than in invariant elution conditions (Figure 1.1).

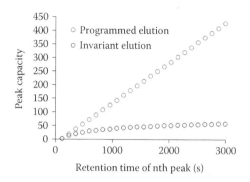

FIGURE 1.1 Comparison of peak capacity of programmed and time invariant elution methods. $N = 5000$, $t_o = 120$ s for both types of elution.

1.3   LIMITATIONS OF THE PEAK CAPACITY CONCEPT

In developing and quantifying the concept of peak capacity several simplifying assumptions are made:

1. All analytes have the same N value.
2. Under time-varying elution all peaks have the same peak width (σ).

It is well known that this second assumption is really only true for species that are strongly retained under the initial elution conditions (see Equation 1.24). However, neither of these limitations is particularly serious in comparison to the assumption that the analytes are essentially uniformly spaced by an interval equal to $2*R(\sigma_i + \sigma_{i+1})$. In practice, one will never see peaks spread so uniformly. The peak capacity concept is fundamentally hypothetical. It sets the *upper limit* to the number of peaks that can be observed [7]. As Giddings [2] said "One has no real hope of resolving 100 components on a column with $n = 100$." Questions about the number of peak maxima (singlets and multiplets) that one will observe *on average* with a sample containing *m randomly spaced* analytes were first answered in a truly classic paper by Davis and Giddings [8] in their study of statistical overlap theory (SOT).

In their model, one assumes that the probability that a peak maximum falls in a retention interval dx (which could be time or volume) is a constant λdx. This assumption corresponds to a Poisson random distribution model. The interval is assumed to be large enough, so that it holds a number of components and is independent of other zones. This corresponds to local randomness that can differ from region to region in the chromatogram. A final but important preliminary point is that the peak sizes are assumed to be uniform. As is well known the minimum resolution needed to see two peak maxima of equal size is 0.5; however, as the peak size ratio increases, the necessary minimum resolution needed to see two peaks for two components increases. This effect is neglected in the Davis–Giddings paper [8].

A major consequence of SOT is the fact that the expected value of the number of singlet (pure component) peaks (s) that will be seen on average is a much smaller

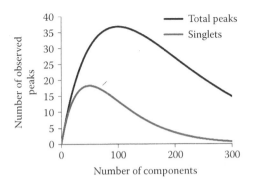

FIGURE 1.2 Davis–Giddings SOT prediction of the average number of observed peaks and singlets for a separation with peak capacity of 100.

number than the peak capacity. The total number of peaks (p)—that is, singlets, doublets, and so on—is larger but still vastly less than the peak capacity of the system. Under the best circumstances one will, *on average*, only see *singlet* peaks numbering no more than 18% of the peak capacity, or a *total* number of peaks that is 37% of the peak capacity. The maximum number of peaks is observed when m is equal to n and the maximum number of singlets occurs when m is equal to $n/2$. Obviously, as the chromatogram becomes more crowded, the number of maxima must decrease because more peaks become fused and, in fact, both s and p decreases very rapidly as shown in Figure 1.2. These results bode very poorly for doing accurate quantitative work with minimal *method development* by adjusting the relative band spacing. Method development really is a battle to lessen the entropy of the separation by imposing order on the relative band spacings. As stated above at some point a mixture becomes so crowded that this approach results in a loosing fight. This point evidently corresponds to an m/n ratio rather less than unity.

The results of the Davis–Giddings SOT work have strongly prompted the use of two-dimensional separation methods in many areas of analytical chemistry, including proteomics [9], metabolomics [10], polymer analysis/characterization [11], and bioanalytical samples [12]. As explained by Giddings [5] when one is dealing with mixtures that have hundreds and even thousands of components, as is the case in bottom-up proteomics, one has no course but to increase the peak capacity. One way to think of increasing the peak capacity is to envision it to be the space (distance or time) available for a separation. A one-dimensional (1D) separation can be thought of as a line of boxes each of which can hold one peak.

In contrast, a 2D separation is a matrix of boxes lying in a plane. Clearly there is a lot more space, that is, more boxes in the 2D separation than in the 1D separation (Figure 1.3) [13].

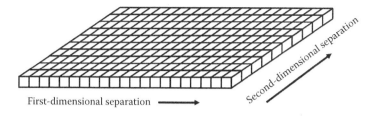

First-dimensional separation ⟶

Second-dimensional separation

FIGURE 1.3 Diagrammatic representation of the spatial increase in moving from 1D to 2D chromatography. (Courtesy of Ron Majors.)

Another way to think of a 2D separation is that if two components fail to separate in the first dimension they have another opportunity to separate in the second dimension provided that the second dimension is *chemically different* from the first in a way that is recognized by the analytes. Obviously in a 2D separation if two components are separated in the first dimension they do not need to be reseparated in the second dimension. This is not the case with two tandem columns; it is required that the two components not be recombined in the second dimension. Tandem separations only increase the separation space marginally, whereas under ideal circumstances Giddings says that the peak capacity of a 2D separation is the product of the peak capacities of the two separations.

$$n_{2D,ideal} = n_1 \times n_2 \qquad (1.15)$$

The question naturally arises in doing a 2D separation: What is the cost of increasing the separation space in this fashion? Let us assume that the first dimension is done on roughly the same timescale as a 1D separation then the time cost is minimal as long as each second-dimension separation is done in a small fraction of the first-dimension time. In fact, generally speaking, high speed in the second dimension is essential to gain a high 2D peak capacity [14,15]. There is an additional cost in 2D separations and that is the additional dilution of the sample during the second-dimension separation [16]. However, the major thrust of this chapter is how one can maximize the peak capacity of the first-dimension separation in a given time. In view of Figure 1.1 we will be talking about gradient elution and focus our attention on reversed-phase chromatography (RPC) as it is the most important area in such fields as proteomics.

1.4 GRADIENT ELUTION REVERSED-PHASE LIQUID CHROMATOGRAPHY

Many eminent chromatographers have contributed markedly to our understanding of gradient elution chromatography [17–31]; however, most relevant to the present manuscript is the work of the late Uwe Neue [25–27], which is a very important combination of peak capacity theory with the theory of RPC. First, we need to give some background on RPC. It is widely agreed that the logarithm of the retention factor (k) is at least a *quasi-linear* function of the volume fraction of the organic modifier in the eluent (ϕ).

$$\ln k = \ln k_w - S \cdot \phi \qquad (1.16)$$

Here, $\ln k_w$ is a hypothetical k factor in pure water assuming that the equation is perfectly correct and S is the solvent sensitivity coefficient that controls how much k varies as ϕ is changed. In linear solvent strength theory one varies ϕ as a linear function of time:

$$\phi(t) = \phi_o + (\phi_f - \phi_o)\frac{t}{t_G} = \phi_o + \Delta\phi\frac{t}{t_G} \tag{1.17}$$

The terms ϕ_o, ϕ_f, and t_G represent the initial and final mobile phase compositions, and the gradient time. Further it is assumed that the gradient profile propagates through the column at the same rate as the mobile phase moves through the column. This means that the strong component of the mobile phase does not sorb into the stationary phase. This is an important assumption and it has been challenged for fast gradients [29,30]. Assuming that the gradient enters the column at exactly the same time as the sample the solute retention time will be

$$t_R = t_o\left(1 + \frac{1}{b}\ln(bk_o + 1)\right) \tag{1.18}$$

where k_o is the retention factor at time equal to zero and b is the *dimensionless gradient slope*.

$$b = \frac{S\Delta\phi t_o}{t_G} \tag{1.19}$$

Defining a *gradient retention factor* (k_G):

$$k_G \equiv \frac{t_R - t_o}{t_o} = \frac{1}{b}\ln(b \cdot k_o + 1) \tag{1.20}$$

and then substituting b from Equation 1.19 we get

$$k_G \equiv \frac{t_R - t_o}{t_o} = \frac{t_G}{S\Delta\phi t_o}\ln\left(\frac{S\Delta\phi t_o}{t_G} \cdot k_o + 1\right) \tag{1.21}$$

Now that we have gradient retention under control we have to consider the peak width. This subject is a bit less certain. The following equation is deemed theoretically correct by most authors [27,28]:

$$\sigma = \frac{t_o}{\sqrt{N}}(1 + k_e)G(p) \tag{1.22}$$

The term $t_o/\sqrt{N}$ corresponds to σ for a peak eluting at the column dead time, and k_e is the local retention factor of a solute as it elutes from the column. Under all of the aforementioned conditions often termed linear solvent strength theory (LSST), it can be shown that

$$k_e = \frac{k_o}{bk_o + 1} \tag{1.23}$$

An important limiting case occurs when k_o is large.

$$k_e \approx \frac{1}{b} \tag{1.24}$$

As solute-to-solute differences in k_o are generally much larger than those in S, variations in k_e are often said to be minor when k_o is large, so for mathematical simplicity S is held constant when the integration in Equation 1.11 is done in all the theoretical work reported to date. This is a significant weakness in the generality of the theory because S often tends to increase as solute retention increases.

Taken together the first two terms in Equation 1.22 correspond to what happens in isothermal/isocratic chromatography. The last term $G(p)$ is the *gradient* compression factor. This accounts for the fact that the front of a peak is always in a weaker eluent than the rear of a peak and thus the peak will be somewhat compressed. Poppe [28] showed that this term is

$$G(p) = \sqrt{\frac{1 + p + \frac{p^2}{3}}{(1+p)^2}} \tag{1.25}$$

where p is defined as

$$p = \frac{k_o \cdot b}{1 + k_o}; \text{ note, when } k_o \gg 1, \ p \approx b \tag{1.26}$$

Obviously, if $b = 0$ then $p = 0$ and one is doing isocratic chromatography. $G(p)$ must then be equal to 1.0 (no zone compression). At the other extreme as p becomes exceedingly large (corresponding to very fast gradients) $G(p)$ approaches $\sqrt{1/3}$, which is about 0.58. Consequently, a peak can be narrowed at most by a factor of almost 2 by the gradient compression factor. Several studies disputed the need for $G(p)$ because of several failures of the basic assumptions of LSST and extra-column broadening act to almost cancel out the gradient-induced zone compression factor but the issue now seems to be resolved [27] and the theoretically derived result—that is, Equation 1.25—is correct given the numerous assumptions of LSST.

In his theoretical work [25,26] on gradient peak capacity Neue started with Grushka's integral (Equation 1.11), but used the value of σ for gradient elution obtained from Equation 1.22 with $G(p)$ set equal to unity, with N and S assumed to be the same for all solutes and the retention time of the last peak set equal to $t_o + t_G$. This result in Equation 1.22 with $G(p)$ set equal to 1.0 (Equation 1.27):

$$n_c \approx 1 + \frac{\sqrt{N}}{4} \frac{1}{1+b} \ln\left(\frac{1+b}{b} e^{bk_G} - \frac{1}{b}\right) = 1 + \frac{\sqrt{N}}{4} \frac{1}{1+b} \ln\left(\frac{1+b}{b} e^{S\Delta\phi} - \frac{1}{b}\right) \tag{1.27}$$

We will refer to this equation as *Neue's exact equation*. In deriving Equation 1.27 Neue used Equation 1.23; however, a much simpler equation is obtained when one

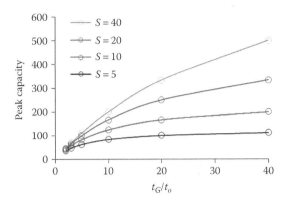

FIGURE 1.4 Peak capacity of reversed-phase gradient elution liquid chromatography according to Neue's approximate equation. In all cases, $\Delta\phi = 1$. The last peak has a retention time of $t_G + t_o$.

assumes that all the solutes are well retained in the initial eluent, so that k_e is given by Equation 1.24. In this case, the integral simplifies nicely:

$$n_c = 1 + \frac{\sqrt{N}}{4R} \frac{t_{R,n} - t_o}{t_o} \frac{1}{1+k_e} \approx 1 + \frac{\sqrt{N}}{4R} \frac{t_{R,n} - t_o}{t_o} \frac{b}{1+b} \tag{1.28}$$

We will return later to see what happens when Equation 1.25 is used for $G(p)$ in the recent work by Gritti et al. [29] and Blumberg and Desmet [21–24]. As done previously, Equation 1.28 is put in final form by taking the retention time of the last peak equal to be $t_G + t_o$ (Figure 1.4).

$$n_c \approx 1 + \frac{\sqrt{N}}{4R} \frac{t_G}{t_o} \frac{b}{1+b} \approx 1 + \frac{\sqrt{N}}{4R} \frac{S\Delta\phi t_G}{t_G + S\Delta\phi t_o} \tag{1.29}$$

We will refer to both of the forms in Equation 1.29 as *Neue's approximate equation*. Clearly, the peak capacity increases monotonically with gradient time and is higher for analytes with larger values of S and samples that require a wide range in eluent composition. The question arises as to how much error is made when one uses Equation 1.24 for k_e rather than Equation 1.23. As shown in Table 1.1 the error is rather small for fast gradients but gets to be more substantial for slow gradients. Fortunately, as $S\Delta\phi$ increases the error decreases. A value of $S = 5$ is actually pretty small and would be typical of a molecule of the size of benzene ($MW = 78$); however, S gets bigger as the solute MW increases. Clearly, Equation 1.29 gives a reasonable estimate that is always conservative, as the exact Equation 1.28 gives higher peak capacities (Table 1.1).

Although these papers are useful from a theoretical perspective, when the analyst has a *real* sample in hand one needs a way to optimize the peak capacity for that sample. One of the first approaches to this problem was based on adopting Poppe's

TABLE 1.1
Error Resulting from the Use of the Neue's
Approximate versus Exact Equations

t_o/t_G	% Difference[a]			
	5[b]	10[b]	20[b]	40[b]
0.5	6.1	1.7	0.5	0.1
0.2	12.0	3.9	1.1	0.3
0.1	17.8	6.4	2.0	0.6
0.05	24.2	9.8	3.3	1.0
0.025	30.3	13.8	5.2	1.7

[a] %Difference = 100 (exact–approximate)/exact.
[b] $S\Delta\phi$.

method of optimizing the isocratic plate count per unit system dead time [32] to optimize gradient elution peak capacity per unit gradient time [33]. Such methods require a great deal of information about the sample's retention and dynamic properties as well as the column and chromatographic system.

1.5   SPEED IN LIQUID CHROMATOGRAPHY AND OPTIMIZATION OF PEAK CAPACITY

The past decade has witnessed several critical developments in column technology and LC instrumentation. These include the introduction of fully porous sub-2 μm particles along with the pumps and other hardware improvements needed to take full advantage of these developments. In addition, smaller (less than 3 micron) superficially porous particles (also called core–shell particles) with an outstanding performance at reasonable pressures have become commercially available from a number of vendors [34–37].

The basic principles of the optimization of isocratic performance as measured by column efficiency and time have been known for quite some time. The observation of the strong dependence of the minimum analysis time on particle size and system pressure dates to the pioneering work of Knox [38–40], Purnell [41], and Halasz [42]. More recent work that has clarified how one can achieve maximum performance in HPLC includes that of Horvath [43] and Guiochon [44], as well as Poppe [32], Desmet [45], and this laboratory [46,47].

The early work by Knox and Saleem showed that *the best possible plate count* in a given amount of time requires that *the eluent velocity, column length, and the particle size be simultaneously adjusted at the maximum allowable instrument pressure.* Only if this is done do the optimum reduced velocity and plate height correspond to those at the minimum in a van Deemter plot [40,42,44,46,47]. Optimizing according to the method of Knox and Saleem leads to values for the optimum column length (L^*), optimum interstitial velocity (u_e^*), and optimum particle size (d_p^*):

$$L^* = \left[\frac{P_{max}B/C}{\Phi\eta} \right]^{1/4} D_m^{1/2} (\lambda t_o)^{3/4} \tag{1.30}$$

$$u_e^* = \left[\frac{P_{max}B/C}{\Phi\eta} \right]^{1/4} D_m^{1/2} (\lambda t_o)^{-1/4} \tag{1.31}$$

$$d_p^* = \left[\frac{\Phi\eta B/C}{P_{max}} \right]^{1/4} D_m^{1/2} (\lambda t_o)^{1/4} \tag{1.32}$$

P_{max} is the maximum system pressure, t_o is the dead time at the optimum eluent velocity and optimum column length, Φ is the dimensionless column resistance factor (typically 500), λ is the ratio of the interstitial porosity (ε_e) to the total column porosity (ε_{tot}), and h_{min} is the reduced plate height at the optimum reduced velocity.

These in turn lead to the maximum possible plate count at t_o.

$$N_{max} = \frac{L^*}{h_{min}d_p^*} = \sqrt{\frac{P_{max}\lambda t_o}{\Phi\eta}} \frac{1}{h_{min}} = \Psi\sqrt{t_o} \tag{1.33}$$

On the other hand, if one merely optimizes the eluent velocity and column length at some *arbitrary but available particle size* one will sacrifice some efficiency. Recent work from this laboratory [48] showed that if one changes particle size from one commercially available particle size to the next larger available size *at the appropriate time* the lost performance will only amount to about 10%; thus, it makes sense to use the results of the Knox–Saleem optimization (Equation 1.33) to estimate the maximum value of N to use in the gradient peak capacity equations as done by Meyer [20].

We now substitute Equation 1.33 in Equation 1.29 to get

$$n_c \approx 1 + \frac{\sqrt{\Psi}(t_o)^{1/4}}{4R} \frac{S\Delta\phi t_G}{t_G + S\Delta\phi t_o} \tag{1.34}$$

The presence of the dead time in both the numerator and denominator of Equation 1.34 suggests that there is an optimum value for t_o. Differentiation of Equation 1.34 leads to the result that there is a maximum peak capacity (Figure 1.5) when

$$t_G = 3S\Delta\phi t_o \tag{1.35}$$

Of course, Equation 1.35 tells us that an optimum value of the dimensionless gradient slope parameter (b) must exist:

$$b_{opt} = \frac{S\Delta\phi t_o}{t_{G,opt}} = \frac{1}{3} \tag{1.36}$$

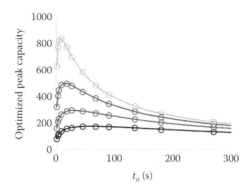

FIGURE 1.5 Dependence of optimized peak capacity versus dead time at fixed gradient time. Conditions: $t_G = 900$ s, curves (black) $S\Delta\phi = 5$, curve (red) $S\Delta\phi = 10$, curve (blue) $S\Delta\phi = 20$, and curve (green) $S\Delta\phi = 40$. Conditions $P_{max} = 400$ bar, $T = 40°C$, and viscosity = 0.69 with acetonitrile at volume fraction 0.25 in water. This is the composition of maximum viscosity. The diffusion coefficient is assumed to be 9.0×10^{-6} cm²/s. The reduced A, B, and C terms in the van Deemter equation are 0.95, 7, and 0.04, respectively. $\Phi = 450$, $\lambda = 0.67$.

Substitution of b_{opt} in Equation 1.34 gives the amazingly simple result for the optimum peak capacity:

$$n_{max} \approx 1 + \frac{3\sqrt{\Psi}}{16R} S\Delta\phi t_o^{1/4} \approx 1 + \frac{0.1875\sqrt{\Psi}}{R} S\Delta\phi t_o^{1/4} \tag{1.37}$$

Alternatively, in terms of t_G we get

$$n_{max} \approx 1 + \frac{\sqrt{\Psi}}{16R}(3S\Delta\phi)^{3/4} t_G^{1/4} \approx 1 + 0.1424\frac{\sqrt{\Psi}}{R}(S\Delta\phi)^{3/4} t_G^{1/4} \tag{1.38}$$

There seems to be some disagreement on the value of the maximum possible peak capacity or the existence and value of the optimum dimensionless gradient slope [20,49]. The studies of Desmet and Blumberg [24] also arrive at the conclusion that there *must be an optimum gradient rate*. However, because they include the gradient compression factor $G(p)$ in their calculation of the peak width, the optimum b is a function of k_o.

The most important thing that Equations 1.37 and 1.38 teach is that the peak capacity under optimum conditions only increases with the one-fourth power of time whether it is gauged by the column dead time or the gradient time. It should now be clear that Equation 1.29 is somewhat misleading in two regards. First, the initial rate of increase in the optimum peak capacity with t_G is not linear with t_G as Equation 1.29 suggests and second, the optimized peak capacity does not approach a horizontal asymptote but rather it increases indefinitely with the amount of time invested. These issues result because Equation 1.29 assumes that the column length, velocity, and particle size are all held fixed as t_G is increased, whereas it is well known that as t_G is increased, one needs to increase the column length to maximize the peak capacity [33]. These results, as a function of the dead time, are shown

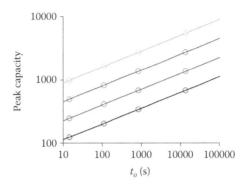

FIGURE 1.6 Optimum peak capacity according to Equation 1.33. The points marked on the curve correspond to 1.8, 3.0, 5.0, and 10 micron particle diameters from left to right. All conditions as in Figure 1.5. $S\Delta\phi = 5$ (black), $S\Delta\phi = 10$ (red), $S\Delta\phi = 20$ (blue), and $S\Delta\phi = 40$ (green).

TABLE 1.2

Maximum Possible Peak Capacity[a-c] as a Function of Gradient Time

	$S\Delta\phi$			
t_G (min)	5	10	20	40
5	140	230	385	640
15	180	300	500	840
30	210	360	600	1000
60	250	420	710	1200
120	300	500	850	1400

[a] Based on Equation 1.38.
[b] Requires $t_G = 3*S\Delta\phi*t_o$.
[c] All conditions are as in Figure 1.5.

---

in Figure 1.6. The slope of these log-log plots is essentially 0.25 as unity is much smaller than the second time-dependent term in Equation 1.37. Plots of the maximum possible peak capacity versus the gradient time are more complex in that the value of t_G corresponding to a given t_o depends on $S\Delta\phi$. Instead, we give results versus t_G in tabular form (Tables 1.2 and 1.3).

The aforementioned results tell us that the optimized peak capacity increases almost exactly in proportion to the one-fourth power of both the column dead time and the gradient time. Further, we see that both pressure and temperature increase the peak capacity (see Table 1.3). In addition, increasing both temperature and pressure has a cumulative effect. As Neue's approximate equation gives lower peak capacities than his more exact equation, the numerical values given here are somewhat conservative; however, the maximum difference between the two equations is only about 10% (Table 1.1) when $S\Delta\phi = 10$ and gets smaller as $S\Delta\phi$ increases. Finally, neither Equation 1.36

TABLE 1.3

Effect of Temperature and Pressure on Maximum Possible Peak Capacity[a]

Temperature (°C)	40		120	
Pressure (bar)	400	1200	400	1200
t_G (min)		Peak Capacity		
5	230	300	290	385
15	300	390	385	505
30	360	470	460	600
60	420	560	540	715
120	500	660	645	850

[a] All conditions are as in Figure 1.5 except the viscosity and diffusion coefficient are adjusted for the elevated temperature; $S\Delta\phi = 10$.

nor 1.37 include the gradient compression term $G(p)$ that was included in the work of Desmet and Blumberg [23,24]. Since compression reduces the peak width for some peaks it follows that even higher results can be expected (see Table 1.4).

It is often stated that the peak capacity under gradient elution varies with the square root of the column length. This is not true under optimized conditions. Jorgenson [50] rightly points out that this assumes that all of the conditions are held constant in the developing Equation 1.29. Keep in mind that Knox–Saleem optimization of N requires that the column length, particle diameter, and eluent velocity must be continuously and simultaneously varied as the analysis timescale is increased (Equations 1.30 through 1.32). The equation based on Knox–Saleem optimization relating N_{max} to L^* is easily derived:

$$N_{max} = \frac{1}{h_{min}} \sqrt[3]{\frac{P_{max}}{\Phi \eta D_m}} \left[\frac{C}{B}\right]^{1/6} L^{*2/3} \tag{1.39}$$

TABLE 1.4

Comparison of Peak Capacities with and without the Gradient Compression Factor[a]

$S\Delta\phi$	t_G min	n, with $G(p) = 1$	n, with $G(p)$ = Equation 1.26	Ratio
5	15	121.6	136.5	1.122
10	30	216.5	243.4	1.124
20	60	403.9	456.7	1.131
30	90	591.3	669.9	1.133
40	120	778.7	883.6	1.135

[a] $b = 1/3$, $N = 10,000$, $t_o = 1$ min.

Since the peak capacity depends on $\sqrt{N_{max}}$ (Equation 1.29), it follows that the optimized peak capacity varies with $L^{*1/3}$; thus, to merely double the peak capacity the column length needs to be increased eight-fold, whereas increasing the analysis time 16-fold. Interestingly, this result agrees with that of Blumberg and Desmet's Equation 1.12 in Reference 24.

It is also interesting to examine the dependence of N_{max} on the optimum particle size (d_p^{*2}) and optimum interstitial eluent velocity (u_e^{*2}):

$$N_{max} = \frac{1}{h_{min}} \left[\frac{P_{max}}{\Phi \eta D_m} \right] \left[\frac{C}{B} \right]^{1/2} d_p^{*2} \tag{1.40}$$

$$N_{max} = \frac{1}{h_{min}} \left[\frac{P_{max} D_m}{\Phi \eta} \right] \left[\frac{B}{C} \right]^{1/2} \frac{1}{u_e^{*2}} \tag{1.41}$$

These two equations tell us that as t_o is increased we must simultaneously increase both the column length and particle size, while decreasing the eluent velocity to achieve the maximum possible plate count. As the peak capacity increases with the square root of N_{max} it follows that, to double the peak capacity the plate count must be quadrupled, consequently the particle size must be doubled. Similarly, to double the peak capacity the eluent velocity must be decreased two-fold.

It is clearly very *pricey* to buy peak capacity by increasing the analysis time in 1D gradient elution by increasing time. This underscores the importance of 2D methods when high peak capacities are needed. Another important point is that we assume all solutes have the same diffusion coefficient. A somewhat surprising result of Knox–Saleem optimization is that while all the three optimization variables (eluent velocity, column length, and particle size) vary with the solute's diffusion coefficient, N_{max} (see Equation 1.33) does not.

1.6 EFFECT OF THE GRADIENT COMPRESSION FACTOR ON THE PEAK CAPACITY

As mentioned earlier, Blumberg and Desmet included $G(p)$ in their approach to computing the peak capacity. Again based on Grushka's integral formulation (Equation 1.11) the peak capacity will now be given by

$$\int_1^n dn = \int_{t_o}^{t_{R,n}} \frac{dt}{4R\sigma} = \frac{\sqrt{N}}{4Rt_o} \int_{t_o}^{t_{R,n}} \frac{dt}{(1+k_e)G(p)} \tag{1.42}$$

As both k_e and $G(p)$ depend on k_o, which we expect to vary substantially as time progresses across the gradient, the actual integral is rather complex. Table 1.5 shows the influence of the initial retention factor and the dimensionless gradient slope on the extent of gradient compression.

TABLE 1.5
Dependence of the Gradient Compression Factor on k_o and b

b	0.01	0.05	0.1	0.333	0.5	1
k_o			$G(p)$			
1	0.9975	0.988	0.976	0.930	0.902	0.839
2	0.9967	0.984	0.969	0.911	0.878	0.808
5	0.9959	0.980	0.962	0.894	0.857	0.784
10	0.9955	0.978	0.959	0.886	0.849	0.774
30	0.9952	0.977	0.956	0.881	0.842	0.767
100	0.9951	0.977	0.955	0.879	0.840	0.765
10,000	0.9951	0.976	0.955	0.878	0.839	0.764

At the optimum value of b (=1/3) as given by Equation 1.36 we see that $G(p)$ only varies by about 7.7% as k_o varies from 1.0 to an essentially infinitely large value. The limiting value of $G(p)$ as k_o becomes very large is

$$G(p) \rightarrow \sqrt{\frac{1+b+b^2/3}{(1+b)^2}} \tag{1.43}$$

If we also use the limiting value of k_e as is done to get Neue's approximate peak capacity (Equation 1.24) then the denominator in the integral of Equation 1.42 becomes independent of k_o and consequently of time, so we can easily do the integral to obtain

$$n_c \approx 1 + \frac{\sqrt{N}}{4R} \frac{t_G}{t_o} \frac{b}{\sqrt{1+b+b^2/3}} \approx 1 + \frac{\sqrt{N}}{4R} \frac{S\Delta\phi}{\sqrt{1+b+b^2/3}} \approx 1 + \frac{\sqrt{N}}{4R} \frac{S\Delta\phi}{\sqrt{1+b+b^2/3}} \tag{1.44}$$

Now, when we substitute Equation 1.33 for N we get

$$n \approx 1 + \frac{\sqrt{\psi}}{4R} \frac{S\Delta\phi t_o^{1/4}}{\sqrt{1+b+b^2/3}} \tag{1.45}$$

Once again there is an optimum value of b that maximizes n; in this case it is

$$b_{\text{opt}} = 0.618 \tag{1.46}$$

And thus:

$$n_{\max} \approx 1 + \frac{0.1894\sqrt{\psi}}{R} S\Delta\phi t_o^{1/4} \tag{1.47}$$

Equation 1.47 should be compared with Equation 1.37. The difference in the two approaches (based on Neue [25]) in comparison to Desmet–Blumberg [24] is barely greater than 1%. It should be recalled that according to Table 1.5 the compression effect is at a maximum when k_o is very large. Thus, we do not anticipate any big difference between Neue's more approximate method and the exact treatment of Desmet and Blumberg when k_o is large. Combining Equation 1.45 with Equation 1.46 gives the analog to Equation 1.38:

$$n \approx 1 + \frac{0.1679\sqrt{\psi}}{R}(S\Delta\phi)^{3/4} t_G^{1/4} \qquad (1.48)$$

In terms of t_G the Desmet–Blumberg treatment at the limiting value of k_o gives an increased peak capacity of about 17% when Equation 1.48 is compared to Equation 1.38.

The aforementioned approach assumes that k_o is always very large, so that $G(p)$ is given by Equation 1.43. The exact closed form integral is quite complex but Equation 1.42 is easily integrated numerically. One can then compare Neue's exact equation with $G(p) = 1$ to the exact numerical integral with k_e and p given by Equations 1.23 and 1.26, respectively. The comparison is most significant at the optimum value of b given by using Neue's approximate equation (i.e., $b_{opt} = 1/3$).

We see that there is only a 10%–13% error between the Desmet–Blumberg result and Neue's exact result. Obviously, when gradient compression is included, the peaks are narrower and higher peak capacities will be obtained. Nonetheless, we see that what we have called Neue's exact equation (Equation 1.27) and the exact treatment of Desmet and Blumberg (Equation 1.42) are quite similar.

It must be understood that Neue's integrations, which lead to Equations 1.27 and 1.28, and that of Desmet and Blumberg, which gives Equation 1.40 all assume that N and b (and thus S) are numerically the same for all solutes in a given sample. Furthermore, Desmet and Blumberg [23] assume a value of S of about 10 for a *typical* solute of MW 100–500. As shown in Figure 1.7 these assumptions are at best rather approximate. If we first consider the five alkyl benzenes varying from

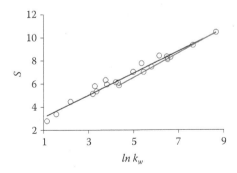

FIGURE 1.7 Relationship between S and $ln\ k_w$. A C_8 column with water–acetonitrile eluents was used. Solute set A: 23 polar and nonpolar solutes (black). Solute set B: alkylbenzenes from benzene to n-pentylbenzene (red). (Data from Mao, Y., Selectivity Optimization in Liquid Chromatography Using the Thermally Tuned Tandem Column (T³C) Concept, PhD dissertation, University of Minnesota (2001).)

benzene ($MW = 78$) to n-butyl benzene ($MW = 134$) we see that S varies from 5.9 to 8.7 for a typical C_8 type phase using acetonitrile/water mobile phases at 40°C. If we were to take this linear relationship out to n-pentadecyl benzene (with $MW = 289$) S would be about 17.4, which is more than a three-fold increase in both S and b. The solute's polarity also has a considerable effects. Consider solutes ranging from n-benzyl formamide ($S = 2.83$, $MW = 144$) to benzophenone ($S = 8.45$, $MW = 183$). It must be admitted that the correlation between S and $ln\ k_w$ for an unrelated collection of molecules is seldom as good as shown in Figure 1.7; nonetheless, the errors caused by the assumption that N and b are constant under the integration are very likely considerably bigger than when we ignore gradient compression; consequently, we feel that k_e and p can reasonably be approximated as $1/b$ and b, respectively (Equations 1.24 and 1.28).

REFERENCES

1. E. Grushka, Chromatographic peak capacity and the factors influencing it, *Anal. Chem.* 42 (1970) 1142–1147. doi:10.1021/ac60293a001.
2. J.C. Giddings, Maximum number of components resolvable by gel filtration and other elution chromatographic methods, *Anal. Chem.* 39 (1967) 1027–1028. doi:10.1021/ac60252a025.
3. J.C. Giddings, Generation of variance, theoretical plates, resolution, and peak capacity in electrophoresis and sedimentation, *Sep. Sci.* 4 (1969) 181–189. doi:10.1080/01496396908052249.
4. C.G. Horvath, S.R. Lipsky, Peak capacity in chromatography, *Anal. Chem.* 39 (1967) 1893. doi:10.1021/ac50157a075.
5. J.C. Giddings, Two-dimensional separations: Concept and promise, *Anal. Chem.* 56 (1984) 1258A–1270A. doi:10.1021/ac00276a003.
6. E. Grushka, Chromatographic peak shapes. Their origin and dependence on the experimental parameters, *J. Phys. Chem.* 76 (1972) 2586–2593. doi:10.1021/j100662a020.
7. J.M. Davis, M.R. Schure, Is the number of peaks in a chromatogram always less than the peak capacity? in: N. Grinberg, P.W. Carr (Eds.), *Advances in Chromatography*, 2017.
8. J.M. Davis, J.C. Giddings, Statistical theory of component overlap in multicomponent chromatograms, *Anal. Chem.* 55 (1983) 418–424. doi:10.1021/ac00254a003.
9. Q. Wu, H. Yuan, L. Zhang, Y. Zhang, Recent advances on multidimensional liquid chromatography–mass spectrometry for proteomics: From qualitative to quantitative analysis—A review, *Anal. Chim. Acta.* 731 (2012) 1–10. doi:10.1016/j.aca.2012.04.010.
10. N.L. Kuehnbaum, P. Britz-McKibbin, New advances in separation science for metabolomics: Resolving chemical diversity in a post-genomic era, *Chem. Rev.* 113 (2013) 2437–2468. doi:10.1021/cr300484s.
11. E. Uliyanchenko, S. van der Wal, P.J. Schoenmakers, Challenges in polymer analysis by liquid chromatography, *Polym. Chem.* 3 (2012) 2313–2335. doi:10.1039/c2py20274c.
12. D.R. Stoll, Recent advances in 2D-LC for bioanalysis, *Bioanalysis.* 7 (2015) 3125–3142. doi:10.4155/bio.15.223.
13. J.C. Giddings, Concepts and comparisons in multidimensional separation, *J. High Resolut. Chromatogr.* 10 (1987) 319–323. doi:10.1002/jhrc.1240100517.
14. X. Li, D.R. Stoll, P.W. Carr, Equation for peak capacity estimation in two-dimensional liquid chromatography, *Anal. Chem.* 81 (2009) 845–850. doi:10.1021/ac801772u.
15. L.W. Potts, D.R. Stoll, X. Li, P.W. Carr, The impact of sampling time on peak capacity and analysis speed in on-line comprehensive two-dimensional liquid chromatography, *J. Chromatogr. A.* 1217 (2010) 5700–5709. doi:10.1016/j.chroma.2010.07.009.

16. M.R. Schure, Limit of detection, dilution factors, and technique compatibility in multi-dimensional separations utilizing chromatography, capillary electrophoresis, and field-flow fractionation, *Anal. Chem.* 71 (1999) 1645–1657. doi:10.1021/ac981128q.

17. L.R. Snyder, J.W. Dolan, *High-Performance Gradient Elution: The Practical Application of the Linear-Solvent-Strength Model*, John Wiley, Hoboken, NJ, 2007.

18. P.J. Schoenmakers, H.A.H. Billiet, R. Tussen, L. De Galan, Gradient selection in reversed-phase liquid chromatography, *J. Chromatogr. A.* 149 (1978) 519–537. doi:10.1016/S0021-9673(00)81008-0.

19. P. Jandera, J. Churáček, Gradient elution in liquid chromatography, *J. Chromatogr. A.* 91 (1974) 223–235. doi:10.1016/S0021-9673(01)97902-6.

20. V.R. Meyer, How to generate peak capacity in column liquid chromatography, *J. Chromatogr. A.* 1187 (2008) 138–144. doi:10.1016/j.chroma.2008.02.019.

21. L.M. Blumberg, Theory of gradient elution liquid chromatography with linear solvent strength: Part 1. Migration and elution parameters of a solute band, *Chromatographia.* 77 (2014) 179–188. doi:10.1007/s10337-013-2555-y.

22. L.M. Blumberg, Theory of gradient elution liquid chromatography with linear solvent strength: Part 2. Peak width formation, *Chromatographia.* 77 (2014) 189–197. doi:10.1007/s10337-013-2556-x.

23. L.M. Blumberg, G. Desmet, Metrics of separation performance in chromatography: Part 3: General separation performance of linear solvent strength gradient liquid chromatography, *J. Chromatogr. A.* 1413 (2015) 9–21. doi:10.1016/j.chroma.2015.07.122.

24. L.M. Blumberg, G. Desmet, Optimal mixing rate in linear solvent strength gradient liquid chromatography, *Anal. Chem.* 88 (2016) 2281–2288. doi:10.1021/acs.analchem.5b04078.

25. U.D. Neue, Theory of peak capacity in gradient elution, *J. Chromatogr. A.* 1079 (2005) 153–161. doi:10.1016/j.chroma.2005.03.008.

26. U. Neue, Peak capacity in unidimensional chromatography, *J. Chromatogr. A.* (2008) 107–130. doi:10.1016/j.chroma.2007.11.113.

27. U. Neue, D. Marchand, L. Snyder, Peak compression in reversed-phase gradient elution, *J. Chromatogr. A.* 1111 (2006) 32–39. doi:10.1016/j.chroma.2006.01.104.

28. H. Poppe, J. Paanakker, M. Bronckhorst, Peak width in solvent-programmed chromatography, *J. Chromatogr. A.* 204 (1981) 77–84. doi:10.1016/S0021-9673(00)81641-6.

29. F. Gritti, M.-A. Perdu, G. Guiochon, Gradient HPLC of samples extracted from the green microalga *Botryococcus braunii* using highly efficient columns packed with 2.6 µm Kinetex-C_{18} core–shell particles, *J. Chromatogr. A.* 1229 (2012) 148–155. doi:10.1016/j.chroma.2012.01.013.

30. F. Gritti, G. Guiochon, The distortion of gradient profiles in reversed-phase liquid chromatography, *J. Chromatogr. A.* 1340 (2014) 50–58. doi:10.1016/j.chroma.2014.03.004.

31. F. Gritti, G. Guiochon, Separations by gradient elution: Why are steep gradient profiles distorted and what is their impact on resolution in reversed-phase liquid chromatography, *J. Chromatogr. A.* 1344 (2014) 66–75. doi:10.1016/j.chroma.2014.04.010.

32. H. Poppe, Some reflections on speed and efficiency of modern chromatographic methods, *J. Chromatogr. A.* 778 (1997) 3–21. doi:10.1016/S0021-9673(97)00376-2.

33. X. Wang, D.R. Stoll, P.W. Carr, P.J. Schoenmakers, A graphical method for understanding the kinetics of peak capacity production in gradient elution liquid chromatography, *J. Chromatogr. A.* 1125 (2006) 177–181. doi:10.1016/j.chroma.2006.05.048.

34. R. Hayes, A. Ahmed, T. Edge, H. Zhang, Core–shell particles: Preparation, fundamentals and applications in high performance liquid chromatography, *J. Chromatogr. A.* 1357 (2014) 36–52. doi:10.1016/j.chroma.2014.05.010.

35. N. Tanaka, D.V. McCalley, Core–shell, ultrasmall particles, monoliths, and other support materials in high-performance liquid chromatography, *Anal. Chem.* 88 (2016) 279–298. doi:10.1021/acs.analchem.5b04093.
36. S. Fekete, E. Olh, J. Fekete, Fast liquid chromatography: The domination of core-shell and very fine particles, *J. Chromatogr. A.* 1228 (2012) 57–71. doi:10.1016/j. chroma.2011.09.050.
37. J. De Vos, K. Broeckhoven, S. Eeltink, Advances in ultrahigh-pressure liquid chromatography technology and system design, *Anal. Chem.* 88 (2016) 262–278. doi:10.1021/ acs.analchem.5b04381.
38. J.H. Knox, High speed liquid chromatography, *Annu. Rev. Phys. Chem.* 24 (1973) 29–49. doi:10.1146/annurev.pc.24.100173.000333.
39. J.H. Knox, The speed of analysis by gas chromatography, *J. Chem. Soc. Resumed.* (1961) 433. doi:10.1039/jr9610000433.
40. J.H. Knox, M. Saleem, Kinetic conditions for optimum speed and resolution in column chromatography, *J. Chromatogr. Sci.* 7 (1969) 614–622. doi:10.1093/chromsci/7.10.614.
41. J. Purnell, *Gas Chromatography*, New York, John Wiley & Sons, 1962.
42. I. Halász, G. Görlitz, Optimal parameters in high speed liquid chromatography(HPLC), *Angew. Chem. Int. Ed. Engl.* 21 (1982) 50–61. doi:10.1002/anie.198200501.
43. H. Chen, C. Horvath, High-speed high-performance liquid chromatography of peptides and proteins, *J. Chromatogr. A.* 705 (1995) 3–20. doi:10.1016/0021-9673(94)01254-C.
44. G. Guiochon, Optimization in liquid chromatography, in: C. Horváth (Ed.), *High-Performance Liquid Chromatography: Advances Perspectives*, Academic Press, New York, 1982, pp. 1–56.
45. G. Desmet, D. Clicq, P. Gzil, Geometry-independent plate height representation methods for the direct comparison of the kinetic performance of LC supports with a different size or morphology, *Anal. Chem.* 77 (2005) 4058–4070. doi:10.1021/ac050160z.
46. P.W. Carr, X. Wang, D.R. Stoll, Effect of pressure, particle size, and time on optimizing performance in liquid chromatography, *Anal. Chem.* 81 (2009) 5342–5353. doi:10.1021/ ac9001244.
47. P.W. Carr, D.R. Stoll, X. Wang, Perspectives on recent advances in the speed of high-performance liquid chromatography, *Anal. Chem.* 83 (2011) 1890–1900. doi:10.1021/ ac102570t.
48. A.J. Matula, P.W. Carr, Separation speed and power in isocratic liquid chromatography: Loss in performance of Poppe vs Knox-Saleem optimization, *Anal. Chem.* 87 (2015) 6578–6583. doi:10.1021/acs.analchem.5b00329.
49. L.R. Snyder, J.J. Kirkland, J.W. Dolan, Gradient elution, in: Introduction to Modern Liquid Chromatography, 3rd ed, Wiley, Hoboken, NJ, 2010, p. 453.
50. K.M. Grinias, J.M. Godinho, E.G. Franklin, J.T. Stobaugh, J.W. Jorgenson, Development of a 45 kpsi ultrahigh pressure liquid chromatography instrument for gradient separations of peptides using long microcapillary columns and sub-2 μm particles, *J. Chromatogr. A.* 1469 (2016) 60–67. doi:10.1016/j.chroma.2016.09.053.
51. Y. Mao, Selectivity Optimization in Liquid Chromatography Using the Thermally Tuned Tandem Column (T^3C) Concept, PhD dissertation, University of Minnesota (2001).

2 Laser Applications in Chromatography

Anneli Kruve and Israel Schechter

CONTENTS

2.1 INTRODUCTION

Light amplification by stimulated emission of radiation (Lasers) becomes more and more popular in analytical applications. Lasers have several properties that allow their utilization in which other light sources fail. Their most important properties for analytical applications are highly directional beam, excellent monochromaticity (spectral purity), high irradiation power, short pulse option, and photon coherence. In analytical chromatography, laser is mainly used for analyte excitation in detectors and for analyte desorption and ionization, which are needed for sampling and for coupling chromatography to mass spectrometry (MS).

The high directionality means that the laser beam diverges only very little (it may be below 10^{-5} radians). This high directionality also allows focusing the narrow laser beam to a very small area (e.g., a few μm^2), close to the diffraction limit. This property is beneficial in scanning applications, such as reading a developed thin-layer chromatography (TLC) plate, in which high spatial resolution is needed for an improved analytical performance. This focusing ability is also needed for achieving very high power densities, which are needed when the concentrations are very low (e.g., while combining nano-LC with fluorescence detection). The high power densities also results in local heating, which can be used for sample desorption or ablation. These properties have been used for reading TLC plates with MS, in which producing gas phase ions is essential.

Due to these unique properties lasers are used in various ways in all fields of chromatography. This review is mainly focused on detection systems in chromatography that are based on lasers. The most widely used detection systems for gas chromatography (GC), high-performance liquid chromatography (HPLC), and TLC covering both optical and MS-based units and utilization of lasers are discussed. Optical detection systems based on Raman spectroscopy, laser-induced fluorescence (LIF), and multiphoton spectroscopy are reviewed.

There are also other laser-based detection methods of minor use in chromatography, such as photoacoustic spectroscopy, fluorescence excitation spectroscopy, and so on. Lasers are also used in preparation of chromatographic materials, fragmentation of compounds in GC, and so on. These methods are out of the scope of this chapter. This chapter is also limited to the techniques that were popular in recent scientific publications and to new promising technologies.

2.2 RAMAN AND SURFACE-ENHANCED RAMAN SPECTROSCOPIES

Raman scattering is affected by the vibrational energy levels of the analyte; therefore, Raman spectroscopy provides significant molecular information. It is useful in both compound identification and quantitation. However, the Raman spectral features are not always sufficient to differentiate between similar compounds, especially in complicated mixtures. Therefore, in such cases, chromatography is needed for separating the compounds for reducing the complexity of the mixture. Raman spectroscopy became a powerful method for detection in various chromatographic techniques, as it also allows gaining structural information. The first realizations in Raman detection in chromatography were already made in the 1980s. Nevertheless, the applications of Raman spectroscopy in chromatography have not yet reached commercialization; and therefore, are not being accessible to a wide community.

The major drawback of Raman spectroscopy that hinders its widespread applications in GC, HPLC, and TLC is its rather low sensitivity. The limits of detection (LODs) are high due to the high quantum yield of Raman scattering (ca. 10^{-5}), and are usually not compatible with other common detectors. Therefore, regular Raman scattering is hardly used in chromatography applications. This issue is solved using surface-enhanced Raman spectroscopy (SERS).

Resonance Raman spectroscopy (RRS) or more recently SERS[1] significantly enhances the sensitivity. In recent applications of this method in chromatography, the LODs have been reduced by many orders of magnitude (usually by a factor of 10^4, but improvements up to 10^8 have also been demonstrated).

The surface-enhancement effect has been achieved by the adsorption of the analyte on the surface of a substrate, usually a metal surface of a fine nanometric structure or colloid particles. Most commonly, Ag and Au nanoparticles have been used in SERS applications, whereas occasional use of Cu may also be found. In SERS, the laser wavelength used for exciting the molecules is in resonance with the surface-plasmon vibration of the used metal. The plasmon band depends on the metal of choice and the actual size of the nanoparticles. Aggregation of the colloid particles causes a red spectral shift. The surface-plasmon bands of the aggregated Ag and Au particles are typically in the visible to near-infrared (NIR) region.

SERS has been coupled with different chromatographic techniques used since the 1980s. In the beginning SERS was mostly coupled with TLC but later also to GC and HPLC. In the recent years, TLC–SERS has shown real advantages and several important applications for the analysis of real samples have been demonstrated.[1] Due to the significant technological differences between HPLC–SERS and TLC–SERS, the following discussion will be separated accordingly.

Although the sensitivity problem in traditional Raman spectroscopy has been solved using SERS, several new issues are related to this technique. These include fluorescence interferences, difficulties in online coupling of the detector to the eluate, mobile-phase background, sensitivity to pH, analyte degradation, and so on. These effects will be addressed in the following sections.

2.2.1 HIGH-PERFORMANCE LIQUID CHROMATOGRAPHY–SURFACE-ENHANCED RAMAN SPECTROSCOPY

2.2.1.1 Fluorescence Interferences

In traditional Raman spectroscopy, the signals are severely interfered by fluorescence, because the fluorescence emission is broad and often of a quantum yield that is higher than that of Raman scattering. One main advantage of SERS is the quenching of fluorescence.[1] Accelerated nonradiative deactivation takes place, because the excited state of the molecule can relax (through energy transfer) to the metallic surface.[1] Thus, molecules adsorbed to the surface undergo fluorescence quenching and the interfering fluorescence is mainly from free molecules. Partial solution is obtained by time-resolved measurements, because usually the fluorescence is delayed relative to the Raman emission. Nevertheless, the temporal resolution significantly increases the instrumental costs.

2.2.1.2 Mobile-Phase Interferences

It would be highly useful and convenient in HPLC if the mobile phase used for separation is compatible with the detection system. This requires a good match between the solvent providing good HPLC separation and good SERS sensitivity. The first and foremost complication of matching these solvents is that common reversed-phase (RP) HPLC eluents produce significant background. In HPLC, the most common organic modifier is acetonitrile. Unfortunately, this solvent yields a significant background spectrum in SERS causing increased detection limits.

Moreover, it has been observed that acetonitrile adsorbs to the surface of silver SERS substrate; and therefore, it interferes with the interaction of the analyte with the nanoparticles. In order to address this problem, Carrillo-Carrion et al.[2] and Cabalin et al.[3] replaced the acetonitrile with methanol in the HPLC separation. Cooper et al.[4] used fully deuterated solvents (deuterated methanol and heavy water) as HPLC solvents, as the Raman peaks for these solvents are shifted toward lower frequencies, resulting in less interferences.

In addition, other mobile-phase components may affect SERS sensitivity. It has been observed that some ions from common buffers may adsorb to the SERS substrate hotspots and reduce the SERS signals.[5]

2.2.1.3 pH Interferences

Duo et al.[6] have shown that the solution pH may alter the analyte's orientation when adsorbed onto the surface of the nanoparticles. They observed that as the ζ potential of gold nanoparticles is negative, basically no lysine was absorbed on the particles at pH of 12.0, as lysine is fully deprotonated in this pH and carries a negative charge. As a result, the SERS signal was almost completely lost at that pH.

Related to this effect, Huang et al.[7] observed a different optimal pH in the analysis of cotinine and *trans*-3′-hydroxycotinine. These effects were attributed to the different adsorption geometry on the silver surface at low pH compared to neutral pH values.

Moreover, the pH also influences the stability of the colloidal particles in SERS substrates.[8] Ni et al.[9] observed that the direction of adsorption of the analyte on the silver electrode surface affect the relation between the intensities of the Raman peaks and the analyte concentration. This all makes evident that the mobile phase is very important for achieving low LODs and superior selectivity.

2.2.1.4 Online Coupling of Surface-Enhanced Raman Spectroscopy to High-Performance Liquid Chromatography

It seems that the popularity of HPLC–SERS is limited by the realization of online coupling. This coupling requires addition or generation of colloid particles after HPLC separation but before SERS detection. Up to date, one of the largest problems for online coupling is the realization of a proper detection cell. In the early works, closed detection cells, similar to commercial UV and fluorescence cells, have been used. However, colloid particles used for SERS may flocculate and contaminate the flow cell, causing a memory effect and peak tailing.[3] To avoid or reduce these effects it is necessary to wash the flow cell with nitric acid, which must be followed by profound washing. This significantly increases the analysis time and also influences the next analyses.

In order to solve the above problems, several groups have used different designs of windowless flow cells (e.g., Figure 2.1).[10,11] Similar flow cells were previously used in fluorescence measurements.[12,13] With such cells a very good repeatability, up to 1% for banned drugs,[11] and LODs below nanomole for purine bases[10] were achieved.

In addition to dealing with carry over and background effects, it is also important to use the conditions that provide the best sensitivity. These include the power of the laser used for SERS measurements and the signal collection time. Cabalin et al.[3] showed that increasing the signal integration time improves both the signal-to-noise ratio and the repeatability. Therefore, long integration times are advantageous. However, increasing the integration time in online HPLC–SERS is very limited, because it starts influencing the chromatographic separation. Therefore, one clear advantage of carrying out off-line measurements is the possibility to increase spectral collection times.

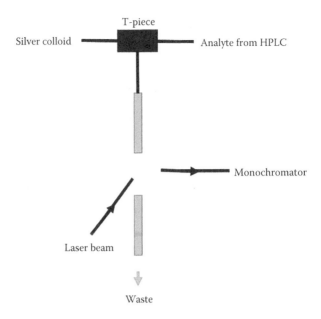

FIGURE 2.1   Schematic diagram of a windowless flow cell used by Cabalin et al.[11]

2.2.1.5   Off-Line Coupling of Surface-Enhanced Raman Spectroscopy to High-Performance Liquid Chromatography

The main motivation for using off-line coupling is the ability to perform long integration times. However, this coupling mode has additional advantages. Good SERS spectra depend not only on proper average size of the nanoparticles, but also their size distribution has to be relatively narrow. A narrow size distribution significantly improves method sensitivity.[14] These factors are much easier to control in an off-line mode.

The colloid particles used in off-line coupling setups are often prepared by reduction of $AgNO_3$ or $HAuCl_4$ solution using a reducing agent.[15] Two eluate collection methods have been mainly used for off-line coupling: (1) collection of the eluate on the colloid particles in microtitration plates[2,8] and (2) collection on TLC plates.[16] Rarely measurements on roughened metal surfaces have also been reported.[9]

The main drawback of the off-line coupling is that the collection step reduces the resolution; thus, losing information already gathered with HPLC. Collecting the effluent on TLC plates allows for the removal of the mobile phase[16]; therefore, reducing the problems related to the mobile-phase background.

The main advantage of the off-line coupling mode, namely longer integration times, has its own limitations. Nirode et al.[14] have shown that increasing the spectrum accusation time using high laser power, may result in loss of information. They observed broadening of Raman bands and increase of background fluorescence, possibly due to degradation of the analyte (riboflavin).

2.2.1.6 Applications of High-Performance Liquid Chromatography–Surface-Enhanced Raman Spectroscopy

Currently, HPLC–SERS is not in routine use, mostly due to technical problems in the coupling of the detector. Nevertheless, several promising applications have been reported: Most of them have focused on (illegal) drugs,[3,5,8,11] including amphetamine, cocaine, heroin, papaverine, procaine, and so on. In addition, basic bases (cytosine, xanthine, hypoxanthine, guanine, thymine, and adenine) have been determined by Cowcher et al.[10] and Carrillo-Carrion et al.[2] using this method. Also similar analysis of test compounds, such as nitrophenols[9] and Rhodamine 6G,[14] have been reported. Interestingly, also the metal cations Pb^{2+} and Hg^{2+} have been analyzed with HPLC–SERS, making use of the Raman spectra of the reaction product of these ions with 4-mercaptobenzoic acid.[17]

2.2.1.7 Chemometric Methods for High-Performance Liquid Chromatography–Surface-Enhanced Raman Spectroscopy

In order to deal with the background problems, chemometric approaches have also been used. Similar background problems exist for both Raman and IR; therefore, the background correction methods proposed for one could also be used for the other.[18] This kind of corrections requires modeling of the background spectrum. However, this is complicated due to (a) change of the background spectrum during elution (especially in case of gradient) and (b) during analyte elution the intensity of the eluent spectrum is reduced.[19] Therefore, the simple subtraction of the background spectrum before and after the analyte peak does not yield best results. In addition, due to extreme concentration difference, the intensity of the background spectrum may be much higher than that of the analytes. Under such conditions, it is possible to miss small peaks. Recently, several new data treatment algorithms have been proposed for background corrections.[1,18,19]

2.2.2 THIN-LAYER CHROMATOGRAPHY–SURFACE-ENHANCED RAMAN SPECTROSCOPY

Compared to HPLC, TLC with SERS detector has already gained popularity, as reflected by the growing number of publications in the recent years. The main advantage of this method over HPLC–SERS is that it does not face the problems related to the eluent background, because the TLC plates are usually dried before the detection.

2.2.2.1 Thin-Layer Chromatography Plates Used in Thin-Layer Chromatography—Surface-Enhanced Raman Spectroscopy

In the beginning mainly silica TLC plates were used in TLC–SERS. In recent years, other plate types have also been introduced, which increase the variety of this technique. For example, Zheng et al.[20] and Gao et al.[21] have introduced the usage of methacrylate-based TLC plates for TLC–SERS analyses. Gao et al.[21] have used molecularly imprinted polymers as TLC plates for the analyses of Sudan I dye in paprika powder. Zheng et al.[20] used a monolithic hydrophobic–hydrophilic dual phase plates for the separation of dyes that could not be separated only in one phase.

2.2.2.2 Application of Nanoparticles to Thin-Layer Chromatography Plates

A proper application of the nanoparticles to the TLC plates is crucial for achieving good performance. In principle, it would be convenient to carry out the elution on a TLC plate already containing the SERS substrate. However, mainly due to the high cost of these substrates and due to their unknown effect on the separation performance, in most cases, the colloids are applied only to the spots obtained in TLC separation.

Takei et al.[22] introduced the TLC plates with a built-in SERS layer of silver and gold nanoparticles. These plates were successfully used for the analyses of three Raman active dyes (rhodamine 6G, crystal violet, and 1,2-di(4-pyridyl)ethylene) and melamine in skim milk. A simpler possibility was used by Herman et al.[23]: They pretreated the silica plates with silver nitrate allowing it to undergo photoreduction and therefore, generate SERS substrate. Separation for mixtures of cresyl violet, bixine, crystal violet, and Cu(II) complex with 4-(2-pyridylazo)-resorcinol was performed on these plates and selective Raman spectra were obtained for each compound. Chen et al.[24] have also used polyvinyl pyrrolidone modified filter paper with preaggregated colloidal silver for SERS measurements. However, the analytes were not separated on the same filter paper, though, in principle, this could be applicable.

2.2.2.3 Dissolving of the Thin-Layer Chromatography Spots

Sometimes dissolving the TLC spot has been required in order to achieve the best sensitivity. After dissolving the spots, SERS substrate is added to the solution and the Raman spectrum is acquired. For example, Chieli et al.[25] observed that the quality of spectra was reduced when the colloids were applied to the TLC plate. They assumed that this was caused by the immediate adsorption of the colloids to the TLC plate. Therefore, they preferred dissolving the TLC spots and placing them on a covered glass slide, in which SERS could be easily measured. Wang et al.[26] also used this approach and in this way they reduced the limit of quantitation (LOQ) by more than two orders of magnitude. A somewhat different approach was suggested by Freye et al.[27] They transferred the TLC spots to a nanocomposite of silver in polydimethylsiloxane with conformal blotting wetted TLC plate and applying pressure.

2.2.2.4 Dynamic Surface-Enhanced Raman Spectroscopy

A modification of TLC–SERS has been lately introduced, named as dynamic SERS (DSERS). In this method, Raman signals are collected from the processed TLC plates (including SERS substrate) during the transformation from wet state to dry state. For example, Fang et al.[28] used 50/50 mixture of water and glycerol and Zheng et al.[20] used mixtures of ethanol and water for wetting the TLC plate before DSERS measurements. In addition, Pozzi et al.[29] obtained a good SERS signal of alkaloids while measuring the TLC plate before complete drying. It has been claimed that the dynamic data provide more information and allow for selecting the optimal measurement conditions.[30,31]

2.2.2.5 Applications of Thin-Layer Chromatography—Surface-Enhanced Raman Spectroscopy

In the recent years, TLC–SERS has been used in a large variety of applications, from cultural heritage studies to metabolomics. Both qualitative and quantitative applications have been presented; usually, the quantitative results have not been at the level of trace analysis.

Chieli et al.[25] and Woodhead et al.[32] have used TLC–SERS for studies of dye components. In conservation science, the information on the dyes used for coloring different materials is essential for conservation procedures, dating, and so on. However, the compounds may be structurally similar and in case of historical objects, contaminants, and also decomposition products may be present.

Chiel et al.[25] have used TLC–SERS for identification of *ammoniacal cochineal* dyes and pigments. Woodhead et al.[32] have analyzed purple dyes used for coloring of four historic silk dresses from the nineteenth century. Using SERS, they were able to distinguish between different synthetic dyes that cannot be distinguished purely on TLC, and also to determine the dyes used for coloring of the dresses. Brosseau et al.[33] analyzed red natural dyes from artworks. Using SERS, they were able to distinguish between alizarin and purpurin that could not be separated on TLC and to detect dyes from a small quantity (1 mg sample size) of wool sample.

TLC–SERS has also proven useful for artificial dye analysis in foodstuff, such as Sudan dyes in paprika powder[21] and Rhodamine B in chili oil.[26] Geiman et al.[34] also analyzed synthetic dyes in ballpoint inks.

Several applications in the determination of illegal adulterants in various mixtures have been published. The group of Feng Lu has presented studies on antidiabetic chemicals,[35] diet aid, the sports-enhancing drug ephedrine,[36] and antitussive and antiasthmatic drugs[28] in botanic dietary supplements. Li et al.[37] also analyzed structurally similar adulterants in botanic dietary supplements. The separation of structurally similar adulterants was achieved for the first time, by applying two-dimensional correlation spectroscopy (2DCOS). In this method, the change in the spectral features is registered as a function of the duration of exposure to the laser.

In addition, biomarkers have been determined with TLC–SERS.[7,38] They determined cotinine and trans-3′-hydroxycotinine in urine samples and have obtained LODs (10 and 200 nM) lower than the expected concentrations in such samples.[7] The method was also successfully used for analysis of smokers' urine samples with very high recoveries. Lucotti et al.[38] attempted developing the TLC–SERS method for determination of apomorphine (short-acting dopamine agonist) in the therapeutically relevant range from human blood plasma. However, due to the matrix interferences, the desired detection limits in plasma were not achieved.

Li et al.[39] was able to detect low level substituted aromatic pollutants in water samples and demonstrated that TLC–SERS is a suitable method for on-site analyses of real samples containing such compounds. They obtained LOD values below 0.2 ppm and the lowest obtained LOD was 8 ppb for benzidine. The disagreement between the TLC–SERS method and the reference method was below 15%, demonstrating both good sensitivity and good accuracy.

Yao et al.[40] has proposed a TLC–SERS method for the analysis of organophosphorus pesticide methidathion in tea leaves, referring to the necessity to monitor this compound on-site, whereas conventional LC/MS and GC/MS methods have been used only in laboratories.

In addition, quantitative monitoring of a reaction progress using TLC–SERS has been reported by Zhang et al.[41] They monitored the Suzuki coupling of phenylboronic acid and 2-bromopyridine and were able to separate the product from 2-bromopyridine based on the SERS spectra only, as these compounds had very close retention factors on TLC plates. In addition, they identified a by-product due to its SERS spectrum that could not be detected purely with TLC.

2.2.2.6 Quantification and Detection Limits

Quantification of the main components and illegal contaminants in various samples using TLC–SERS has been reported.[26,40] The concentration detection limits range from nanomolar to several percents. Usually, quantitative analyses have been performed without the usage of internal standard; however, Wang et al.[26] observed significant improvement in repeatability, from RSD 98.1% to 24.0%, on using melamine as an internal standard. Though, further improvement is required to meet the criteria of most legal bodies and validation guides.[42]

It seems that application of TLC–SERS for rapid on-site screening of contaminants and illegal substances has a great potential. For many other forensic and trace analyses, the current LODs are not sufficient and need to be decreased in order to compete with other techniques, such as MS.

2.2.2.7 Chemometric Methods for Thin-Layer Chromatography—Surface-Enhanced Raman Spectroscopy

Due to the problems related to the background and to low signals in TLC–SERS, several groups have investigated the benefits of various chemometric tools. Gao et al.[21] proposed the use of principal component analysis (PCA) to study the spectral differences between different samples. In addition, they used partial least square regression (PLSR) for quantitative determination of Sudan I dye. However, the linearity obtained was comparable to the one obtained with conventional linear regression.

Lv et al.[36] used partial least squares discriminant analysis (PLSDA) to study spectral differences between ephedrine, pseudoephedrine, methylephedrine, and norephedrine, and to choose the appropriate SERS peak as a marker for ephedrine. Gao et al.[21] have also developed a fast method for screening of Sudan I dye in paprika powder.

2.2.3 Alternative Methods

A very interesting approach has been proposed by Joon Lee et al.[17] for separation of metal ions on a surface-enhanced Raman active medium. They used a capillary column for which the walls were covered with a dense layer of gold nanoparticles treated with 4-mercaptobenzoic acid. This stationary phase simultaneously acts as a separation medium and an efficient SERS substrate. The measurements were performed inside the column by scanning the column with the laser beam, rather than measuring the effluent at the end of the column.

2.3 HIGH-PERFORMANCE LIQUID CHROMATOGRAPHY/ CAPILLARY ELECTROPHORESIS–LASER-INDUCED FLUORESCENCE

Conventional fluorescence detectors for HPLC or capillary electrophoresis (CE) are based on xenon lamps for excitation. LIF has been used as detectors in HPLC and CE instruments, when enhanced sensitivity is needed.

2.3.1 MEASUREMENT SETUPS

A schematic overview of a typical measurement setup for HPLC–LIF is shown in Figure 2.2. These detectors are used especially when small sample amounts, for example, in nano- or micro-HPLC, are available.

HPLC–LIF instruments are usually homemade with various different laser types and wavelengths. The main reason for this is that commercial LIF detectors for HPLC are rare and expensive. Moreover, the sensitivity of this system is high and not needed for most routine applications. In addition, the required maintenance for LIF is much more complicated than for xenon lamp-based detector.

Obviously, LIF detectors can only be applied to fluorescing compounds. However, they can also be applied to detection of nonfluorescing compounds, after their derivatization with suitable reagents to become fluorescent.

2.3.2 DERIVATIZATION

The applications of LIF are very similar to applications of conventional fluorescence detectors in HPLC. The compounds that can successfully be analyzed have to emit fluorescence, or they could be reacted with proper reagents that result in fluorescing products. Therefore, the applications are not limited to the small number of fluorescent compounds.

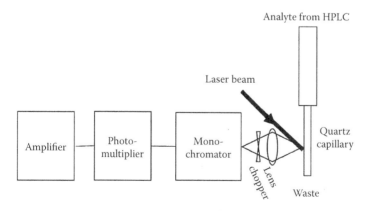

FIGURE 2.2 Block diagram of HPLC–LIF setup, based on the example by Bhat et al.[43]

A large variety of derivatization agents are used in HPLC–LIF. For amino acids and peptide determination, conventional fluorescent reagents such as ortho-phthalaldehyde (OPA),[44] fluorenylmethyloxycarbonyl chloride (FMOC-Cl),[45] 2,3-naphthalenedicarboxaldehyde (NDA),[46] or fluorescein isothiocyanate (FITC)[47,48] have been used. OPA and FMOC-Cl have fluorescence excitation and emission maxima in the UV range. Also more specialized tags such as FITC, and so on, have been used. For derivatization of carboxylic acids, Wu et al.[49] have used 4-nitro-7-piperazinobenzofurazan (NBD-PZ).

Derivatization has also allowed LIF detection of quite unexpected compounds. For example, Wang et al.[50] managed to analyze saccharides by using hydrazine-based derivatization reagents. This analysis is considered complicated for HPLC, because there is neither a good HPLC technique nor a good detector for saccharides. However, labeling with an intensive fluorescence tag helps to overcome both problems simultaneously, making the saccharides more hydrophobic for separation in reversed-phase chromatography and easily detected. Ruhaak et al.[51] have derivatized N-glycans (oligosaccharides) for LIF detection with 8-aminopyrene-1,3,6-trisulfonic acid (APTS). Similarly, also glycan analysis has been achieved with APTS derivatization.[52]

Oyama et al.[53] introduced a new derivatization agent, 4-fluoro-7-nitro-2,1,3-benzoxadiazole (NBD-F), for the analysis of saccharides. Using the example of glycoprotein-derived oligosaccharides, they demonstrated that it is possible to achieve sensitivity approximately 10 times better than LC/MS. Xiong et al.[54] proposed a simultaneous HLPC–LIF/UV/MS method for detection of glycosamine, which made use of derivatization with rhodamine B isothiocyanate (RBITC).

Lately, Zhang et al.[55] introduced a new dual-labeling strategy for simultaneous peptide detection with LIF and inductively coupled plasma MS(ICP-MS); though, ICP–MS still has two orders of magnitude lower detection limits.

2.3.3 Applications of High-Performance Liquid Chromatography/ Capillary Electrophoresis–Laser-Induced Fluorescence

Straightforward applications of HPLC/LIF for naturally fluorescent compounds are possible. One such group of compounds is natural dyes. For example, constituents of red dye from madder plant have been analyzed with CE–LIF technique, differentiating between alizarin, purpurine, carmine, and morine. Also synthetic dyes, such as Safranin T in foodstuff[56] and Phloxine B (used both as a dye and as a pesticide) in coffee beans,[57] have been successfully analyzed.

Also toxins, biomarkers, and so on, have been successfully analyzed with LIF-based chromatographic systems. Arroyo-Manzaneres et al.[58,59] have determined mycotoxine achratoxin A in wine samples. For doing this they have combined a thorough dispersive liquid–liquid extraction procedure with HPLC–LIF analysis.[58] Though several matrix compounds remained in the blank chromatograms after sample preparation, a sufficient separation from matrix peaks for convincing detection was achieved. Ferrari et al.[60] combined LIF with a different type of liquid chromatography: micellar electrokinetic chromatography—for analysis of two biomarkers desmosine and isodesmosine in urine samples. Wu et al.[49] used pressurized capillary electrochromatography in combination with LIF to detect metabolites of cholesterol catabolism: free bile acids, with good repeatability (5.6%) and recovery (above 90%).

A very common application of LIF detection is protein profile determination. As three of the naturally occurring amino acids (tryptophan, tyrosine, and phenylalanine) are also naturally fluorescent, it is possible to determine this kind of profiles both with and without derivatization. The profiles themselves were not very informative, as they have been recorded at one excitation wavelength and one emission wavelength. But the profiles of various tissue samples have been related to indication of different illnesses (mostly cancer).[61–64] The profiles were afterward treated with multivariate statistical approaches based on PCA. The HPLC/LIF protein profile libraries and the concrete classification scheme of such analyses are unfortunately useless for other users worldwide, because they are dependent on the specific chromatographic system, column, eluents, and so on. Therefore, each user must develop a new library and a classification scheme. However, the value of these works relies on demonstrating the ability of differentiating between stages of illness based on HPLC/LIF protein profiles.

2.3.4 COMPARISON OF LASER-INDUCED FLUORESCENCE WITH OTHER CHROMATOGRAPHIC DETECTORS

The performance of LIF detectors have also been compared with other detectors, such as UV, conventional fluorescence, and MS. Ferrari et al.[60] proposed substituting LC/MS method for determination of desmosine and isodesmosine with micellar electrokinetic chromatography in combination with LIF detection. The obtained LODs were approximately two orders of magnitude higher than that for the LC/MS method. However, due to the differences in the chromatographic systems, the results cannot be fully attributed to the detectors. A part of the improved performance is due to the significantly narrower chromatographic peaks in the LC/MS system.[42] Katzenmeyer et al.[65] have also observed that LIF detection may be complementary. They combined LIF and MS detection and observed that LIF provided more reliable quantification, whereas the superior selectivity of the MS allowed characterization of the metabolites (of doxorubicin).

LIF is also advantageous in fast analysis of protein mixtures using HPLC/MS. Post column collection is performed, and the fractions are inspected by UV or LIF for combining them prior to MS identification. This way, the nonrelevant fractions are also removed and not transferred to the MS detector. Chan et al.[66] showed that in this method, LIF provides much more information than UV-based detection and its usage results in more accurate combining of the fractions.

2.4 SIZE EXCLUSION CHROMATOGRAPHY WITH MULTIANGLE LIGHT SCATTERING

Size exclusion chromatography (SEC) is a common method for monitoring the size (molecular weight) of both synthetic and natural polymers. Unfortunately, SEC with UV or similar detectors needs calibration. In this calibration, a function between the retention volume (or time) and the molecular mass of the polymer is established. As the interaction of the stationary phase and the analyte are undesirable in the SEC and the retention time is actually a function of the hydrodynamic volume of the polymer (not its molecular mass), it is very important that the calibration compound would be extremely similar to the analytes. However, achieving this task is almost impossible (because branching significantly influences the calibration function).

Therefore, commonly a multiangle light scattering (MALS) device is used as a detector in the SEC. MALS requires a very narrow wavelength range, and laser sources are ideal for this task. With MALS detector it is possible to accurately measure the molecular weight, without using calibration compounds. In addition, MALS allows simultaneous measurement of the hydrodynamic volume of the polymer, therefore, also indicating the packing density as well as branching of the polymer.[67]

Today, MALS is a common detector in SEC and it has been applied for various sample types. Ammar et al.[68] used SEC–MALS for the analysis of sulfated polysaccharides. They also simultaneously determined the hydrodynamic radii of the polymers and observed that these parameters were significantly dependent on the algae type from which the polymer was obtained. Ono et al.[69] have similarly used SEC–MALS for characterization of cellulose, chitin, and cellulose triacetate. Dahesh et al.[70] have used SEC–MALS to study gluten proteins and to indicate the relation between their hydrodynamic radii and molecular weight. This revealed a much more opened system than expected. Islam et al.[71] have pointed out that MALS is an essential detector for SEC in dendrimer analysis, because for this type of polymer there are no calibration compounds available.

2.5   LASERS FOR CONNECTING CHROMATOGRAPHY AND MASS SPECTROMETRY

Today, MS is an essential detector for all types of chromatography. Thus, considering chromatography detectors without MS would be incomplete. LC/MS is probably the second most common HPLC system used today.[42] Based on the last year's trend, it can be speculated that in the future LC/MS will become the primary solution for routine analyses. However, LC/MS itself is time-consuming and not well compatible with on-site analysis of considerable interest.

TLC method is a good alternative for on-site analysis, because it is compact, fast, producing less waste, and allowing simultaneous monitoring of several samples, while still providing reasonable separations and purification of samples.[72] However, best selectivity is achieved with MS or MS/MS methods.

Lasers are being used for coupling TLC (in which the analyte is in solid form, adsorbed on a surface) and MS (in which the analyte is in the gas phase, in an ionized form). The lasers perform both desorption and ionization of the analyte. In the following section, we shall address these two roles of the lasers.

2.5.1 Laser Ionization in Gas Chromatography/Mass Spectrometry Analysis

Gas chromatography is an extensively used for analysis of volatile compounds and a selective detector such as MS is often desirable for reliable identification of the peaks. However, MS is usually coupled to GC through electron ionization (EI). The kinetic energy of the electrons used in EI is much larger than that required purely for ionization; therefore, the excess of this energy goes for breaking covalent bonds in the molecular ion.[73] This result in large fragmentation of the molecular ion. In some cases, the extensive fragmentation leads to a complete loss of the molecular ions, thus reducing the selectivity and sensitivity.[74] In order to overcome this problem, soft ionization sources, such as single photon ionization (SPI) and multiphoton ionization (MPI), have been proposed and used.[75] They will be discussed next.

2.5.1.1 Single Photon Ionization

SPI is based on irradiating the target molecules with photons of energy larger than their ionization energy. As the ionization energy of most molecules is in the range of 5–15 eV, SPI requires lasers in the vacuum UV (VUV) range. Lasers and also high energy lamps (such as electron beam pumped excimer lamps) have been used for SPI. The application of lasers for SPI/MS is not limited to GC and they were also used in thermal desorption instruments,[76] thermogravimetric instruments,[73] and so on. In all these cases, the softness of SPI is advantageous. This has been demonstrated by Mühlberger et al.,[77] who showed that in cigarette smoke analysis, the EI/MS spectrum is dominated by inorganic oxides and hardly identifiable fragments, whereas several organic smoke components could be easily identified using the SPI/MS spectrum.

GC/SPI/MS has been extensively used, especially in environmental and health applications. Streibel et al.[76] used thermal desorption with SPI/MS (and also resonant multiphoton ionization/MS) for characterization of urban aerosols and ash from spruce wood. The use of soft ionization sources allowed analysis of both aliphatic and carbonylic hydrocarbons, alkanoic acids, and different esters. They succeeded to identify PAH compounds, phenols, guiacol, retene, and other wood combustion indicators in urban air. Even more interesting, chrysene, phenanthrene, and pyrene were also identified. Nowadays, ca. 40% of the urban aerosol compounds are still not identified and coupling SPI to GCMS may contribute toward solving this problem.

Eschner et al.[78] analyzed 14 compounds from cigarette smoke using fast GC/SPI/MS. This challenging analytical issue required both SPI and the combination of GC/MS, because the resolution of one-dimensional analysis is not sufficient. Even more sophisticated approach has been proposed by Welthagen et al.[79] who utilized not only GS/SPI/MS but also GC/GC/SPI/MS for analysis of diesel samples.

The softness of SPI, compared to EI, is especially relevant in MS/MS analyses. If a too large part of the molecular ions undergo fragmentation during ionization, the left parent ions may not be sufficient for sensitive MS/MS analysis. Schramm et al.[80]

demonstrated that SPI can also be used in analysis of warfare compounds with MS/MS, as well as in online monitoring of coffee-roasting process.

2.5.1.2 Multiphoton Ionization

In addition to single photon ionization, it is also possible to use multiple photons, usually two, for ionization. The process is called MPI. The simultaneous absorption of several photons is a rare nonlinear process, which is enhanced by the laser flux and strongly depends on the laser pulse duration (short pulses are preferred). In the classical picture, the first photon excites the molecule to either real or virtual electronic state, and absorption of additional photons may accumulate and reach the ionization threshold. The efficiency of MPI is significantly higher if one of the photons excites the molecule to a real electronic state and these cases are called resonance multiphoton ionization (REMPI). Therefore, the ionization efficiency is higher for compounds that have absorption bands in the laser wavelength region. Usually, lasers from 200 to 350 nm have been used for MPI. In this region, many aromatic compounds absorb light, allowing for their sensitive and selective detection. Many studies indicated the analytical potential and applicability of MPI processes.[81-95] Similarly to SPI, MPI is also considered a soft ionization method.

The majority of the GC/MPI/MS applications have been performed for analysis of PAH compounds. The ionization potential of these compounds is rather low, such that often only two photons in the UV are sufficient for their ionization. The first PAH analysis using GC/MPI/MS already appeared in the 1980s. Sack et al.[75] used the fourth harmonic of a Nd:YAG laser (266 nm) for multiphoton ionization of gasoline samples. They have shown the 3 primary advantages of MPI: (a) moderate selectivity, (b) high ionization efficiency (up to 100%), and (c) reduced fragmentation. They also indicated that MPI method is sensitive to compounds that cannot be detected with EI (e.g., phenanthren).

PAH compounds have also been analyzed by Li et al.[96] who reported subpicogram LOD values for most of the 16 PAHs in the EPA priority list. Watanabe-Ezoe et al.[97] have analyzed polychlorinated dibenzo-p-dioxins and polychlorinated dibenzofurans and soil samples using GC/MPI/MS. They demonstrated that this method provides results that are as good as HRGC/EI/HRMS, even with significantly reduced sample preparation. Li et al.[98] also demonstrated the applicability of GC/MPI/MS for analysis of 49 pesticides in fruits and vegetables. They also observed better LODs using GC/MPI/MS compared to GC/EI/MS. Only the pesticides that do not possess an absorption band at the laser wavelength provide higher LODs.

2.5.1.3 Comparison of Electron Ionization, Single Photon Ionization, and Multiphoton Ionization

Mitschke et al.[99] compared SPI and MPI in MS analysis of diesel samples. They observed that varying the ionization mode and laser wavelength can help in tuning the selectivity of the method. Using a VUV laser for SPI they observed that the spectra were dominated by aliphatic hydrocarbons, but when using MPI, mostly aromatic compounds were observed (Figure 2.3). Tuning the laser wavelength in MPI changes the MS spectra. When irradiating with a 250 nm laser, all aromatic compounds

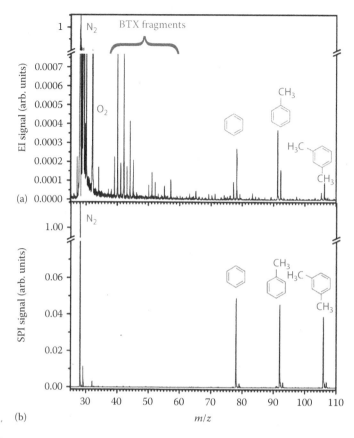

FIGURE 2.3 (a) Electron impact ionization (70 eV) mass spectrum (EI-TOFMS) of a calibration gas mixture containing 10 ppm benzene, toluene, and m-xylene, respectively, in nitrogen (analysis time 1000 ms). The observed detection limit for toluene at the base peak (91 m/z fragment) is 700 ppb. (b) Single photon ionization mass spectrum (EBEL SPI–oaTOFMS, Ar-excimer radiation) of the same calibration gas mixture (analysis time 650 ms). The observed detection limit for toluene at the molecular peak (92 m/z) is 35 ppb. (From Mühlberger, F. et al., *Anal. Chem.*, 79, 8118–8124, 2007.)

were observed. When using 275 nm, benzene and toluene was not observed, and at 300 nm irradiation, only compounds with three or more rings could be identified.

2.5.2 LASER METHODS IN TLC/MALDI/MS

The most common ionization sources for TLC/MS are matrix-assisted laser desorption/ionization (MALDI) and electrospray ionization (ESI). In the following section, we focus on MALDI approach to connect TLC with MS. The resulted method, named TLC/MALDI/MS has a significant impact in analytical chemistry and such instruments are commercially available.

2.5.2.1 Instrumental Methods

MALDI has been originally designed to analyze samples crystallized together with matrix components on a surface.[100] The matrix absorbs laser radiation (usually in the UV range) and a part of this energy is transferred to the analyte. In turn, the analyte desorbs from the solid state to the gas phase and becomes available for MS detection. This method can be utilized for scanning TLC plates, after the chromatographic separation is completed. However, the TLC plates are usually made of a layer of silica particles and do not consist of any matrix components that absorb radiation. Moreover, the laser radiation does not penetrate the TLC plate material.

Therefore, coupling of TLC plates to MS detection is not straightforward. The most laborious solution is dissolving spots from the developed TLC plate and crystallizing them together with matrix components.[72] In addition to being laborious, this approach sacrifices the resolution and is time-consuming. Thus, the advantages of TLC separation are lost. Solving this problem requires direct coupling of TLC and MALDI/MS.

Several strategies have been suggested and examined for direct coupling of TLC and MALDI/MS. In these methods, the analytes are extracted from the stationary phase and brought to the surface, where they are crystallized with matrix components. One popular automatic technique is the spray coating, where the TLC plate is homogeneously sprayed with a solution that contains the matrix components. The solution extracts the analytes from the stationary phase and brings them in contact with the matrix components.[72]

This method works well and is commercially available. Nevertheless, applying the solvent to the TLC plates always results in some degree of peak broadening. Reducing of this peak-broadening effect was achieved by coating the TLC plates with already crystallized matrix.[72]

After exposing the analytes on the TLC plate to the matrix compounds, their scanning with the laser beam is necessary. Different scanning modes have been proposed: (a) scanning at specific previously chosen spots, (b) 1D scanning of the plate in the relevant direction, and (c) 2D scanning of all the plate. The more detailed the scanning is, the more time-consuming it is; however, it provides a more detailed analytical information from the plates.[72]

2.5.2.2 Applications of Thin-Layer Chromatography/Matrix-Assisted Laser Desorption/Ionization/Mass Spectrometry

TLC/MALDI/MS has been applied in a large variety of fields. One example is the analysis of phospholipids, which is very challenging because their MALDI/MS spectrum is usually rich in various adducts, that complicate the interpretation. Batubara et al.[101] and Rohlfing et al.[102] have achieved excellent analysis of phospholipids using TLC/MALDI/MS. Batubara et al. have added chemometric treatment in order to differentiate between metabolomics changes occurring after drug treatment of the phospholipids. Rohlfing et al.[102] succeeded in developing a semiquantitative method for phospholipid analysis in both positive and negative ionization modes. They reached LODs of 10–150 pmol and correlation coefficients of more than 0.9.

Salo et al.[103] have demonstrated the bioanalytical applicability of TLC/MALDI/ MS by analyzing 7 benzodiazepines in human urine and achieving LODs in pico- molar range. They used solid phase extraction (SPE) sample cleanup prior to TLC analysis, in order to keep TLC zones narrow; however, the total analysis time was not significantly shortened. The developed method was also successfully applied to analyze urine from patients who had taken diazepam and two metabolites temazepam and N-desmethyldiazepam were identified. In order to reduce the interfering analyte fragmentation in MALDI/MS, vibrational cooling has been suggested for TLC plate scanning.[104] With this technique, the LOD for gangliosides was reduced to 120 fmol.

Due to the problems with the matrix application, also matrix-free laser desorption ionization (LDI) has been suggested and used[105] for TLC plate analysis. However, this approach is limited to a small range of analytes.

2.5.3 Ambient Laser Ionization for Thin-Layer Chromatography

A set of new ionization sources, called ambient ionization has been introduced and is now often used to minimize or avoid chromatographic separation.[106] A laser beam is used for desorption and ionization of the analyses. Usually, the ionization step follows desorption; however, these processes may take place simultaneously.

The ambient ionization has first been evaluated for a variety of imaging applica- tions, such as imaging of different tissues, biological samples, and so on.[106] These ionization sources are also suitable for TLC plate scanning, as discussed in the following.

Lasers are used in most of these ambient ionization sources. The common methods include electrospray laser desorption and ionization (ELDI [Figure 2.4]), laser-induced acoustic desorption–electrospray ionization (LIAD–ESI), laser ablation electrospray ionization (LAESI), laser desorption spray postionization (LDSPI), laser electrospray mass spectrometry (LEMS), and laser desorption atmospheric pressure chemical ionization (LDAPCI). Although different in name, the principle behind these methods is similar.

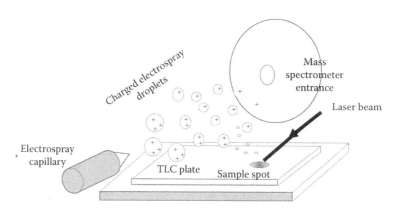

FIGURE 2.4 Schematic overview of ELDI–MS instrumentation for scanning TLC plates.

Compared to MALDI these ionization sources do not require addition of matrix to the TLC plate, which contributes to peak broadening. Therefore, high resolution scanning of the TLC plates is possible, when only the diameter of the laser spot determines the spatial resolution.

Lin et al.[107] used laser desorption coupled to ESI for analysis of dyes, amines, and extracts of drug tablets from both normal and reversed-phase TLC plates. To achieve sufficient selectivity, they also applied MS/MS analysis. LODs were quite high, in the micromolar range. They also showed that TLC/ELDI/MS also helps in identifying compounds even if not fully separated on TLC. However, ionization suppression in the coelution region may occur. Cheng et al.[108] have presented a very effective way to automate analysis of TLC plates, using TLC–ELDI–MS and plate transportation with Lego bricks.

In 2013 Herdering et al.[109] introduced the combination of laser ablation and atmospheric pressure chemical ionization (APCI) source and used it for scanning TLC plates. The method was applied to detect caffeine and acetaminophen in Thomapyrin tablets. Similarly Peng et al.[110] coupled laser desorption (using an IR laser) with APCI source and successfully analyzed two phospholipids, lecithin and sphingomyelin, on TLC plates. They used plain and graphite-coated TLC plates and showed that this coating allowed for reduced laser power. This is of importance because the lower laser power results in less fragmentation of the analytes and also allows for using smaller and less expensive lasers.

2.6 CONCLUSION

In general, lasers possess unique properties that are advantageous in analytical chemistry, and in chromatography, in particular. Most applications are in chromatographic detectors, in which their abilities to induce fluorescence and Raman radiation, as well as soft ionization through multiphoton process, are utilized. However, lasers are also instrumental in sampling of surfaces and ablation and ionizing materials for chromatographic methods coupled with MS. Some of the best LODs in chromatography have been achieved due to the usage of lasers.

It seems that the application of lasers in chromatography always follows new technological development in lasers. Therefore, one can anticipate that two such recent advances in laser technology will also be applied in chromatography in the near future: One is the development of rigid and stable optical parametric oscillators (OPO) lasers, which are now commercially available. They allow for convenient and stable tuning of laser radiation in a wide spectral range, with subnanometer spectral resolution. The other laser technology that recently became ready for analytical applications is the ultrashort pulse lasers. Although further analytical research is needed for the proper introduction of these technologies in chromatography, feasibility has already been proven and success is foreseen.

REFERENCES

1. Dijkstra, R., Ariese, F., Gooijer, C., and Brinkman, U. (2005) Raman spectroscopy as a detection method for liquid-separation techniques. *TrAC Trends Anal. Chem. 24*, 304–323.

2. Carrillo-Carrión, C., Armenta, S., Simonet, B. M., Valcárcel, M., and Lendl, B. (2011) Determination of pyrimidine and purine bases by reversed-phase capillary liquid chromatography with at-line surface-enhanced Raman spectroscopic detection employing a novel SERS substrate based on ZnS/CdSe silver–quantum dots. *Anal. Chem. 83*, 9391–9398.

3. Cabalín, L. M., Rupérez, A., and Laserna, J. J. (1996) Flow-injection analysis and liquid chromatography: Surface-enhanced Raman spectrometry detection by using a windowless flow cell. *Anal. Chim. Acta 318*, 203–210.

4. Cooper, S. D., Robson, M. M., Batchelder, D. N., and Bartle, K. D. (1997) Development of a universal Raman detector for microchromatography. *Chromatographia 44*, 257–262.

5. Sägmüller, B., Schwarze, B., Brehm, G., and Schneider, S. (2001) Application of SERS spectroscopy to the identification of (3,4-methylenedioxy)amphetamine in forensic samples utilizing matrix stabilized silver halides. *The Analyst 126*, 2066.

6. Dou, X., Jung, Y. M., Yamamoto, H., Doi, S., and Ozaki, Y. (1999) Near-infrared excited surface-enhanced Raman scattering of biological molecules on gold colloid I: Effects of pH of the solutions of amino acids and of their polymerization. *Appl. Spectrosc. 53*, 133–138.

7. Huang, R., Han, S., and Li, X. (2013) Detection of tobacco-related biomarkers in urine samples by surface-enhanced Raman spectroscopy coupled with thin-layer chromatography. *Anal. Bioanal. Chem. 405*, 6815–6822.

8. Sägmüller, B., Schwarze, B., Brehm, G., Trachta, G., and Schneider, S. (2003) Identification of illicit drugs by a combination of liquid chromatography and surface-enhanced Raman scattering spectroscopy. *J. Mol. Struct. 661–662*, 279–290.

9. Ni, F., Thomas, L., and Cotton, T. M. (1989) Surface-enhanced resonance Raman spectroscopy as an ancillary high-performance liquid chromatography detector for nitrophenol compounds. *Anal. Chem. 61*, 888–894.

10. Cowcher, D. P., Jarvis, R., and Goodacre, R. (2014) Quantitative online liquid chromatography-surface-enhanced Raman scattering of purine bases. *Anal. Chem. 86*, 9977–9984.

11. Cabalin, L. M., Ruperez, A., and Laserna, J. J. (1993) Surface-enhanced Raman spectrometry for detection in liquid chromatography using a windowless flow cell. *Talanta 40*, 1741–1747.

12. Voigtman, E., Jurgensen, A., and Winefordner, J. D. (1981) Comparison of laser excited fluorescence and photoacoustic limits of detection for static and flow cells. *Anal. Chem. 53*, 1921–1923.

13. Xu, J., Yang, B.-C., Tian, H.-Z., and Guan, Y.-F. (2006) A windowless flow cell-based miniaturized fluorescence detector for capillary flow systems. *Anal. Bioanal. Chem. 384*, 1590–1593.

14. Nirode, W. F., Devault, G. L., Sepaniak, M. J., and Cole, R. O. (2000) On-column surface-enhanced Raman spectroscopy detection in capillary electrophoresis using running buffers containing silver colloidal solutions. *Anal. Chem. 72*, 1866–1871.

15. Lee, P. C. and Meisel, D. (1982) Adsorption and surface-enhanced Raman of dyes on silver and gold sols. *J. Phys. Chem. 86*, 3391–3395.

16. Roth, E. and Kiefer, W. (1994) Surface-enhanced Raman spectroscopy as a detection method in gas chromatography. *Appl. Spectrosc. 48*, 1193–1195.

17. Lee, S. J. and Moskovits, M. (2011) Visualizing chromatographic separation of metal ions on a surface-enhanced Raman active medium. *Nano Lett. 11*, 145–150.

18. Kuligowski, J., Quintás, G., Garrigues, S., Lendl, B., de la Guardia, M., and Lendl, B. (2010) Recent advances in on-line liquid chromatography–infrared spectrometry (LC-IR). *TrAC Trends Anal. Chem. 29*, 544–552.

19. Boelens, H. F. M., Dijkstra, R. J., Eilers, P. H. C., Fitzpatrick, F., and Westerhuis, J. A. (2004) New background correction method for liquid chromatography with diode array detection, infrared spectroscopic detection and Raman spectroscopic detection. *J. Chromatogr. A 1057*, 21–30.
20. Zheng, B., Liu, Y., Li, D., Chai, Y., Lu, F., and Xu, J. (2015) Hydrophobic-hydrophilic monolithic dual-phase layer for two-dimensional thin-layer chromatography coupled with surface-enhanced Raman spectroscopy detection: Other Techniques. *J. Sep. Sci. 38*, 2737–2745.
21. Gao, F., Hu, Y., Chen, D., Li-Chan, E. C. Y., Grant, E., and Lu, X. (2015) Determination of Sudan I in paprika powder by molecularly imprinted polymers–thin layer chromatography–surface enhanced Raman spectroscopic biosensor. *Talanta 143*, 344–352.
22. Takei, H., Saito, J., Kato, K., Vieker, H., Beyer, A., and Gölzhäuser, A. (2015) TLC-SERS plates with a built-in SERS layer consisting of cap-shaped noble metal nanoparticles intended for environmental monitoring and food safety assurance. *J. Nanomater. 2015*, 1–9.
23. Herman, K., Mircescu, N. E., Szabo, L., Leopold, L. F., Chiş, V., and Leopold, N. (2013) In situ silver spot preparation and on-plate surface-enhanced Raman scattering detection in thin layer chromatography separation. *J. Appl. Spectrosc. 80*, 311–314.
24. Chen, Y., Cheng, H., Tram, K., Zhang, S., Zhao, Y., Han, L., Chen, Z., and Huan, S. (2013) A paper-based surface-enhanced resonance Raman spectroscopic (SERRS) immunoassay using magnetic separation and enzyme-catalyzed reaction. *The Analyst 138*, 2624.
25. Chieli, A., Sanyova, J., Doherty, B., Brunetti, B. G., and Miliani, C. (2016) Chromatographic and spectroscopic identification and recognition of ammoniacal cochineal dyes and pigments. *Spectrochim. Acta A. Mol. Biomol. Spectrosc. 162*, 86–92.
26. Wang, C., Cheng, F., Wang, Y., Gong, Z., Fan, M., and Hu, J. (2014) Single point calibration for semi-quantitative screening based on an internal reference in thin layer chromatography-SERS: The case of Rhodamine B in chili oil. *Anal. Methods 6*, 7218.
27. Freye, C. E., Crane, N. A., Kirchner, T. B., and Sepaniak, M. J. (2013) Surface enhanced Raman scattering imaging of developed thin-layer chromatography plates. *Anal. Chem. 85*, 3991–3998.
28. Fang, F., Qi, Y., Lu, F., and Yang, L. (2016) Highly sensitive on-site detection of drugs adulterated in botanical dietary supplements using thin layer chromatography combined with dynamic surface enhanced Raman spectroscopy. *Talanta 146*, 351–357.
29. Pozzi, F., Shibayama, N., Leona, M., and Lombardi, J. R. (2013) TLC-SERS study of Syrian rue (*Peganum harmala*) and its main alkaloid constituents: TLC-SERS study of alkaloids from Syrian rue. *J. Raman Spectrosc. 44*, 102–107.
30. Yang, L., Li, P., Liu, H., Tang, X., and Liu, J. (2015) A dynamic surface enhanced Raman spectroscopy method for ultra-sensitive detection: From the wet state to the dry state. *Chem. Soc. Rev. 44*, 2837–2848.
31. Yan, X., Li, P., Yang, L., and Liu, J. (2016) Time-dependent SERS spectra monitoring the dynamic adsorption behavior of bipyridine isomerides combined with bianalyte method. *The Analyst 141*, 5189–5194.
32. Woodhead, A. L., Cosgrove, B., and Church, J. S. (2016) The purple coloration of four late 19th century silk dresses: A spectroscopic investigation. *Spectrochim. Acta. A. Mol. Biomol. Spectrosc. 154*, 185–192.
33. Brosseau, C. L., Gambardella, A., Casadio, F., Grzywacz, C. M., Wouters, J., and Van Duyne, R. P. (2009) Ad-hoc surface-enhanced Raman spectroscopy methodologies for the detection of artist dyestuffs: Thin layer chromatography-surface enhanced Raman spectroscopy and in situ on the fiber analysis. *Anal. Chem. 81*, 3056–3062.

34. Geiman, I., Leona, M., and Lombardi, J. R. (2009) Application of Raman spectroscopy and surface-enhanced Raman scattering to the analysis of synthetic dyes found in ballpoint pen inks. *J. Forensic Sci. 54*, 947–952.

35. Zhu, Q., Cao, Y., Cao, Y., Chai, Y., and Lu, F. (2014) Rapid on-site TLC–SERS detection of four antidiabetes drugs used as adulterants in botanical dietary supplements. *Anal. Bioanal. Chem. 406*, 1877–1884.

36. Lv, D., Cao, Y., Lou, Z., Li, S., Chen, X., Chai, Y., and Lu, F. (2015) Rapid on-site detection of ephedrine and its analogues used as adulterants in slimming dietary supplements by TLC-SERS. *Anal. Bioanal. Chem. 407*, 1313–1325.

37. Li, H., xia Zhu, Q., sian Chwee, T., Wu, L., feng Chai, Y., Lu, F., and fang Yuan, Y. (2015) Detection of structurally similar adulterants in botanical dietary supplements by thin-layer chromatography and surface enhanced Raman spectroscopy combined with two-dimensional correlation spectroscopy. *Anal. Chim. Acta 883*, 22–31.

38. Lucotti, A., Tommasini, M., Casella, M., Morganti, A., Gramatica, F., and Zerbi, G. (2012) TLC–surface enhanced Raman scattering of apomorphine in human plasma. *Vib. Spectrosc. 62*, 286–291.

39. Li, D., Qu, L., Zhai, W., Xue, J., Fossey, J. S., and Long, Y. (2011) Facile on-site detection of substituted aromatic pollutants in water using thin layer chromatography combined with surface-enhanced Raman spectroscopy. *Environ. Sci. Technol. 45*, 4046–4052.

40. Yao, C., Cheng, F., Wang, C., Wang, Y., Guo, X., Gong, Z., Fan, M., and Zhang, Z. (2013) Separation, identification and fast determination of organophosphate pesticide methidathion in tea leaves by thin layer chromatography–surface-enhanced Raman scattering. *Anal. Methods 5*, 5560.

41. Zhang, Z.-M., Liu, J.-F., Liu, R., Sun, J.-F., and Wei, G.-H. (2014) Thin layer chromatography coupled with surface-enhanced Raman scattering as a facile method for on-site quantitative monitoring of chemical reactions. *Anal. Chem. 86*, 7286–7292.

42. Kruve, A., Rebane, R., Kipper, K., Oldekop, M.-L., Evard, H., Herodes, K., Ravio, P., and Leito, I. (2015) Tutorial review on validation of liquid chromatography–mass spectrometry methods: Part I. *Anal. Chim. Acta 870*, 29–44.

43. Bhat, S., Patil, A., Rai, L., Kartha, V. B., and Santhosh, C. (2010) Protein profile analysis of cellular samples from the cervix for the objective diagnosis of cervical cancer using HPLC-LIF. *J. Chromatogr. B 878*, 3225–3230.

44. Sánchez, B. Á., Capote, F. P., and de Castro, M. L. (2011) Targeted analysis of sphingoid precursors in human biofluids by solid-phase extraction with in situ derivatization prior to μ-LC-LIF determination. *Anal. Bioanal. Chem. 400*, 757–765.

45. Chan, K. C., Janini, G. M., Muschik, G. M., and Issaq, H. J. (1993) Laser-induced fluorescence detection of 9-fluorenylmethyl chloroformate derivatized amino acids in capillary electrophoresis. *J. Chromatogr. A 653*, 93–97.

46. Vyas, C. A., Rawls, S. M., Raffa, R. B., and Shackman, J. G. (2011) Glutamate and aspartate measurements in individual planaria by rapid capillary electrophoresis. *J. Pharmacol. Toxicol. Methods 63*, 119–122.

47. He, Y., Zhao, L., Yuan, H., Xu, Z., Tang, Y., Xiao, D., and Choi, M. M. F. (2011) HPLC with in-capillary optical fiber laser-induced fluorescence detection of picomolar amounts of amino acids by precolumn fluorescence derivatization with fluorescein isothiocyanate. *Chromatographia 74*, 541–547.

48. Jiang, Y., Nikolau, B., and Ma, Y. (2010) Separation and quantification of short-chain coenzyme A in plant tissues by capillary electrophoresis with laser-induced fluorescence detection. *Anal. Methods 2*, 1900.

49. Wu, Y., Wang, X., Wu, Q., Wu, X., Lin, X., and Xie, Z. (2010) Separation and determination of structurally related free bile acids by pressurized capillary electrochromatography coupled to laser induced fluorescence detection. *Anal. Methods 2*, 1927.

50. Wang, C., Gao, M., Huang, Z., and Zhang, X. (2013) Characterization of saccharide using high fluorescent 5-(((2-(carbohydrazino)methyl)thio)acetyl)-aminofluorescein tag by Capillary-HPLC-LIF and MALDI-TOF-MS. *Talanta 117*, 229–234.
51. Ruhaak, L. R., Hennig, R., Huhn, C., Borowiak, M., Dolhain, R. J. E. M., Deelder, A. M., Rapp, E., and Wuhrer, M. (2010) Optimized workflow for preparation of APTS-labeled N-glycans allowing high-throughput analysis of human plasma glycomes using 48-channel multiplexed CGE-LIF. *J. Proteome Res. 9*, 6655–6664.
52. Hamm, M., Wang, Y., and Rustandi, R. (2013) Characterization of N-linked glycosylation in a monoclonal antibody produced in NS0 cells using capillary electrophoresis with laser-induced fluorescence detection. *Pharmaceuticals 6*, 393–406.
53. Oyama, T., Yodohsi, M., Yamane, A., Kakehi, K., Hayakawa, T., and Suzuki, S. (2011) Rapid and sensitive analyses of glycoprotein-derived oligosaccharides by liquid chromatography and laser-induced fluorometric detection capillary electrophoresis. *J. Chromatogr. B 879*, 2928–2934.
54. Xiong, B., Wang, L.-L., Li, Q., Nie, Y.-T., Cheng, S.-S., Zhang, H., Sun, R.-Q., Wang, Y.-J., and Zhou, H.-B. (2015) Parallel microscope-based fluorescence, absorbance and time-of-flight mass spectrometry detection for high performance liquid chromatography and determination of glucosamine in urine. *Talanta 144*, 275–282.
55. Zhang, Z., Yan, X., Xu, M., Yang, L., and Wang, Q. (2011) A dual-labelling strategy for integrated ICPMS and LIF for the determination of peptides. *J. Anal. At. Spectrom. 26*, 1175.
56. Su, Z., Zhai, H., Chen, Z., Zhou, Q., Li, J., and Liu, Z. (2014) Molecularly imprinted solid-phase extraction monolithic capillary column for selective extraction and sensitive determination of safranine T in wolfberry. *Anal. Bioanal. Chem. 406*, 1551–1556.
57. Zhai, H., Su, Z., Chen, Z., Liu, Z., Yuan, K., and Huang, L. (2015) Molecularly imprinted coated graphene oxide solid-phase extraction monolithic capillary column for selective extraction and sensitive determination of phloxine B in coffee bean. *Anal. Chim. Acta 865*, 16–21.
58. Arroyo-Manzanares, N., Gámiz-Gracia, L., and García-Campaña, A. M. (2012) Determination of ochratoxin A in wines by capillary liquid chromatography with laser induced fluorescence detection using dispersive liquid–liquid microextraction. *Food Chem. 135*, 368–372.
59. Arroyo-Manzanares, N., García-Campaña, A. M., and Gámiz-Gracia, L. (2011) Comparison of different sample treatments for the analysis of ochratoxin A in wine by capillary HPLC with laser-induced fluorescence detection. *Anal. Bioanal. Chem. 401*, 2987–2994.
60. Ferrari, F., Fumagalli, M., Piccinini, P., Stolk, J., Luisetti, M., Viglio, S., Tinelli, C., and Iadarola, P. (2012) Micellar electrokinetic chromatography with laser induced detection and liquid chromatography tandem mass-spectrometry-based desmosine assays in urine of patients with chronic obstructive pulmonary disease: A comparative analysis. *J. Chromatogr. A 1266*, 103–109.
61. Patil, A., Prabhu, V., Choudhari, K. S., Unnikrishnan, V. K., George, S. D., Ongole, R., Pai, K. M. et al. (2010) Evaluation of high-performance liquid chromatography laser-induced fluorescence for serum protein profiling for early diagnosis of oral cancer. *J. Biomed. Opt. 15*, 67007.
62. Patil, A., Bhat, S., Pai, K. M., Rai, L., Kartha, V. B., and Chidangil, S. (2015) Ultra-sensitive high performance liquid chromatography–laser-induced fluorescence based proteomics for clinical applications. *J. Proteomics 127*, 202–210.
63. Bhat, S., Patil, A., Rai, L., Kartha, V. B., and Chidangil, S. (2012) Application of HPLC combined with laser induced fluorescence for protein profile analysis of tissue homogenates in cervical cancer. *Sci. World J. 2012*, 1–7.

64. Kumar, K. K., Chowdary, M. V. P., Mathew, S., Rao, L., Krishna, C. M., and Kurien, J. (2009) Protein profile study of breast-tissue homogenates by HPLC-LIF. *J. Biophotonics 2*, 313–321.
65. Katzenmeyer, J. B., Eddy, C. V., and Arriaga, E. A. (2010) Tandem laser-induced fluorescence and mass spectrometry detection for high-performance liquid chromatography analysis of the in vitro metabolism of doxorubicin. *Anal. Chem. 82*, 8113–8120.
66. Chan, K. C., Veenstra, T. D., and Issaq, H. J. (2011) Comparison of fluorescence, laser-induced fluorescence, and ultraviolet absorbance detection for measuring HPLC fractionated protein/peptide mixtures. *Anal. Chem. 83*, 2394–2396.
67. Podzimek, S. (2014) Truths and myths about the determination of molar mass distribution of synthetic and natural polymers by size exclusion chromatography. *J. Appl. Polym. Sci. 131*, 40111–40120.
68. Ammar, H. H., Lajili, S., Said, R. B., Le Cerf, D., Bouraoui, A., and Majdoub, H. (2015) Physico-chemical characterization and pharmacological evaluation of sulfated polysaccharides from three species of Mediterranean brown algae of the genus Cystoseira. *DARU J. Pharm. Sci. 23*, 1.
69. Ono, Y., Ishida, T., Soeta, H., Saito, T., and Isogai, A. (2016) Reliable dn/dc values of cellulose, chitin, and cellulose triacetate dissolved in LiCl/N,N-dimethylacetamide for molecular mass analysis. *Biomacromolecules 17*, 192–199.
70. Dahesh, M., Banc, A., Duri, A., Morel, M.-H., and Ramos, L. (2014) Polymeric assembly of gluten proteins in an aqueous ethanol solvent. *J. Phys. Chem. B 118*, 11065–11076.
71. Islam, M. T., Shi, X., Balogh, L., and Baker, J. R. (2005) HPLC separation of different generations of poly(amidoamine) dendrimers modified with various terminal groups. *Anal. Chem. 77*, 2063–2070.
72. Srivastava, M. (Ed.). (2011) *High-Performance Thin-Layer Chromatography (HPTLC)*. Springer Science & Business Media, Berlin, Germany.
73. Geissler, R., Saraji-Bozorgzad, M. R., Gröger, T., Fendt, A., Streibel, T., Sklorz, M., Krooss, B. M. et al. (2009) Single photon ionization orthogonal acceleration time-of-flight mass spectrometry and resonance enhanced multiphoton ionization time-of-flight mass spectrometry for evolved gas analysis in thermogravimetry: Comparative analysis of crude oils. *Anal. Chem. 81*, 6038–6048.
74. Skoog, D. A., Holler, F. J., and Crouch, S. R. (2007) *Principles of Instrumental Analysis* 6th ed. Thomson Brooks/Cole, Belmont, CA.
75. Sack, T. M., McCrery, D. A., and Gross, M. L. (1985) Gas chromatography multiphoton ionization Fourier transform mass spectrometry. *Anal. Chem. 57*, 1290–1295.
76. Streibel, T., Weh, J., Mitschke, S., and Zimmermann, R. (2006) Thermal Desorption/ Pyrolysis coupled with photoionization time-of-flight mass spectrometry for the analysis of molecular organic compounds and oligomeric and polymeric fractions in urban particulate matter. *Anal. Chem. 78*, 5354–5361.
77. Mühlberger, F., Saraji-Bozorgzad, M., Gonin, M., Fuhrer, K., and Zimmermann, R. (2007) Compact ultrafast orthogonal acceleration time-of-flight mass spectrometer for on-line gas analysis by electron impact ionization and soft single photon ionization using an electron beam pumped rare gas excimer lamp as VUV-light source. *Anal. Chem. 79*, 8118–8124.
78. Eschner, M. S., Selmani, I., Gröger, T. M., and Zimmermann, R. (2011) Online comprehensive two-dimensional characterization of puff-by-puff resolved cigarette smoke by hyphenation of fast gas chromatography to single-photon ionization time-of-flight mass spectrometry: Quantification of hazardous volatile organic compounds. *Anal. Chem. 83*, 6619–6627.

79. Welthagen, W., Mitschke, S., Mühlberger, F., and Zimmermann, R. (2007) One-dimensional and comprehensive two-dimensional gas chromatography coupled to soft photo ionization time-of-flight mass spectrometry: A two- and three-dimensional separation approach. *J. Chromatogr. A 1150*, 54–61.

80. Schramm, E., Kürten, A., Hölzer, J., Mitschke, S., Mühlberger, F., Sklorz, M., Wieser, J. et al. (2009) Trace detection of organic compounds in complex sample matrixes by single photon ionization ion trap mass spectrometry: Real-time detection of security-relevant compounds and online analysis of the coffee-roasting process. *Anal. Chem. 81*, 4456–4467.

81. Schechter, I., Schroeder, H., and Kompa, K.L. (1992) Quantitative laser mass-analyzer by time resolution of the ion-induced voltage in multi-photon-ionization processes. *Anal. Chem. 64*, 2787–2796.

82. Gridin, V. V., Korol, A., Bulatov, V., and Schechter, I. (1996) Laser two-photon ionization of pyrene on contaminated soils. *Anal. Chem. 68*, 3359–3363.

83. Gridin, V. V., Bulatov, V., Korol, A., and Schechter, I. (1997) Particulate material analysis by a laser ionization fast conductivity method. Water content Effects. *Anal. Chem. 69*, 478–484.

84. Gridin, V. V., Litani-Barzilai, I., Kadosh, M., and Schechter, I. (1997) A renewable liquid droplet method for on-line pollution analysis by multi-photon ionization. *Anal. Chem. 69*, 2098–2102.

85. Gridin, V. V., Litani-Barzilai, I., Kadosh, M., and Schechter, I. (1998) Determination of aqueous solubility and surface adsorption of polycyclic aromatic hydrocarbons by laser multiphoton ionization. *Anal. Chem. 70*, 2685–2692.

86. Inoue, T., Gridin, V. V., Ogawa, T., and Schechter, I. (1998) Simultaneous laser-induced multiphoton ionization and fluorescence for analysis of polycyclic aromatic hydrocarbons. *Anal. Chem. 70*, 4333–4338.

87. Spurný, K. (Ed.). (1999) Analytical Chemistry of Aerosols. Lewis, Boca Raton, FL.

88. Gridin, V. V., Inoue, T., Ogawa, T., and Schechter, I. (2000) On-line screening of airborne PAH contamination by simultaneous multiphoton ionization and laser induced fluorescence. *Instrum. Sci. Technol. 28*, 131–141.

89. Litani-Barzilai, I., Fisher, M., Gridin, V. V., and Schechter, I. (2001) Fast filter-sampling and analysis of PAH aerosols by laser multiphoton ionization. *Anal. Chim. Acta 439*, 1–8.

90. Litani-Barzilai, I., Bulatov, V., Gridin, V. V., and Schechter, I. (2004) Detector based on time-resolved ion-induced voltage in laser multiphoton ionization and laser-induced fluorescence. *Anal. Chim. Acta 501*, 151–156.

91. Vinerot, N., Chen, Y., Gridin, V. V., Bulatov, V., Feller, L., and Schechter, I. (2010) A novel method for direst nondestructive diagnosis of caries affected tooth surfaces by laser multiphoton ionization. *Instrum. Sci. Technol. 38*, 143–150.

92. Chen, Y., Bulatov, V., Vinerot, N., and Schechter, I. (2010) Multiphoton ionization spectroscopy as a diagnostic technique of surfaces under ambient conditions. *Anal. Chem. 82*, 3454–3456.

93. Vinerot, N., Chen, Y., Bulatov, V., Gridin, V. V., Fun-Young, V., and Schechter, I. (2011) Spectral characterization of surfaces using laser multi-photon ionization. *Opt. Mater. 34*, 329–335.

94. Tang, S., Vinerot, N., Fisher, D., Bulatov, V., Yavetz-Chen, Y., and Schechter, I. (2016) Detection and mapping of trace explosives on surfaces under ambient conditions using multiphoton electron extraction spectroscopy (MEES). *Talanta 155*, 235–244.

95. Tang, S., Vinerot, N., Bulatov, V., Yavetz-Chen, Y., and Schechter, I. (2016) Multiphoton electron extraction spectroscopy and its comparison with other spectroscopies for direct detection of solids under ambient conditions. *Anal. Bioanal. Chem. 408*, 8037–8051.

96. Li, A., Uchimura, T., Tsukatani, H., and Imasaka, T. (2010) Trace analysis of polycyclic aromatic hydrocarbons using gas chromatography-mass spectrometry based on nanosecond multiphoton ionization. *Anal. Sci. 26*, 841–846.

97. Watanabe-Ezoe, Y., Li, X., Imasaka, T., Uchimura, T., and Imasaka, T. (2010) Gas chromatography/femtosecond multiphoton ionization/time-of-flight mass spectrometry of dioxins. *Anal. Chem. 82*, 6519–6525.

98. Li, A., Imasaka, T., Uchimura, T., and Imasaka, T. (2011) Analysis of pesticides by gas chromatography/multiphoton ionization/mass spectrometry using a femtosecond laser. *Anal. Chim. Acta 701*, 52–59.

99. Mitschke, S., Welthagen, W., and Zimmermann, R. (2006) Comprehensive gas chromatography–time-of-flight mass spectrometry using soft and selective photoionization techniques. *Anal. Chem. 78*, 6364–6375.

100. Cole, R. B. (Ed.). (2010) *Electrospray and MALDI Mass Spectrometry: Fundamentals, Instrumentation, Practicalities, and Biological Applications* 2nd ed. Wiley, Hoboken, NJ.

101. Batubara, A., Carolan, V. A., Loadman, P. M., Sutton, C., Shnyder, S. D., and Clench, M. R. (2015) Thin-layer chromatography/matrix-assisted laser desorption/ionisation mass spectrometry and matrix-assisted laser desorption/ionisation mass spectrometry imaging for the analysis of phospholipids in LS174T colorectal adenocarcinoma xenografts treated with: TLC/MALDI-MS and MALDI-MSI of DMXAA-treated xenografts. *Rapid Commun. Mass Spectrom. 29*, 1288–1296.

102. Rohlfing, A., Müthing, J., Pohlentz, G., Distler, U., Peter-Katalinić, J., Berkenkamp, S., and Dreisewerd, K. (2007) IR-MALDI-MS analysis of HPTLC-separated phospholipid mixtures directly from the TLC plate. *Anal. Chem. 79*, 5793–5808.

103. Salo, P. K., Vilmunen, S., Salomies, H., Ketola, R. A., and Kostiainen, R. (2007) Two-dimensional ultra-thin-layer chromatography and atmospheric pressure matrix-assisted laser desorption/ionization mass spectrometry in bioanalysis. *Anal. Chem. 79*, 2101–2108.

104. Ivleva, V. B., Elkin, Y. N., Budnik, B. A., Moyer, S. C., O'Connor, P. B., and Costello, C. E. (2004) Coupling thin-layer chromatography with vibrational cooling matrix-assisted laser desorption/ionization fourier transform mass spectrometry for the analysis of ganglioside mixtures. *Anal. Chem. 76*, 6484–6491.

105. Shariatgorji, M., Spacil, Z., Maddalo, G., Cardenas, L. B., and Ilag, L. L. (2009) Matrix-free thin-layer chromatography/laser desorption ionization mass spectrometry for facile separation and identification of medicinal alkaloids. *Rapid Commun. Mass Spectrom. 23*, 3655–3660.

106. Huang, M.-Z., Cheng, S.-C., Cho, Y.-T., and Shiea, J. (2011) Ambient ionization mass spectrometry: A tutorial. *Anal. Chim. Acta 702*, 1–15.

107. Lin, S.-Y., Huang, M.-Z., Chang, H.-C., and Shiea, J. (2007) Using electrospray-assisted laser desorption/ionization mass spectrometry to characterize organic compounds separated on thin-layer chromatography plates. *Anal. Chem. 79*, 8789–8795.

108. Cheng, S.-C., Huang, M.-Z., Wu, L.-C., Chou, C.-C., Cheng, C.-N., Jhang, S.-S., and Shiea, J. (2012) Building blocks for the development of an interface for high-throughput thin layer chromatography/ambient mass spectrometric analysis: A green methodology. *Anal. Chem. 84*, 5864–5868.

109. Herdering, C., Reifschneider, O., Wehe, C. A., Sperling, M., and Karst, U. (2013) Ambient molecular imaging by laser ablation atmospheric pressure chemical ionization mass spectrometry: Laser ablation atmospheric pressure chemical ionization for MS imaging. *Rapid Commun. Mass Spectrom. 27*, 2595–2600.

110. Peng, S., Edler, M., Ahlmann, N., Hoffmann, T., and Franzke, J. (2005) A new interface to couple thin-layer chromatography with laser desorption/atmospheric pressure chemical ionization mass spectrometry for plate scanning. *Rapid Commun. Mass Spectrom. 19*, 2789–2793.

3 Hafnia and Zirconia Chromatographic Materials for the Enrichment of Phosphorylated Peptides

Stefan Vujcic, Lisandra Santiago-Capeles,
Karina M. Tirado-González,
Amaris C. Borges-Muñoz, and Luis A. Colón

CONTENTS

3.1 PREAMBLE BY LUIS A. COLÓN

This volume of the *Advances in Chromatography* book series is dedicated to the memory of Prof. Eli Grushka, who sadly passed early in the spring of 2016. We join the other authors of this volume in tribute to Eli's life and his immense contributions to the field of separation science, which include the coeditorship of this book series for many years. After his postdoctoral work with J. Calvin Giddings, Eli joined the faculty in the Department of Chemistry at the State University of New York at Buffalo (UB), where he stayed for about 10 years. He was a full professor before leaving UB to join The Hebrew University of Jerusalem.

Coincidentally, I currently hold the faculty line that Eli occupied (i.e., separation scientist) in the Department of Chemistry at UB. I met Eli Grushka years later, when he was on a sabbatical at Stanford University in the lab of Prof. Richard N. Zare at a social gathering in a scientific conference of former members of the Zare research group (I conducted my postdoctoral research with Zare). We had great discussions, including separation science, of course, and about the times Eli had in Buffalo, NY and the people he knew at UB. He told me then that Buffalo was his *alma mater* regarding his independent research in academia. Although sporadically, we kept in contact with each other since then.

Eli's contributions to the advancement of chromatographic and electrophoretic separations were many. One of his latest interests was on the evaluation of the adsorptive properties of new separation media [1–3]. In his honor, my research group provides a contribution to this volume of *Advances in Chromatography* that relates to the adsorptive properties of Group IV metal oxides toward phosphorylated peptides. Eli was an outstanding separation scientist. He will be greatly missed.

3.2 INTRODUCTION

The major challenge in proteomics is to isolate and identify a protein from a given organism, and then evaluate its post-translational modifications to gain insight into the protein's function [4]. A difficulty in correlating protein expression with a given biological function is that biological activity is a function of protein phosphorylation state. The phosphorylation modification works similar to an *on/off* switch in a protein, changing its conformation, activity, and modes of interactions [5–6], making protein phosphorylation a key process in many biological pathways [5]. Phosphorylation is reversible and is known to occur on serine, threonine, and tyrosine amino acid residues in a protein [4–5]. This process is one of the most prevalent intracellular protein modifications responsible for numerous cellular processes, such as cell differentiation, signaling, metabolism, migration, and apoptosis [7–8]. Approximately 30%–50% of proteins in a cell are phosphorylated at any given time and space [5–6]; however, determining the site of phosphorylation is a challenging task [9]. The high complexity of the proteome and the very low concentration of the phosphoproteins in complex samples make their detection very difficult [10]. In many instances, it is critical to develop and include an enrichment step in the phosphoproteomic workflow to separate phosphorylated from nonphosphorylated peptides (or proteins) [5,10] prior to chemical analysis, which is commonly performed through mass spectrometry (MS).

Different approaches for the selective isolation and/or enrichment of phosphopeptides have been developed. Traditional methods include labeling of the protein with the radioisotope ^{32}P and Edman degradation chemistry on phosphopeptides to localize the site of phosphorylation [7]. Other common isolation techniques include immobilized metal ion affinity chromatography (IMAC) [8,11], immunoprecipitation using phosphoprotein-specific antibodies [12], or specific chemical modification strategies targeted for phosphorylated amino acids [13]. Of these methods, IMAC purification has been used most frequently. Very recently, Lin and Deng reported the successful use of immobilized Hf^{4+} in IMAC for phosphopeptide enrichment [14]. The IMAC approach, however, has several disadvantages. The enrichment and recovery of phosphopeptides are strongly dependent on the type of metal ion used, column material, and the loading/eluting procedures utilized. It may have low molecular mass contaminants, and often it is a labor-intensive procedure [15–16]. Identification and localization of protein phosphorylation are mostly performed by means of MS, as it can be appreciated from the immense literature. MS offers high sensitivity, selectivity, and speed of analysis as compared to most biochemical techniques [4,17–19]. Electrospray ionization (ESI) and matrix-assisted laser desorption/ionization (MALDI) are ionization techniques that have had the greatest impact on biological MS and in the field of proteomics [5]. However, difficulties in MS analysis of phosphopeptides still remain due to the low ionization and detection efficiencies in MALDI and ESI [20]. Ion suppression effects could have an adverse effect in proteomics investigations [21]. Therefore, isolation and enrichment of phosphorylated peptides prior to MS are very important to increase signal intensities and obtain more accurate results.

3.2.1 Enrichment of Phosphopeptides by Group IV Metal Oxides

A metal oxide affinity chromatography (MOAC) utilizing metal oxides from the Group IV of the periodic table of the elements is an alternative way to purify/isolate phosphopeptides. The metal oxide materials titania (TiO_2) and zirconia (ZrO_2) have been utilized for this purpose [22–28]. Titania and zirconia have been used in both off-line [22,24,27] and online [23,29–31] chromatographic applications. The Lewis acid sites at the surface of the metal oxides are the key for unique affinity in the isolation and enrichment of phosphorylated compounds. Although it has been reported that zirconia complements titania by having more affinity toward monophosphorylated peptides [22], there is also evidence that zirconia behaves very similar to titania, with less affinity toward monophosphorylated species [31].

Hafnia is another Group IV metal oxide, which has neither been explored much for chromatography nor for the potential enrichment/isolation of phosphorylated peptides in the field of proteomics until recently. Our research group has synthesized chromatographic monoliths of hafnia [32] and tested the material for the enrichment of phosphopeptides [28]. The hafnia material was synthesized in-house as a monolithic structure using sol gel processing by reacting $HfCl_4$ with N-methylformamide and propylene oxide to form a gel. The gel was subsequently aged and submitted to a temperature treatment of 700°C to form the solid piece of metal oxide. After fabricating the hafnia material, it was ground and adapted for use in a micropipette tip format. Initial testing using ESI–MS showed that hafnia was able to successfully enrich both mono- and tetra-phosphorylated peptides without any apparent bias. Zirconia from a commercial source, however, appeared to show bias toward tetraphosphorylated peptides.

Others have also explored the use of hafnia material for the isolation of phosphorylated peptides. Blacken et al. [26], coated plasma-treated stainless steel surfaces with the oxides of zirconium, titanium, and hafnium by reactive landing of gasphase ions produced through ESI from the metal alcoxides in 1-propanol solutions. The oxide-coated stainless steel surfaces were then used for *in situ* enrichment of phosphopeptides prior to analysis by MALDI–MS. Enrichment of mono- and diphos-phorylated peptides was achieved using the prepared materials. Under their experi-mental conditions, zirconia proved to be more effective than hafnia and showed that the enrichment is proportional to the thickness of zirconia on the plate.

In the work by Nelson, et al. [27], mesoporous hafnia and zirconia materials were synthesized and analyzed by linear trap quadropole (LTQ) MS and LTQ/FTICR hybrid MS. The metal oxide materials were synthesized by adding Pluronic F127, ethanol, $HfCl_4$, or $ZrCl_4$ and NH_4OH under stirring and allowed to age for 3 days at 40°C. The sols were heated in a furnace (in air) ramped from room temperature to 360°C over 6 h and held 2 h at 360°C. The mesoporous material was then crushed using mortar and pestle and used for enrichment. It was reported that choosing the pH and additives of the solutions used during the binding, washing, and elution steps are essential to obtain specific enrichment of phosphorylated peptides to maximize the removal of interfering species. They achieved detection of 20 and 21 phosphopeptides after enrichment with zirconia and hafnia, respectively. Under the tested conditions, hafnia appeared to be more effective for the enrichment of multiphosphorylated peptides when compared to zirconia. Both metal oxides, however, had comparable affinity toward monophosphorylated peptides.

Graphene–hafnia, graphene–titania, and graphene–zirconia composites have been synthesized by means of a hydrothermal and/or sol gel reactions followed by a temperature treatment [25]. The synthesized composite materials were tested for the enrichment of phosphorylated peptides by MALDI–MS. It was reported that the graphene–hafnia composite was more effective for enrichment than the other composites, not only on the amount of phosphopeptides but also exhibited better selectivity toward mono- and multiphosphorylated peptides, particularly for tetraphosphorylated ones.

As a general observation, it appears that the enrichment properties of the metal oxides toward phosphorylated peptides are influenced by how the materials are synthesized. In Sections 3.4 and 3.5, we evaluate the isolation and enrichment characteristics of hafnia and zirconia materials synthesized in our laboratory toward phosphorylated peptides. Using β-casein as a model phosphorylated protein, it is shown that the enrichment of phophopeptides is influenced by the heat treatment used during the synthesis/processing of the metal oxides.

3.3   TRYPTIC DIGEST AND PHOSPHOPEPTIDE ENRICHMENT PROCEDURES

β-Casein (from bovine) has been used as a model protein in phosphoproteomics because it is well known that it contains five phosphorylated serine residues [33]. When the protein is digested with trypsin, both monophosphorylated and tetraphos-phorylated peptides are released. This makes it a very good model to use in the study of the adsorptive characteristics of metal oxides toward mono- and multiphosphory-lated peptides. In our laboratory, the tryptic digestion has been performed using a readily available procedure described previously [34]. The tryptic digest (8 nmol) is diluted in 25 μL of a solution consisting of 300 mg/mL dihydroxybenzoic acid (DHB) in ACN/0.1% TFA (80/20). The phosphopeptides are then enriched using micropipette tips containing approximately 1 mg of zirconia or hafnia materials. After sample uptake, elution of nonphosphorylated peptides is accomplished by rinses with 300 mg/mL DHB in acetonitrile/0.1%TFA (80/20), 20 μL (2 times), and acetonitrile/0.1% TFA (80/20), 20 μL (5 times); the enriched phosphopeptides are then eluted with 0.25% ammonium hydroxide in water (pH 11). After the phosphorylated species are eluted, they are dried under a flow of nitrogen.

The dried peptides are reconstituted in 75 μL of 0.1% ammonium acetate solution (1:1:2, NH$_4$OH:H$_2$O:MeOH, v/v/v) prior to ESI–MS analysis in the negative ion mode. For MALDI-TOF-MS analysis, the dried peptides have been reconstituted in 20 μL of sterile water and desalted according to manufacturer instructions using PerfectPure C-18 Tip (Eppendorf, Hamburg, Germany). Eluted peptides are directly deposited onto the MALDI plate and subsequently spotted with the matrix. The samples are dried at room temperature. Different matrices have been used for phosphopeptides analysis, including α-cyano-4-hydroxycinnamic acid, 2,5-dihydroxybenzoic acid, and 2′,4′,6′-trihydroxyacetophenone, among many others [35–37]. We have used 20 mg/mL DHB in 50:50 (v/v) ACN:H$_2$O with 1% phosphoric acid because it has been reported that phosphoric acid as an additive enhances the phosphopeptide ions signal [38].

3.4 HAFNIA AND ZIRCONIA ADSORPTIVE MATERIALS

Commercial products of zirconia and titania are readily available but not for hafnia. A simple approach to synthesize hafnia and zirconia oxides is through sol-gel processing, which we have reported in the literature [28,32]. Briefly, an aqueous solution of the metal chloride (~1.3 M) is reacted with eight molar equivalents of N-methylformamide. Seven molar equivalents of propylene oxide are added to induce the polycondensation reactions leading to polymerization that renders a gel within 30–120 s. Typically, the gels are allowed to age at 50°C before undergoing solvent exchanges with water and acetone. The solids are dried at room temperature and then submitted to a desired temperature treatment at 300°C or 1100°C (see below) for at least 18 hours. The materials can be easily crushed and a determined quantity adapted to a micropipette tip.

3.5 ENRICHMENT OF BOVINE β-CASEIN PHOSPHOPEPTIDES ON ZIRCONIA AND HAFNIA MATERIALS

In a study comparing the enrichment capabilities of titania (TiO$_2$ NuTip^{TM}) and zirconia (ZrO$_2$ NuTip^{TM}) microtips for isolation of phosphopeptides, it was concluded that the titania microtips showed higher selectivity toward multiphosphorylated peptide species, whereas zirconia tips enriched primarily monophosphorylated peptides [22]. The authors suggested that the selectivity of zirconia for monophosphorylated peptides can be attributed to either the higher acidity of zirconia or the higher coordination number as compared to titania. There have also been contradicting reports suggesting that zirconia behaves more similar to titania [16,31]. We hypothesized that the zirconia material utilized in these reports may have been different in nature, possibly due to the processing of the material during synthesis that can alter material's surface properties. This hypothesis is explored herein by evaluating the isolation/enrichment properties of zirconia and hafnia materials prepared in our laboratory as a function of preparation conditions; specifically, subjecting the metal oxides to various temperature treatments during their preparation process and observing the effects on the enrichment patterns using trypsin proteolytic digests of bovine β-casein and MS analysis.

3.5.1 ELECTROSPRAY IONIZATION—MASS SPECTROMETRIC ANALYSIS

An ESI–MS spectrum obtained from a β-casein tryptic digest sample that was not submitted to metal oxide enrichment is shown in Figure 3.1. ESI–MS was operated in negative ion mode with a source voltage of 3.5 kV and capillary voltage of −17 V using nitrogen as the sheath gas. Product ion spectra were recorded using a scan range between m/z 500 and 2000 with collision energy of 35% using helium. Due to the ionization mechanism of ESI, multiply charged peptide molecular ions are produced and detected through MS. The spectrum in Figure 3.1 is predominated by peaks corresponding to nonphosphorylated compounds with some low-intensity phosphorylated peaks present. The mass of various phosphopeptide sequences from β-Casein tryptic digest observed by ESI–MS are shown in Table 3.1; these were identified by means of the proteolytic peptide database generated with the *Peptide Mass Tool* on the Expasy website (http://ca.expasy.org/) [39]. Each peptide has been labeled as α_x (where x is a designated number for each charged peptide ion). The same tryptic digest sample mixture submitted to the enrichment procedure on the metal oxides produced cleaner spectra with prominent intensities for the phosphorylated peptides (see Figures 3.2 and 3.3).

Zirconia materials prepared at two different temperatures during the final stages of their synthetic procedure (i.e., 300°C and 500°C) were used to adsorb peptides from a β-casein tryptic digest sample. The MS spectra of the species retained and eluted

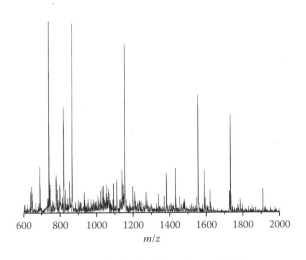

FIGURE 3.1 ESI-mass spectrum of a bovine β-casein tryptic digest.

TABLE 3.1

Molecular Masses of Phosphopeptides in Bovine β-Casein Identified in the Proteolytic Peptide Database Using the ESI–MS Technique

m/z	Charge	Monoisotopic Mass	Phosphorylation: Site	Sequence	ESI Label
1030.4	−2	2060.8	1P:50	48–63	α_1
1040.4	−3	3121.2	4P:30,32,33,34	16–40	α_2
1278.6	−2	2557.2	1P:50	44–63	α_3
1482.6	−2	2965.2	4P:30,32,33,34	17–40	α_4
1560.6	−2	3121.2	4P:30,32,33,34	16–40	α_5

Source: Artimo, P. et al., *Nucleic Acids Res.*, 40, W597, 2012.

from these materials are shown in Figure 3.2a and b. For comparison, a commercially available product (ZrO$_2$ NuTip™) was also used for enrichment of the phosphorylated peptides of the same β-casein tryptic digest sample and the MS spectrum of this analysis is shown in Figure 3.2c. The spectrum in Figure 3.2a, where zirconia treated at 300°C was used, shows peaks corresponding to the monophosphorylated (2060.8 Da) and tetraphosphorylated peptides (3121.2 Da) of β-casein. The similar peak intensities suggest that there is no discrimination toward mono- or tetraphosphorylated peptides during the enrichment process. MS/MS analysis of the known phosphorylated species at m/z of 1030, 1040, and 1560 showed the successive loses of 98 or 49 Da masses depending on the charge of the peptide. Figure 3.2b, which shows the spectrum of the tryptic digest enriched with zirconia treated at 500°C, shows a different pattern. Comparing Figure 3.2a and b, one can notice differences in the peak intensities for the mono- and tetraphosphorylated species. First, the relative abundance of the tetraphosphorylated peptide predominates with respect to monophosphorylated peptides in the spectrum shown in Figure 3.2b. This observation suggests that the heating treatment used during the preparation of the zirconia material has an effect on the enrichment selectivity. The MS spectrum of the sample enriched with the commercially available zirconia material (ZrO$_2$ NuTip™) depicted in Figure 3.2c showed a similar enrichment pattern to the in-house zirconia treated at 300°C. In other words, they both have similar adsorptive properties (i.e., selectivity) toward mono- and tetra phosphorylated species.

Figure 3.3 shows ESI mass spectra of the species from a tryptic digest of bovine β-casein retained and eluted from hafnia materials processed at two different temperatures during synthesis. The spectrum in Figure 3.3a corresponds to the phosphopeptides enriched on hafnia treated at 300°C, whereas Figure 3.3b shows enrichment on hafinia treated at 1100°C. Mono-(α_1 and α_3) and tetra-(α_3, α_4, and α_5) phosphorylated peptides are present in the spectra. However, the spectrum of the sample that used the hafnia treated at the higher temperature clearly displays a selective discrimination in favor of the monophosphorylated peptides. The intensity of the tetraphosphorylated species has been lowered by a factor of more than two, which illustrates the selectivity of hafnia toward monophosphorylated peptides [28].

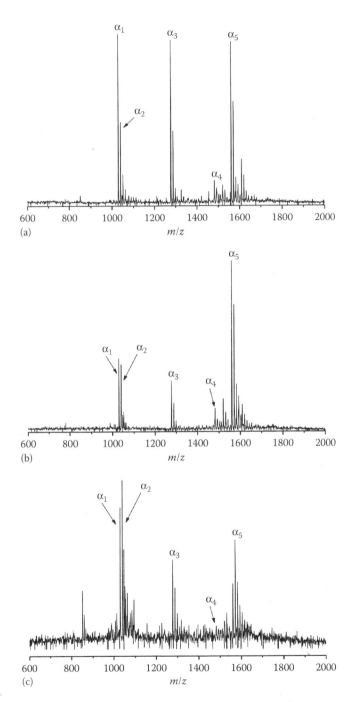

FIGURE 3.2 ESI-mass spectra of the components in a bovine β-casein tryptic digest retained in and eluted from (i.e., enriched) zirconia materials: (a) zirconia treated at 300°C, (b) zirconia treated at 500°C, and (c) commercial NuTip™ ZrO₂.

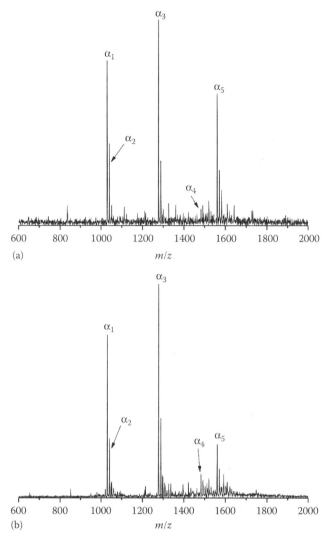

FIGURE 3.3 ESI-mass spectra of the components in a bovine β-casein tryptic digest retained in and eluted from (i.e., enriched) hafnia materials processed at (a) 300°C and (b) 1100°C during their synthesis.

3.5.2 Matrix-Assisted Laser Desorption/Ionization-Time-of-Flight-Mass Spectrometric Analysis

MALDI causes little or no fragmentation of the compounds being analyzed by MS [5], and similar to ESI, it belongs to the soft ionization techniques in MS. MALDI coupled with a time-of-flight (TOF) mass analyzer has been extensively used for proteomics investigations, including the study of posttranslational modifications such as phosphorylation [5,9,35–37,40–44]. MALDI–MS has had such a great impact in

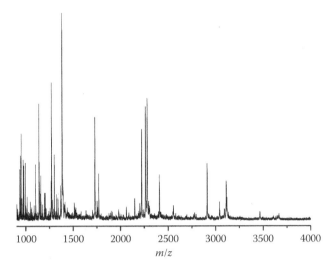

FIGURE 3.4 MALDI-mass spectrum of a β-casein protein tryptic digest.

proteomic analysis, in part, because of its ability to generate mass spectra that are considerably less complex than those obtained by ESI–MS; typically, one peak corresponding to each charge state of the ionized molecule. MALDI combined with TOF–MS is an excellent tool that provides for the detection of a very wide range of masses [45]. We have used MALDI–TOF operated in the linear positive ion mode to study and confirm phosphopeptide enrichment patterns by hafnia and zirconia adsorptive materials. Even when positive ion mode does not favor the detection of highly acidic peptides, it tends to have better signal-to-noise ratio relative to negative ion mode [42]. In addition, the positive ion mode can provide information that may be complementary to ESI–MS.

Figure 3.4 depicts the MALDI–MS spectrum for a bovine β-casein tryptic digest sample without treatment/enrichment by metal oxides. The spectrum is dominated mostly by nonphosphorylated peptides. Table 3.2 contains a list of phosphopeptide sequences for β-casein identified in the proteolitic peptide database as analyzed by MALDI–MS [39]. Three phosphopeptides are observed: a set of two

TABLE 3.2

Molecular Masses of Phosphopeptides in Bovine β-Casein Identified in the Proteolytic Peptide Database Using the MALDI–MS Technique

Monoisotopic Mass	Phosphorylation: Site	Sequence	MALDI Label
2060.8	1P:50	48–63	β_1
2557.2	1P:50	44–63	β_2
3121.2	4P:30,32,33,34	16–40	β_3

Source: Artimo, P. et al., *Nucleic Acids Res.*, 40, W597, 2012.

monophosphorylated peptides of monoisotopic masses 2060.8 Da (β1) and 2557.2 Da (β2), and a tetraphosphorylated peptide with a mass of 3121.2 Da (β3).

Figure 3.5a and b show MALDI-mass spectra of bovine β-casein tryptic digests after enrichment and elution from zirconia materials that had been processed at 300°C (a) and 500°C (b) during their synthesis. Both, monophosphorylated and tetraphosphorylated peptides can be identified on the spectra. The peptides labeled β_1 and β_2 contain one phosphoserine, which is typically straightforward to detection by MALDI–MS [38]. The tetraphosphorylated peptide, labeled β_3, is challenging to detect because of its size (it contains four phosphoserines) and it is slightly hydrophobic in nature, which tends to be lost during sample preparation [38]. Interestingly, this tetraphosphorylated peptide is very abundant in both spectra, indicating the effectiveness of zirconia in isolating/enriching this phosphopeptide. There is an apparent presence of non-phosphorylated species below ~1500 Da, which have also been reported in the literature for other zirconia type of materials [38]. However, the intensity of the peaks corresponding to the phosphorylated peptides is remarkably different between the two spectra in Figure 3.5, indicating that the temperature processing of the material during its preparation has an effect on its adsorptive properties toward Phosphopeptides. Figure 3.5b underscores the selective preference toward the enrichment of the tetraphosphorylated peptide by the zirconia material that had been processed at 500°C. This is in agreement with the findings observed by ESI–MS (see Figure 3.2b).

One can consider the enrichment patterns of the commercially available zirconia (ZrO$_2$ NuTip™) and titania (TiO$_2$ NuTip™) materials and then compare such with our in-house prepared metal oxides. Figure 3.6a and b show the MALDI-mass spectra when using these for enrichment of phosphopeptides in β-casein tryptic digest. All three phosphorylated peptides can easily be identified in each spectrum, indicating that both, zirconia and titania NuTip™ materials, can isolate the monophosphorylated and the tetraphosphorylated peptides from the tryptic digest. The patterns resemble that of Figure 3.5a when the in-house zirconia prepared at 300°C was used for enrichment, although the relative abundance of the β_3 peak is somewhat reduced, which may indicate a slight selectivity toward the monophosphorylated species by the zirconia NuTip™. This suggests that both the commercially available zirconia tip and our in-house zirconia process at 300°C behaved in a similar fashion. As shown in Figure 3.6b, the MALDI-mass spectrum indicates similar phosphopeptide enrichment behaviors between the zirconia NuTip™ and the titania NuTip™. Unfortunately, we do not have information regarding the temperature treatment used to manufacture the commercially available products.

The MALDI-mass spectra of peptides from bovine β-casein tryptic digest enriched by means of hafnia prepared at 300°C (a) and 1100°C (b) are illustrated in Figure 3.7. When using the hafnia produced at 300°C for enrichment (Figure 3.7a), the monophosphorylated species, β_1 and β_2, are present at a higher intensity than the tetraphosphorylated peptide β_3 but all species are easily identified. The hafnia material processed at 1100°C, on the other hand, showed a remarkable degree of discrimination in favor of the monophosphorylated peptides (Figure 3.6b). In fact, the signal intensity of β_3 was almost indistinguishable from the baseline noise. This is a clear indication that hafnia material prepared at 1100°C is exceedingly selective toward monophosphorylated peptides.

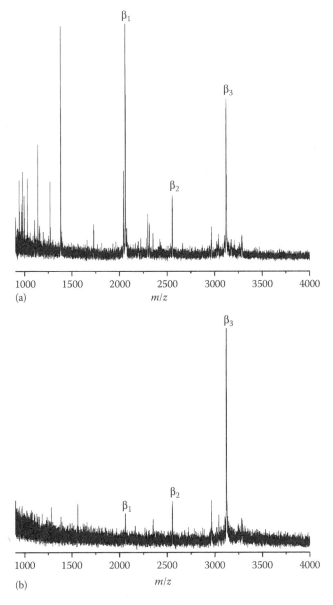

FIGURE 3.5   MALDI-mass spectra of bovine β-casein tryptic digest peptides after enrichment and elution from zirconia materials that had been processed at (a) 300°C and (b) 500°C during their synthesis.

Figure 3.8 illustrates the intensity of the β_3 tetraphosphorylated peptide obtained from the MALDI-mass spectra normalized by the intensities of monophosphorylated peptides (i.e., β_1 and β_2) for the digest samples enriched by various metal oxides. The graph shows the selectivity of zirconia (300°C and 500°C), hafnia (300°C and 1100°C), and commercial NuTip™ materials toward tetraphosphorylated peptides

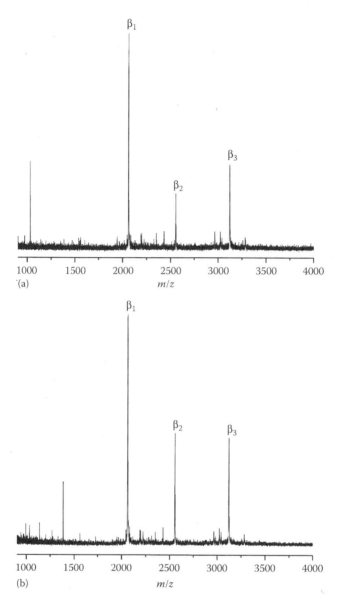

FIGURE 3.6 MALDI-mass spectra of bovine β-casein tryptic digest peptides after enrichment and elution from (a) zirconia NuTip™ and (b) titania NuTip™.

in the tryptic digests of β-casein. From Figure 3.8, one can notice that enrichment microtips containing hafnia prepared at 1100°C and zirconia prepared at 500°C are complementary to each other. Hafnia processed at 1100°C can almost exclusively enrich monophosphorylated species; whereas zirconia processed at 500°C is selective toward tetraphosphorylated species. The zirconia- and titania-enrichment

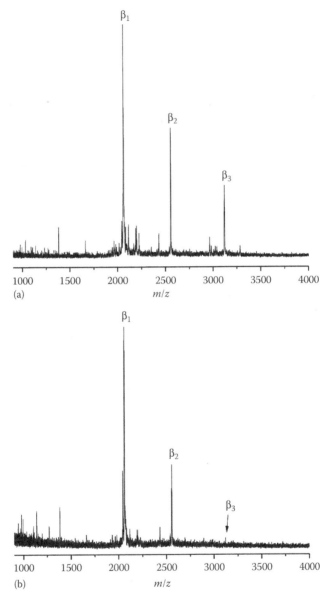

FIGURE 3.7   MALDI-mass spectra of bovine β-casein tryptic digest peptides after enrichment and elution from hafnia materials that had been processed at (a) 300°C and (b) 1100°C
during their synthesis.

microtips from commercial sources showed essentially the same enrichment
patterns for β-casein phosphopeptides, with somewhat similar selectivity to the
in-house materials prepared at 300°C. This is in contrast with a previous study
indicating that zirconia and titania had preferential selectivity in the enrichment of
monophosphorylated and multiphosphorylated peptides, respectively [22].

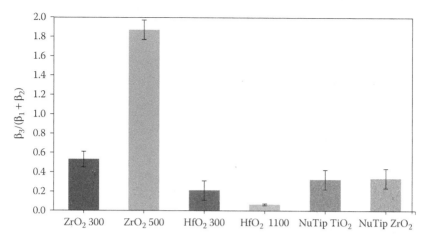

FIGURE 3.8 Peak intensity of β_3 (tetraphosphorylated peptide) normalized by the intensities of β_1 and β_2 (monophosphorylated peptides) from the MALDI-mass spectra obtained after enrichment and elution from various metal oxides. The 300, 500, and 1000 labels refer to the temperature treatment used to prepare the materials. The NuTip™ refers to the commercially available materials.

3.6   FINAL REMARKS

The study of phosphorylation in a given protein is a challenging task, in part, because of the low concentration of phosphorylated peptides, and although major phosphorylation sites might be located easily, minor sites may be very difficult to identify [18]. Isolation/enrichment of phosphopeptides by means of chromatographic retention on Group IV metal oxide materials followed by proper elution and mass spectrometric analysis provide an effective approach to identify phosphorylated peptides in proteomics research. There have been reported discrepancies in the selectivity of metal oxides toward the adsorption of phosphorylated peptides. Such discrepancies, even for the same metal oxide, can be attributed to: (1) the forms of the different materials (e.g., micro-, nano-, and mesoporous), (2) enrichment condition, including different binding, washing, and eluting buffers, and (3) MS ionization and tuning conditions [27]. Regarding this proposition, it should be pointed out that neither the enrichment conditions nor the samples that are used to evaluate the enrichment properties have been standardized [25–28]. Furthermore, expanding on the forms of the material, it is important to highlight that the temperature utilized in the preparation/treatment of the material influences the crystalline structure of the metal oxide, influencing the material surface characteristics [46]. We have evaluated and compared the isolation/enrichment characteristics of several Group IV metal oxide materials using bovine β-casein as a model protein and found that by processing the metal oxides (i.e., hafnia and zirconia) at different temperatures during their synthesis influences the enrichment selectivity toward phosphorylated peptides. When zirconia and hafnia were treated at 300°C (amorphous phase for both materials), the data suggest no discrimination toward mono- or tetraphosphorylated peptides.

Interestingly, when hafnia was processed at 1100°C (monoclinic crystalline phase), it showed an almost exclusive enrichment/isolation characteristic toward monophosphorylated peptides, making it more selective than the commercially available materials tested. On the other hand, zirconia material processed at 500°C (tetragonal crystalline phase) showed to a large extent preferential enrichment toward tetraphosphorylated peptides. Not only the in-house materials complement each other, the data also suggest that using specific processing of the material during synthesis can provide for tunable adsorptive/enrichment characteristics.

ACKNOWLEDGMENT

We acknowledge the financial support by the U.S. National Science Foundation (CHE-1508105 and CHE-0554677) for the research work in our laboratory. Any opinions, findings, and conclusions or recommendations expressed in this material are those of the authors and not necessarily reflect the views of the National Science Foundation.

REFERENCES

1. M. Pumera, J. Wang, E. Grushka, O. Lev, *Talanta, 72*: 711 (2007).
2. D. Benhaim, E. Grushka, *J. Chromatogr. A, 1217*: 65 (2010).
3. D. Benhaim, E. Grushka, *J. Liq. Chromatogr. Related Technol., 31*: 2198 (2008).
4. E. Salih, *Mass Spectrom. Rev., 24*: 828 (2005).
5. A. Kraj, J. Silberring, *Proteomics: Introduction to Methods and Applications.* John Wiley & Sons: Hoboken, NJ, 2008.
6. C. Piggee, *Anal. Chem., 81*: 2418 (2009).
7. J. X. Yan, N. H. Packer, A. A. Gooley, K. L. Williams, *J. Chromatogr. A, 808*: 23 (1998).
8. A. Gruhler, J. V. Olsen, S. Mohammed, P. Mortensen, N. J. Faergeman, M. Mann, O. N. Jensen, *Mol. Cell. Proteomics, 4*: 310 (2005).
9. J. Dong, H. Zhou, R. Wu, M. Ye, H. Zou, *J. Sep. Sci., 30*: 2917 (2007).
10. M. Rainer, H. Sonderegger, R. Bakry, C. W. Huck, S. Morandell, L. A. Huber, D. T. Gjerde, G. K. Bonn, *Proteomics, 8*: 4593 (2008).
11. D. C. A. Neville, C. R. Rozanas, E. M. Price, D. B. Gruis, A. S. Verkman, R. R. Townsend, *Protein Sci., 6*: 2436 (1997).
12. S. Ficarro, O. Chertihin, V. A. Westbrook, F. White, F. Jayes, P. Kalab, J. A. Marto et al., *J. Biol. Chem., 278*: 11579 (2003).
13. D. T. McLachlin, B. T. Chait, *Anal. Chem., 75*: 6826 (2003).
14. H. Lin, C. Deng, *Talanta, 149*: 91 (2016).
15. S. B. Ficarro, M. L. McCleland, P. T. Stukenberg, D. J. Burke, M. M. Ross, J. Shabanowitz, D. F. Hunt, F. M. White, *Nat. Biotechnol., 20*: 301 (2002).
16. S. S. Jensen, M. R. Larsen, *Rapid Commun. Mass Spectrom., 21*: 3635 (2007).
17. R. S. Annan, M. J. Huddleston, R. Verma, R. J. Deshaies, S. A. Carr, *Anal. Chem., 73*: 393 (2001).
18. M. Mann, S.-E. Ong, M. Gronborg, H. Steen, O. N. Jensen, A. Pandey, *Trends Biotechnol., 20*: 261 (2002).
19. D. T. McLachlin, B. T. Chait, *Curr. Opin. Chem. Biol., 5*: 591 (2001).
20. J. Gropengiesser, B. T. Varadarajan, H. Stephanowitz, E. Krause, *J. Mass Spectrom., 44*: 821 (2009).

21. R. Mallet Claude, Z. Lu, R. Mazzeo Jeff, *Rapid Commun. Mass Spectrom., 18*: 49 (2004).
22. H. K. Kweon, K. Hkansson, *Anal. Chem., 78*: 1743 (2006).
23. M. R. Larsen, T. E. Thingholm, O. N. Jensen, P. Roepstorff, T. J. D. Jorgensen, *Mol. Cell. Proteomics, 4*: 873 (2005).
24. H.-C. Hsieh, C. Sheu, F.-K. Shi, D.-T. Li, *J. Chromatogr. A, 1165*: 128 (2007).
25. X. Huang, J. Wang, J. Wang, C. Liu, S. Wang, *RSC Advances, 5*: 89644 (2015).
26. G. R. Blacken, M. Volny, M. Diener, K. E. Jackson, P. Ranjitkar, D. J. Maly, F. Turecek, *J. Am. Soc. Mass Spectrom., 20*: 915 (2009).
27. C. A. Nelson, J. R. Szczech, C. J. Dooley, Q. Xu, M. J. Lawrence, H. Zhu, S. Jin, Y. Ge, *Anal. Chem., 82*: 7193 (2010).
28. J. G. Rivera, Y. S. Choi, S. Vujcic, T. D. Wood, L. A. Colon, *Analyst, 134*: 31 (2009).
29. M. W. H. Pinkse, S. Mohammed, J. W. Gouw, B. van Breukelen, H. R. Vos, A. J. R. Heck, *J. Proteome Res., 7*: 687 (2008).
30. G. T. Cantin, T. R. Shock, S. K. Park, H. D. Madhani, J. R. Yates, III, *Anal. Chem., 79*: 4666 (2007).
31. M. Cuccurullo, G. Schlosser, G. Cacace, L. Malorni, G. Pocsfalvi, *J. Mass Spectrom., 42*: 1069 (2007).
32. D. C. Hoth, J. G. Rivera, L. A. Colon, *J. Chromatogr., A, 1079*: 392 (2005).
33. B. R. Dumas, G. Brignon, F. Grosclaude, J. C. Mercier, *Eur. J. Biochem., 20*: 264 (1971).
34. Y. S. Choi, T. D. Wood, *Rapid Commun. Mass Spectrom., 22*: 1265 (2008).
35. K. Janek, H. Wenschuh, M. Bienert, E. Krause, *Rapid Commun. Mass Spectrom., 15*: 1593 (2001).
36. T. Nabetani, K. Miyazaki, Y. Tabuse, A. Tsugita, *Proteomics, 6*: 4456 (2006).
37. A. Tholey, *Rapid Commun. Mass Spectrom., 20*: 1761 (2006).
38. S. Kjellstroem, O. N. Jensen, *Anal. Chem., 76*: 5109 (2004).
39. P. Artimo, M. Jonnalagedda, K. Arnold, D. Baratin, G. Csardi, E. de Castro, S. Duvaud et al., *Nucleic Acids Research, 40*: W597 (2012).
40. M. Dong, M. Wu, F. Wang, H. Qin, G. Han, J. Dong, R. A. Wu, M. Ye, Z. Liu, H. Zou, *Anal. Chem., 82*: 2907 (2010).
41. J. T. Du, Y. M. Li, Y. F. Zhao, M. Nakagawa, X. R. Qin, T. Nemoto, H. Nakanishi, *Chin. Chem. Lett., 15*: 927 (2004).
42. H. Zhang, C. Zhang, G. A. Lajoie, K. K. C. Yeung, *Anal. Chem., 77*: 6078 (2005).
43. P. M. Bigwarfe, T. D. Wood, *Int. J. Mass Spectrom., 234*: 185 (2004).
44. A. Navare, M. Zhou, J. McDonald, F. G. Noriega, M. C. Sullards, F. M. Fernandez, *Rapid Commun. Mass Spectrom., 22*: 997 (2008).
45. E. De Hoffman, J. Charette, V. Stroobant, (Eds.), *Mass Spectrometry: Principles and Applications.* Wiley: Chichester, 1996, 340 pp.
46. M. Yashima, T.-A. Kato, M. Kakihana, M. A. Gulgun, Y. Matsuo, M. Yoshimura, *J. Mater. Res., 12*: 2575 (1997).

4 Making Top–Down Sequencing of All/Any Proteins a Reality. How Might This Be Accomplished?

Ira S. Krull and Jared R. Auclair

CONTENTS

4.1 INTRODUCTION AND BACKGROUND

In what might be termed Part 1 of this discussion on the current state and future evolvement of top–down sequencing (TDS) for proteins, we described how the methods and instrumentation have evolved in recent years to the current status quo [1]. In brief, TDS has evolved to include many different forms of molecular ion fragmentation from the original, high-energy collision dissociation (HCD) or collision-induced dissociation (CID), to include today's electron transfer dissociation (ETD), electron capture dissociation (ECD), high-energy electron capture dissociation (HecD), infrared multiphoton dissociation (IRMPD), ultraviolet photodissociation (UVPD), and probably others. In addition, the mass spectrometers have included Fourier transform MS (FTMS), Obitraps, Quadrupole-Time of Flight (Q-ToF), and other, perhaps homemade variants.

Though the most recent applications of UVPD appear to have the greatest success in percent sequencing realized for many proteins, it is not yet successful with proteins >50 kDa or monoclonal antibodies (mAbs) or polyclonal antibodies (pAbs). In addition, in many instances, lower molecular weight (MW) proteins, that is, <50 kDa often do not realize sequencing approaches to 100%. On the other hand, in general, the lower the MW, the greater the percent sequencing success. With peptides, those below 10 kDa almost always exhibit very close to 100% success rates [2–9]. What is it about these polyamino acids, that is, as they grow in size and conformations, having secondary and tertiary structures, if not higher order complexes, that they then defy being efficiently (100%) sequenced by one or any combination of the aforementioned fragmentation routes? In addition, why can we not seemingly understand what we are doing that has not led to 100% coverage for any and all proteins of any MWs? Despite the often held, but somewhat unsubstantiated belief that current TDS methods are 100% successful, the facts belie this wishful thinking.

Thus, in order to achieve 100% sequence coverage of any protein, of any size, in TDS, one first must understand the nature of the protein in the gas phase. At first glance, one would expect a protein to unfold in the gas phase, due to the lack of the hydrophobic effect and the enhancement of long-range, Coulombic interactions [10]. However, in practice, it is likely that low charge state proteins remain partially folded, whereas high charge state proteins remain largely unfolded. In fact, it is possible that proteins in the gas phase fold such that polar residues are on the inside and hydrophobic residues reside on the outside (the exact opposite than folding in solution) [10]. In a vacuum, the folding landscape is similar to that in the solution phase (Figure 4.1); however, apparently it is much easier for proteins to get trapped in local energy minima [10]. In particular, the lower charge state (purple) proteins are more likely to get trapped in these energy minima, rather than the higher, charged proteins which lead to the spaghetti ball analogy, described in more details in Section 4.2.

4.2 THE SPAGHETTI BALL ANALOGY

Several years ago, McLafferty began studying the possible application of TDS for all proteins but quickly realized, along with others, that this was not an easy row to hoe [11–12]. He never did really succeed in being able to sequence all the proteins

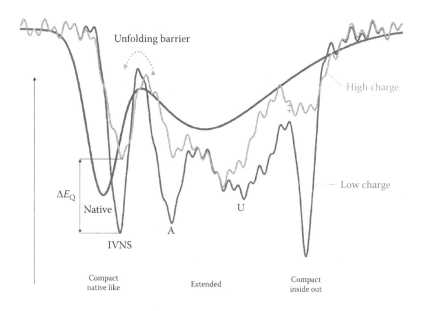

FIGURE 4.1   Folding landscape in the gas phase. Purple is the plot of low, charge state proteins; brown is plot of high, charge state proteins. The low-charge state proteins are more likely to get stuck in an unfolding, local minima. (Reprinted from Meyer, T. et al., *WIRES Comput. Mol. Sci.*, 3, 408–425, 2013. With permission.)

Limitations

1. Proteins more than 30 kDa are difficult to analyze.

   - Tertiary and secondary structures interfere with fragmentation.

   - Spaghetti ball analogy.

2. High detection limit.

3. Does not work on all proteins.

FIGURE 4.2   Limitations already existing in doing successful TDS of larger proteins. (Reprinted from G.T. Hermanson, *Bioconjugate Techniques*, 3rd ed., Academic Press, Elsevier, Oxford, UK, 2013. With permission of the copyright holder, Neil Kelleher, Northwestern University, Evanston, IL.)

100%, regardless of their MWs. However, he speculated that the problem may have resided in the ability of most proteins to retain their higher order structures (HOS), even within the spectrometer under any fragmentation methods. He used the analogy of proteins assuming a *spaghetti ball* conformation (Figure 4.2) once inside the high vacuum regions of the typical MS, regardless of how they were first introduced. Basically, even when proteins are chemically or thermally denatured prior to introduction into any MS, it is reasonable to believe that such proteins rapidly lose denaturing agents (SDS, surfactants, organics, pH, salts, etc.), and reform to varying extents, their native structures *in vacuo*, at room temperatures.

Further, Breuker and McLafferty considered the desolvation of protein molecules in electrospray ionization (ESI), using molecular dynamics simulations [13]. Based on their simulations, Figure 4.3 shows the potential folding states of a protein after electrospray on the order of 10^{-12}–10^2 for side-chain collapse, unfolding, and refolding. Desolvation of the protein occurs globally at first, and then from the charged amino acid side chains, which happen on the nanosecond timescale. If disulfide bonds are reduced in solution, then, in theory, one would have a completely denatured protein. In addition, this would be the *ideal* state for 100% sequence coverage by TDS (this would be another state of the protein, between C and D in Figure 4.3). However, this in and of itself, is the challenge, as picoseconds after this denatured state occurs, if it even exists at all, the amino acid side chains would interact with the neighboring side chains, and then with the peptide backbone, thus creating a more folded state (D–G in Figure 4.3) [13].

Even with UVPD dissociation in the ion trap of an Orbitrap, there is no evidence that proteins remain denatured. In this particular case, even if not fully denatured within the MS, the UV can still get to most/many internal amide bonds and fragment them, though not all. This may well be why UVPD appears to be the most efficient and effective of all fragmentation methods for TDS, thus far [4–6]? Other fragmentation methods, which rely on collisions between electrons or molecules (ETC, ECD), do not appear to be able to reach as many amide bonds, especially for very high MW analytes. This again argues that larger proteins reform, to varying extents, to HOSs, on a very short timescale, once introduced into the high vacuum regions of the MS, now more resistant to complete TDS. How can this, then be prevented? Figure 4.2 suggests some of the more serious, current limitations in performing 100% TDS.

This is potentially an important point, because though most TDS studies have denatured their proteins, reducing disulfide bonds, end capping free thiols, and adding surfactants and inorganic/organic additives, known to denature proteins, there is no guarantee that solution-denatured proteins remain denatured inside the high vacuum regions of the instrument! Most publications tend to disregard this idea—though it may well prove fatal for TDS purposes? In addition, there are no discussions anywhere of how one might keep the solution-denatured proteins denatured once inside the MS? There is very little, if any, real discussion on this topic. Of course, there may well be other reasons beyond this spaghetti ball effect, as to the apparent inability to 100% sequence any and all proteins of any MW? In that which follows, we are only going to deal with how to ensure that all proteins remain denatured and with little or no HOS inside the MS. With this in mind, the most denatured form of the protein would exist at the front of the mass spectrometer or the entry point [13]. Thus, ideally, one would want to denature their protein in solution, and then mimic the denaturing conditions at the entry of the MS, until detection of the molecules through the detector at the end of the MS instrument. This leaves us with two areas for discussion: (1) sample preparation and (2) instrumental parameters.

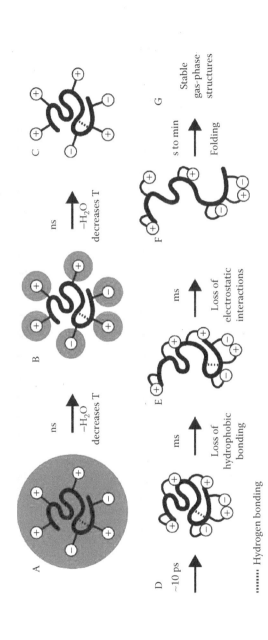

FIGURE 4.3 Unfolding of a globular protein in electrospray ionization. The key is to have an unfolded structure between structures C and D where the disulfide is broken and there are no secondary contacts between amino acid side chains. (From Breuker K. and McLafferty, F.W., *Proc. Natl. Acad. Sci.*, 105, 18145–18152, 2008. With permission.)

4.3 SAMPLE PREPARATION

4.3.1 Can the Right Solvents Help in Achieving 100% Sequence Coverage?

Different solvent conditions can aid in producing a denatured protein. Protein biochemists can easily indicate some of the best protein denaturants, including things such as urea and guanidine hydrochloride. However, any mass spectrometrist will indicate that such solutions are not amenable to MS. The trick would be to find a denaturing agent that is amenable or suitable to MS conditions. Ideally, such denaturing conditions might include a reducing agent (e.g., dithiothreitol [DTT]), high temperature, and an acidified, organic solvent amenable to MS, such as acetic acid or acetonitrile with formic acid [10,14]. One might ask why the solvents matter, because we desolvate on initial entry into the MS, and on desolvation, the protein refolds into a vacuum, as the native state? There is some experimental evidence that even after desolvation, the protein maintains a *memory* of its native state in solution, much similar to a muscle that has a memory [10]. The correct, experimental, solution conditions, may in fact, aid in maintaining a denatured protein's structure.

4.3.2 Do Higher Charge States of Proteins Result in a Higher, Percent Sequence Coverage or Not? Supercharging versus Proton Transfer Reactions in the Gas Phase

There are, at least, two ways to treat the usual, charge state distribution of any protein, and bring that away from its normal state. One is to reduce the protein's molecular ion charge states (lower M+ values), which usually increases the concentration of those ions, using a variety of reagents [15–16]. This is usually termed, proton transfer reactions (PTR). The use of such molecular ions, though now of higher concentrations because there are fewer, lower charged ions, has *not* provided a higher percent sequencing than with the originally formed, molecular ions. Most papers in TDS have tended to use the molecular ion of highest concentration and MS intensity, which is usually *not* that of highest charge state. However, such ions tend to provide the lowest limits of detection and highest sensitivity, ideal for quantitation in bioanalysis or other, quantitative applications.

The alternate approach would be to increase the M+ values, add more charges, a narrower distribution, and generate more ions at higher charge states [17]. This is commonly termed, supercharging. Figure 4.1 provides further evidence as to why supercharging may be advantageous in obtaining a fully denatured protein [10]. A low charge state still maintains the ability for electrostatic interactions to stabilize the protein, whereas at a high charge state, the coulombic repulsion is dominant, and the unfolding barrier can be more easily overcome. In more recent reports, two solution additives, propylene carbonate and ethylene carbonate, were used to form significantly higher protonation states of three, common test proteins. These same proteins have been reported by the use of other methods and additives, such as that obtained with the benchmark *supercharging* additives (m-nitrobenzyl alcohol and sulfolane) [15–16]. By use of these newer reagents, nearly the entire charge state

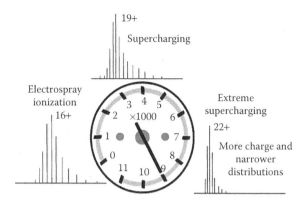

FIGURE 4.4 Supercharging and extreme supercharging of proteins beyond the theoretical, maximum proton transfer limit in ESI–MS. (Reprinted with permission from Teo, C.A. and Donald, W.A., *Anal. Chem.*, 86, 4455–4462, 2014. Copyright 2014 American Chemical Society and the *Journal Analytical Chemistry.*)

distributions of protonated ubiquitin and cytochrome C ions were shifted to higher charge states than the theoretical, maximum protein charging protonation limit in ESI, which was predicted on the basis of proton-transfer reactivity [18].

Perhaps a typical scenario is going from the normal ESI charge state distribution with a maximum intensity (concentration) of 16+ to a supercharging state of 19+ to an extreme, supercharging state of 22+, with more charge and narrower distributions, Figure 4.4 [17]. On account of each supercharged ion that has a higher charge state, there are fewer such ions, and there is also a higher concentration of those species over the normal ESI charge states at lower charge per ion. These may well be desirable properties for more extensive and complete TDS in the future? The mechanism of charging and supercharging molecules in ESI has also been discussed by Loo et al. [15–16], as well as in previous publications [18]. Although it appears that we are able to perform the above changes in the distribution and concentrations of higher charged states, it is unclear if this is a road to improved TDS or not? We have found but a single reference wherein this approach to possibly improving TDS has been discussed [15]. Here, Yin and Loo have studied two supercharging reagents, viz., m-nitrobenzyl alcohol and sulfolane, for a protein–ligand complex in TDS. In addition, though they reported that increasing the analyte charge allows for more effective, tandem MS of protein complexes, there is no mention of any increased, percent coverage, or sequencing by TDS of the protein alone or its complex.

One should differentiate between supercharging, as aforementioned, and proton transfer reactions (ESI charged protein to reagent), which reduces the total charge on each molecular ion in the spectrometer but does not appear to improve TDS sequencing at all [1,19]. It is unclear from the existing literature, whether supercharging, as described earlier, has the ability to improve TDS percent sequencing, just because it leads to a more highly charged, molecular ions at higher levels in the gas phase? The reasoning on our parts has been that molecular ions having higher and higher charge states than normally formed might have more unfolded conformations (ion–ion repulsions) in the gas phase? It is possible that negative ion TDS, described much less than for positive ion TDS might show this desired effect more? It is possible that, more unfolded, molecular ions should or might lead to improved fragmentation because of ion–ion repulsion in the usual, spaghetti ball conformations in the gas phase? And, thereby allowing for improved TDS sequencing, perhaps? The literature does not yet support this possibility.

In addition to, or possibly in combination with, supercharging it may be possible to increase 100% TDS by reducing the protein to one, single charge state of highest charge? Obviously, this would take advantage of the higher charge states, as above-mentioned, and at the same time, produce one form of the protein to be submitted to such top–down techniques as UVPD, ECD, ETD, and others. Common buffers used in charge state reduction are basic salts, such as triethylammonium bicarbonate, *proton sponges*, or imidazole. The major hurdle, of course, would be to find a solvent that reduces the total number of charge states observed but not the overall charge of the protein (e.g., maximized). In other words, for example, instead of having a protonated protein with charge states 22+, 23+, and 24+, one would find a buffer that resulted in only a 24+ charge state protein (or better a buffer that shifted everything to a higher charge state than the highest originally observed) [20]. This would also improve sensitivity and minimum detection limits by having a single, charge state, rather than many, on which to quantitate.

4.3.3   Would Pre-Mass Spectrometry, Chemical Derivatization Improve the Percent Sequence Coverage of Typical Proteins?

In a search for preventing, once solution denatured proteins, from refolding inside the MS, and thus presumably being more amenable to different, fragmentation methods, another possible approach would be to fully tag the protein in solution. There is, of course, a significant literature on using derivatization approaches for proteins, especially in separations such as ultrahigh performance liquid chromatography (UHPLC) or high performance capillary electrophoresis (HPCE) [21–24]. However, none of these tagged proteins have, to the best of our own knowledge, yet been described as to their sequencing abilities through TDS. There is some work reported on using protein tagging and sequential, ion/ion reactions to enhance sequence coverage by ETD in MS [24]. In this study, ion–ion proton transfer reactions were used to simplify the ETD spectra, and to disperse the fragment ions over the entire mass range in a controlled manner. It was also shown that protein derivatization could be used to selectively enhance the sequence information obtained at both termini of a typical protein. Derivatization approaches could, in theory, also lead to more unfolded, denatured states of typical proteins, which should then lead to improved, TDS sequencing. There does not appear to be any discussion in the literature regarding just how fully tagged proteins behave, using recent advances in TDS. There are, however, clear problems in making homogeneous, single derivatives of larger and larger proteins, as well as not having any databases for their successful sequencing in TDS and/or top–down proteomics (TDP) approaches [25–32]. However, it is feasible that fully tagged (all primary amino and carboxyl groups, if not all amides) proteins will be less likely to retain a folded state, once fully ionized, within a typical TDS MS arrangement [24].

There does not appear to be any literature reference wherein such an approach has ever been suggested or described. In addition, depending on the specific tag(s) used, it is possible to form very highly charged, molecular ions that may then fragment much more successfully than the starting, untagged protein? However, interpretation of the MS data with such derivatives may not lend itself to the existing TDS/TDP software for interpretations of the final, fragmentation pattern/species, as none really exists for such derivatives in MS. Thus, the fragmentations obtained may have to be sequenced using *de novo* principles, at least, at first? Another problem may be choosing the best possible tagging reagents to study. One desires tags that will improve ionization efficiencies, forming more highly charged, molecular ions than for the untagged protein, as well as maintaining an unfolded state in the MS. In addition, one also wants fragmentation to occur along the backbone, amide groups in preference to the tagged sites (amide bonds tagged alone may be preferred?).

The difficulties in using this approach for improved TDS or TDP lie in the extra steps required to form 100% tagged proteins, especially those of larger size/MW than others. It also, usually requires sample tagging, optimization steps, removal of excess reagents, cleanup of the now-tagged protein derivative (one derivative, fully tagged), and then determination of the percent purity and tagging before doing any TDS/TDP. Although an increased protein charging may be achieved with tagging, the derivatization process can be relatively laborious (reaction times >1 hr), with additional sample pretreatment and purification steps, it can often result in sample loss, side reactions, and increased analysis times. These are not things that most mass spectrometrists appreciate or relish doing before even starting their MS TDS studies.

4.4 INSTRUMENTAL PARAMETERS

4.4.1 CAN SOURCE PARAMETERS BE REALIZED AND MAINTAINED THROUGHOUT THE ENTIRE MASS SPECTROMETRY TO INCREASE TOP–DOWN SEQUENCING?

Based on simulations, it appears that the majority of the conformational change of a protein in the gas phase occurs in the millisecond timeframe, at the inlet of the MS, suggesting that early intervention in the instrument could prevent refolding of the protein into a rigid, gas phase structure [10,13–14]. Thus, it seems that mimicking the inlet conditions, throughout the entirety of the MS experiment, until detection, might result in the most unfolded protein and therefore, more efficient TDS. The two most basic parameters to consider might be temperature and vacuum. The higher the temperature and the longer under vacuum, should result in a more denatured protein state on fragmentation. Of course, the largest challenge is to maintain a high temperature and vacuum throughout the MS experiment. In fact, with currently available mass spectrometers, it is likely impossible. One way to overcome this restriction may be in using ion mobility spectrometry (IMS), especially at elevated temperatures, as below.

4.4.2 ION MOBILITY SPECTROMETRY AND COLLISIONAL CROSS SECTIONS OF PROTEINS VERSUS TEMPERATURE

There are no simple ways to determine specific conformations of a protein within the high vacuum region of an MS or their collisional cross section (CCS) numbers. Only recently, has IMS begun to be utilized to derive CCS numbers for proteins, and as a function of their temperatures [33–38]. Clearly, temperature has long been recognized as an excellent method to denature proteins, in solution, solid state, or the gas phase, as within an IMS. The size and charge state of ESI generated, protein molecular ions can be resolved using IMS, with a variety of commercial or home-made instruments [33–38]. Though we have only chosen these specific references to discuss the current utility of IMS–MS to study protein, molecular ions in the gas phase, there are numerous others. The intent of such studies has been to resolve all or most of the variously charged, molecular ions coming from any protein, after ESI sample ionization, in a wide variety of IMS instrumentation designs, as a function of drift time, and careful calibration with proteins of known CCS, to calibrate conditions, so as to derive newer CCS versus temperature for other proteins.

CCS is a function of the degree of folding or unfolding of a typical protein, which is a function of the temperature in the IMS, and how the instrument was first calibrated against known proteins of known CCS at known temperatures, of course. In Figure 4.5, we illustrate deriving what is called a *Heatmap* plot of *m/z* versus drift time for a specific protein, native p53 [37]. The *x*-axis is the drift time through this particular IMS, and the *y*-axis is the *m/z* for the various, multiply charged molecular ions of the starting protein. Now, this is for a particular temperature, but as the temperature is varied, as below, the *Heatmap* will change, the number of multiply charged molecular ions may also change, as will the CCS for each of these molecular ions. It is not yet clear if the average molecular charges, as a function of higher temperatures, also increase or not? If supercharging reagents were present, they well might.

It is possible, in a later stage of the MS, to determine the molecular charge on each of these molecular ions (known proteins), and to follow how their CCS may change with increasing temperatures. This becomes important for our purposes, because if

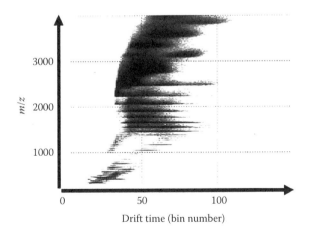

FIGURE 4.5 Synapt HDMS *Heatmap* plot of *m/z* versus drift time (bins) for native p53. (Reprinted from *Int. J. Mass Spectrom.*, 298, Faull, P.A. et al., Utilising ion mobility-mass spectrometry to interrogate macromolecules: Factor H complement control protein modules 10–15 and 19–20 and the DNA-binding core domain of tumour suppressor p53, 99–110, Copyright 2010, with permission from Elsevier.)

we can follow the molecular ion of greatest intensity and abundance, perhaps also of highest charge state ($M+x$, $M+y$, $M+z$), one or more of these may well fragment more than others for TDS purposes. TDS benefits by fragmenting (ECD, ETD, UVPD, etc.) the most abundant, M+ species, of highest charge state, and (presumably) which are most unfolded as a function of temperature. However, such ions may not be present at the highest concentrations, adversely affecting detection and probably, sequencing results. This has never actually been studied to our knowledge, and it does not appear, that anybody has actually ever measured the CCS of any molecular ions used before in TDS studies. It is a commonly held belief that higher charge states of the molecular ion, lead to improved fragmentation under typical TDS conditions by any means [1]. It seems there are no studies, which have used IMS at high temperatures, so as to improve unfolding of the protein before it enters the MS to be sequenced by TDS means.

Quite apart and aside from TDS, Barran and coworkers have described the ability to vary the temperature within an IMS unit before the protein is introduced into the MS itself [35–38]. They have plotted the CCS of several proteins as the temperature is increased, and as one might expect, ex post facto, the gross sizes of most, if not all, proteins studied, show significant increases in their CCSs. The maximum temperature possible in this particular, unique IMS unit was about 570 K. As the temperature was increased, for each protein, its CCS increased but at a particular temperature, it decreased only to continue increasing further until the maximum temperature (possible in the instrumentation available) was reached (ca. 550 K). It was neither clear why there was this break in each protein's CCS at a certain temperature, nor why its size would contract midway between room temperature and 570 K? However, that is not the point for the purposes of improving all TDS sequencing for all proteins, going forward.

There have been several other literature citations on the use of variable temperature IMS–MS in studying changes in protein conformations, but none as seemingly comprehensive as that by Barran et al. [36,39–40]. In addition, such papers really form the basis of suggesting the future use of variable temperature, drift time, ion mobility, mass spectrometry (VT–DT–IM–MS) for improved TDS of all proteins. However, in order to realize higher percent sequencing for all proteins, then in all likelihood, the highest temperature described should be employed. These studies emphasized but three well-known proteins, viz., cyt c, tumor suppressor protein p53 DBD and the N-terminus of the oncoprotein, murine double minute 2 (NTMDM2). The variables involved drift time, temperature, nature of the protein, and the specific MS used, a modified Waters, Q-Tof [36]. The IMS measurements led to well-defined and controlled CCS as a function of temperature (200 K–571 K), residence time in the IMS region, and the nature of the protein. Each protein, however, behaved differently, regarding the number of multiply charged molecular ions, nature of these charged states ($M+x$), presence or absence of monomer versus multimer species, and CCS (sizes). Multiple conformations could be observed over several charge states for all three monomeric proteins. We are making the assumption that the larger the CCS and higher the molecular ion, charge state, the more likely the protein will be sufficiently unfolded to undergo more successful and complete fragmentation and thereby, sequencing by TDS means. At least, that is the rationale.

For example, in the case of NTMDM2, it presented as a single, compact, and conformational family centered on a CCS of 1250 Å at 300 K. This undergoes conformational tightening at higher temperatures, before unfolding completely at the highest temperatures. More extended charge states, presented at several conformers at room temperature, but then underwent thermally induced unfolding before significant, structural collapse to more open structures. However, not all proteins showed significant increases in CCS at the highest temperatures studied, though in this study, most did (2/3). This also depended on the charge states being analyzed for each protein. In general, the most increases in CCS were 2–3X that of the protein at the lowest temperatures in the IMS. Temperatures higher than about 571 K were not attempted. However, bigger and bigger CCSs does not 100% signify that all proteins will be fully denatured with zero secondary or tertiary structures still present but reduced from the native form. There is very little data on how CCS corresponds with the degree of folding or higher order structures of typical proteins, which may well vary from low to very high MWs.

Of course, the above and other studies, do not necessarily guarantee that the increased size of proteins under high temperature IM, will also lead to enough conformational unfolding to permit 100% TDS. This may also very much be protein and charge state dependent, as above for changes in CCS for different, charge state, molecular ions for each protein.

4.4.3   How Can the Above-Mentioned Information Then Be Utilized to Improve All Future Top–Down Sequencing Studies?

The increase in CCS for all or most proteins at elevated temperatures suggests that all or most proteins unfold (perhaps to varying degrees), and lose their original 3D conformation at very high temperatures, as aforementioned. That is, they may become thermally denatured, even at the very high vacuums within the MS or IMS. The vacuums in these two regions, of course, may be vastly different (dependent on the type of IMS or FAIMS utilized), and so there is no guarantee that the loss of conformation induced in the IMS, will remain at the higher vacuum region and lower temperatures within the MS itself? However, given the very short times between the residence in the IMS and introduction into the MS, hope springs eternal that the unfolding suggested by the CSS studies, will remain during the molecular ion selection steps and fragmentation. If that is true, which only actual experimentation might discern, there may well be a much better likelihood that all proteins will now show higher sequencing than ever before. Figure 4.6 illustrates a commonly utilized arrangement for doing IMS–MS studies, and as available from different MS vendors [41–42].

A lingering question revolves around what type of IMS would best serve this purpose? There are, to the best of our understanding, fundamentally three basic types of IMS units described or commercially available. One is most commonly used by researchers, a linear or circular, open tube with a flowing, inert gas, applied voltage, and usually run at room temperature, at atmospheric pressures or higher [41–42]. This has been termed, by some, as a radio frequency (RF)-confining drift cell.

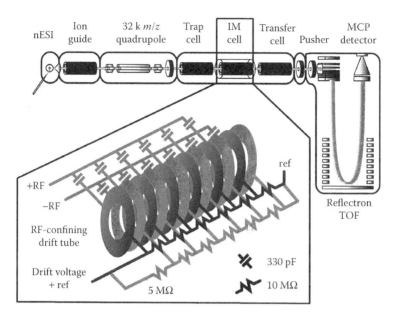

FIGURE 4.6 A specific arrangement of an IMS–MS system, showing the placement of the, at times, high temperature IMS after the ESI and before the other components of the MS instrumentation. Modified quadrupole/ion mobility/time-of-flight hybrid mass spectrometer used for absolute Ω measurements. (Reproduced with permission from Bush, M.F. et al., *Anal. Chem.*, 82, 9557–9565, 2010. Copyright 2010 American Chemical Society and *Analytical Chemistry Journal*.)

In some cases, IMS is performed at a He pressure of about 3.5–4.0 Torr and a cell temperature from room to 571 K [15,18–19]. The nature of the gas used is variable but should be inert to the separations desired, viz., multiply charged, molecular ions of proteins. The rate of He gas flow and applied voltage (15–60 V) are variables dependent on the particular studies being pursued. There are no reports in using this type of IMS–MS for TDS purposes.

The other two, commonly referenced IMS systems are Field Amplified Ion Mobility Spectrometry (FAIMS) and Traveling Wave IMS (TWIMS, Waters). However, it seems highly unlikely and extremely difficult to utilize either of these two IMS configurations for the express purpose of generating very high temperatures to unfold any/all proteins before the entrance to the MS, be that Orbitrap, Q-ToF, or other configurations. This may present a major problem, in that the time required to move the unfolded protein at 571 K from the IMS unit into the ion trap or other regions of the MS, may well allow it to cool and reassume a folded conformation before 100% fragmentation could occur? It is easy to measure CCS within an IMS cell, all at a high temperature but to move such proteins into fragmentation regions of the MS itself is a different requirement. Figure 4.6 illustrates a typical, applied RF-confining arrangement, with a flowing inert gas stream IMS cell, not in common use today in MS [21]. This particular, custom-built model uses an RF-confining drift tube with variable drift voltages, so as to control how fast or slow the molecular ions (or fragment ions from the quadrupole) that are already formed earlier in the nESI part, can move from the selection, quadrupole part to the transfer cell and then into the HRMS Q-ToF portion of this novel, noncommercial MS spectrometer. This does not use a traveling wave in the IM cell but it is really a drift tube with acceleration potentials on the RF-confining rings, so as to control how slow or fast the molecular ions (or other, fragmentation ions) pass through the Q-ToF. It is similar to the inner workings of a Waters Synapt instrument, with some major modifications to the IM cell, notably the removal of any traveling wave. This is in contrast to the standard Synapt HDMS and traveling waves that are not used in this IMS cell.

That cell could be placed before any resolving sectors, in this case, a quadrupole, depending on if one wished to study the intact, molecular ion or fragmentation species derived therefrom, coming out of the quadrupole? There are many other designs for the IM cell, some much simpler than in Figure 4.6. Often it is just a basic, drift cell with voltage potentials at each end, and a flowing stream of inert gas, often with organic additives to improve resolution of the components introduced. This can lead to improved separations of the major ion(s) of interest, usually the molecular ion for TDS purposes. Most of the IMS systems utilized to determine CCS have been home-built, with higher temperature capabilities than all other drift cells on the market today. IMS was not, at first, introduced to study CCS features of proteins but it has evolved into a form that can today perform such studies, quite handily and quickly, as aforementioned. Please note that in all IMS approaches, one must first form the molecular ions of the analytes, usually by nESI or normal ESI, and these total ions can then be fractionated, in this case, by a quadrupole before the IMS cell. However, in most other arrangements, the IMS cell performs the initial separation of intact, molecular ions of multiple charges. And then, each of these molecular ions can be fragmented (alone) and sequenced (if proteins), by what follows in an incredible variety of HRMS instruments in the literature today. Only a very few of those described have actually been commercialized, and rather by a limited number of vendors. However, almost every MS vendor today has at least one model that contains some type of IMS component.

4.4.4 On Combining High Temperature, Differential Ion Mobility Spectrometry with Ultraviolet Photodissociation before MS-MS? How Else Can We Bring about Fragmentation of the Denatured Protein with Maximum Collisional Cross Sections?

What if the unfolded proteins derived from the IMS unit immediately refold, as soon as they enter the much cooler regions of the remaining MS? In addition, they do this before any UVPD, ETD, ECD, or other fragmentation modes have sufficient time to operate on the (presumably) fully unfolded proteins from the IMS? It is very possible that thermally unfolded proteins would not then experience sufficient fragmentation to perform TDS above 50%, for example. How can this be overcome?

One possibility would be to perform the UVPD or ECD immediately after the IMS region, perhaps within the ion trap or HCD parts of the instrument? In the most recent work with UVPD, it was performed within the ion trap (IT) immediately after sample introduction from the ESI ion source [43]. This could allow retaining the initial molecular ions from the IMS within the IT long enough to realize maximum UVPD fragmentation, as to maximize the percent sequencing realized. Perhaps?

4.5 CONCLUSION

Similar to many of the most challenging, scientific problems, some combination of the ideas presented here may well result in the most denatured proteins, thus perhaps leading to ca. 100% TDS of any, and perhaps all, proteins. Experimental optimization of all parameters will be necessary; however, it seems to us that a good starting point might include acidified solvents, with the IMS at high temperatures and using UVPD as the initial source of fragmentation. We might also suggest that the intrinsically disordered proteins (IDP) may provide an interesting starting point (molecule) for these proof-of-concept experiments, as by nature, truly IDP have no inherent structure. Thus, we hope that recent advances in instrumentation and our understanding of protein folding, will lead to 100% TDS, perhaps in the near future? And, thereby reducing our dependence on the more complex, bottom-up sequencing, as currently employed, which seems to have its own, at times, serious drawbacks [44–46].

ACKNOWLEDGMENTS

We have been assisted in numerous ways by the kind exchange of TDS information from several sources, of these the most helpful and meaningful have come from many colleagues, but especially Joe Loo, Neil Kelleher, Jenny Brodbelt, Jeff Agar, and others. We have also had access to all the relevant, scientific literature through the kind resources of Northeastern University's Snell Library system in Boston.

We also appreciate the kind invitation by Nelu Grinberg to contribute to this Memorial Edition to Eli Grushka, who held the role of editor or coeditor, longer than any other person throughout the history of this series of highly recognized contributions in all separations areas, over the course of several decades.

ABBREVIATIONS

CCS	collisional cross section
CID	collision-induced dissociation
DDT	dithiothreitol
ECD	electron capture dissociation
ESI	electrospray ionization
ETD	electron transfer dissociation
FAIMS	field amplified IMS
FTMS	Fourier transform MS
HCD	high-energy collision-induced dissociation
HDMS	high definition MS
HecD	high-energy electron capture dissociation
HOS	higher order structure

IDP	intrinsically disordered proteins
IM	ion mobility
IMS	ion mobility spectrometry or spectrometer
IT	ion trap
IRMPD	infrared multiphoton dissociation
kDa	kilo Daltons
mAb	monoclonal antibody
MS	mass spectrometry or mass spectrometer
MW	molecular weight
nESI	nanospray ESI
PTR	proton transfer reaction
Q-ToF	quadrupole time-of-flight, hybrid mass spectrometer
RF	radio frequency
RF	radiofrequency
TDS	top–down sequencing
TDP	top–down proteomics
VT–DT–IM–MS	variable temperature–drift time–ion mobility–mass spectrometry
TWIMS	traveling wave IMS
UVPD	ultraviolet photodissociation

REFERENCES

1. I.S. Krull, A. Rathore, and J.A. Loo, The current and future state of top-down protein sequencing, *LCGC North America*, 34(7), 492–499 (2016).
2. J.F. Kellie, J.C. Tran, J.E. Lee, D.R. Ahlf, H.M. Thomas, I. Ntai, A.D. Catherman et al. The emerging process of top down mass spectrometry for protein analysis: Biomarkers, protein-therapeutics, and achieving high throughput, *Mol. Biosyst.*, 6(9), 1532–1539 (2010).
3. J.B. Shaw, W. Li, D.D. Holden, Y. Zhang, J. Griep-Raming, R.T. Fellers, B.P. Early, P.M. Thomas, N.L. Kelleher, and J.S. Brodbelt. Complete protein characterization using top-down mass spectrometry and ultraviolet photodissociation, *J. Am. Chem. Soc.*, 135(34), 12646–12651 (2013).
4. D.D. Holden, J.M. Pruet, and J.S. Brodbelt, Ultraviolet photodissociation of protonated, fixed charge, and charge-reduced peptides, *Int. J. Mass Spec.*, 390, 81–90 (2015).
5. J.R. Cannon, D.D. Holden, and J.S. Brodbelt, Hybridizing ultraviolet photodissociation with electron transfer dissociation for intact protein characterization, *Anal. Chem.*, 86, 10970–10977 (2014).
6. M.B. Cammarata, R. Thyer, J. Rosenberg, A. Ellington, and J.S. Brodbelt, Structural characterization of dihydrofolate reductase complexes by top-down ultraviolet photo-dissociation mass spectrometry, *J. Amer. Chem. Soc.*, 137, 9128–9135 (2015).
7. Y.O. Tsybin, L. Fornelli, C. Stoermer, M. Luebeck, J. Parra, S. Nallet, F.M. Wurm, and R. Hartmer, Structural analysis of intact monoclonal antibodies by electron transfer dissociation mass spectrometry, *Anal. Chem.*, 83(23), 8919–8927 (2011).
8. J.J. Wolff, S.L. Van Orden, and J.A. Loo, Native top-down electrospray ionization-mass spectrometry of 158 kDa protein complex by high-resolution Fourier transform ion cyclotron resonance mass spectrometry, *Anal. Chem.*, 86, 317–320 (2014).
9. C.J. DeHart, O.S. Skinner, L. Fornelli, P.D. Compton, P.M. Thomas, G. Lahav, J. Gunawardena, and N.L. Kelleher, Analysis of intact p53 protein by top-down mass spectrometry, *Paper presented 2015 National ASMS Meeting*, St. Louis, MO, June 2015.

10. T. Meyer, V. Gabelica, H. Grubmuller, and M. Orozco, Proteins in the gas phase, *WIRES Comput. Mol. Sci.*, 3, 408–425 (2013).

11. F.W. McLafferty, K. Breuker, M. Jin, X. Han, G. Infusini, H. Jiang, X. Kong, and T.P. Begley, Top-down MS, a powerful complement to the high capabilities of proteolysis proteomics, *FEBS J.*, 274, 6256–6268 (2007).

12. F. Lanucara and C.E. Eyers, Top down mass spectrometry for the analysis of combinatorial post-translational modifications, *Mass Spec. Rev.*, 32(1), 27–42 (2013).

13. K. Breuker and F.W. McLafferty, Stepwise evolution of protein native structure with electrospray into the gas phase, 10^{-12} to 10^2s, *Proc. Natl. Acad. Sci.*, 105(47), 18145–18152 (2008).

14. R.R. Ogorzalek Loo and R.D. Smith, Investigation of the gas-phase structure of electrosprayed proteins using ion-molecule reactions, *J. Am. Soc. Mass Spectrom.*, 5, 207–220 (1994).

15. S. Yin and J.A. Loo, Top-down mass spectrometry of supercharged native protein-ligand complexes, *Int. J. Mass Spectrom.*, 300(2–3), 118–122 (2011).

16. R.R. Ogorzalek Loo, R. Lakshmanan, and J.A. Loo, What protein charging (and supercharging) reveal about the mechanism of electrospray ionization, *J. Am. Soc. Mass Spectrom.*, 25(10), 1675–1693 (2014).

17. C.A. Teo and W.A. Donald, Solution additives for supercharging proteins beyond the theoretical maximum proton-transfer limit in electrospray ionization mass spectrometry, *Anal. Chem.*, 86(9), 4455–4462 (2014).

18. A.T. Iavarone and E.R. Williams, Mechanism of charging and supercharging molecules in electrospray ionization, *J. Am. Chem. Soc.*, 125(8), 2319–2327 (2003).

19. R.R. Ogorzalek Loo, B.E. Winger, and R.D. Smith, Proton transfer reaction studies of multiply charged proteins in a high mass-to-charge ratio quadrupole mass spectrometer, *J. Am. Soc. Mass Spectrom.*, 5, 1064–1071 (1994).

20. J.T.S. Hopper, K. Sokratous, and N.J. Oldham, Charge state and adduct reduction in electrospray ionization-mass spectrometry using solvent vapor exposure, *Anal. Biochem.*, 421, 788–790 (2012).

21. C.F. Poole, Liquid chromatography: Chapter 2, *Derivatization in Liquid Chromatography*, Kindle Edition, Elsevier, Oxford, UK, 2013.

22. R.L. Lundblad, *Chemical Reagents for Protein Modification*, 4th ed., CRC Press, Taylor & Francis Group, Boca Raton, FL, 2014.

23. G.T. Hermanson, *Bioconjugate Techniques*, 3rd ed., Academic Press, Elsevier, Oxford, UK, 2013.

24. L.C. Anderson, A.M. English, W.-H. Wang, D.L. Bai, J. Shabanowitz, and D.F. Hunt, Protein derivatization and sequential ion/ion reactions to enhance sequence coverage produced by electron transfer dissociation mass spectrometry, *Int. J. Mass Spectrom.*, 377, 617 (2015).

25. M.E. Szulc, P. Swett, and I.S. Krull, Size selective derivatizations with polymer immobilized reagents, *Biomed. Chromatogr.*, 11(3), 207 (1997).

26. I.S. Krull, M.E. Szulc, and J. Dai. *Derivatizations in HPCE*. A Primer, Thermo Separation Products, Inc., San Jose, CA, 1997.

27. I.S. Krull, R. Strong, Z. Sosic, B.-Y. Cho, S. Beale, and S. Cohen. Labeling reactions for minute amounts of proteins in HPLC and HPCE, *J. Chromatogr., B, Biomed. Appl.*, 699, 173 (1997).

28. I.S. Krull, A. Sebag, and R. Stevenson, Specific applications of capillary electrochromatography to biopolymers, including proteins, nucleic acids, peptide mapping, antibodies, and so forth. *J. Chromatogr. A*, 887, 137 (2000).

29. I.S. Krull and R. Strong, *Liquid Chromatography-Derivatization*. Invited contribution to Encyclopedia of Separation Science, Elsevier Scientific Publishers, Amsterdam, The Netherlands, 2000, p. 583.

30. H.J. Liu, B.-Y. Cho, R. Strong, I.S. Krull, S. Cohen, K.C. Chan, and H.J. Issaq, Derivatization of peptides and small proteins for improved identification and detection in capillary zone electrophoresis (CZE), *Anal. Chim. Acta*, 400, 181 (1999).
31. H.J. Liu, B.-Y. Cho, I.S. Krull, and S.A. Cohen, Homogeneous, fluorescent derivatization of large proteins, *J. Chromatogr. A*, 927, 77 (2001).
32. H.J. Liu, I.S. Krull, and S.A. Cohen, Femtomole peptide mapping by derivatization, HPLC and fluorescence detection, *Anal. Biochem.*, 294, 7 (2001).
33. E.G. Marklund, M.T. Degiacomi, C.V. Robinson, A.J. Baldwin, and J.L.P. Benesch, Collision cross sections for structural proteomics, *Structure*, 23, 1–9 (2015).
34. M. Sharon, Structural MS pulls its weight, *Science*, 340, 1059–1060 (2013).
35. R. Beveridge, A.S. Phillips, L. Denbigh, H.M. Saleem, C.E. MacPhee, and P.E. Barran, Relating gas phase to solution conformations: Lessons from disordered proteins, *Proteomics*, 15, 2872–2883 (2015).
36. E.R. Dickinson, E. Jurneczko, K.J. Pacholarz, D.J. Clarke, M. Reeves, K.L. Ball, T. Hupp, D. Campopiano, P.V. Nikolova, and P.E. Barran, Insights into the conformation of three, structurally diverse proteins: Cytochrome c, p54 and MDM2, provided by variable-temperature ion mobility spectrometry, *Anal. Chem.*, 87, 3231–3238 (2015).
37. P.A. Faull, H.V. Florance, C.Q. Schmidt, N. Tomczyk, P.N. Barlow, T.R. Hupp, P.V. Nikolova, and P.E. Barran, Utilising ion mobility-mass spectrometry to interrogate macromolecules: Factor H complement control protein modules 10-15 and 19-20 and the DNA-binding core domain of tumour suppressor p53, *Int. J. Mass Spectrom.*, 298, 99–110 (2010).
38. P. Barran, The benefits of performing ion mobility at various temperatures from a biological and a technical perspective, *Paper presented at the 29th Meeting of the Israeli Society for Mass Spectrometry, Joint Meeting of the British and Israeli MS Societies*, February 28–29, 2016, Weizmann Institute of Science, Rehovot, Israel.
39. Y. Mao, J. Woenckhaus, J. Kolafa, M.A. Ratner, and M.F. Jarrold, Thermal unfolding of unsolvated Cytochrome c: Experiment and molecular dynamics simulations, *J. Am. Chem. Soc.*, 121, 2712–2721 (1999).
40. M.F. Jarrold, Unfolding, refolding and hydration of proteins in the gas phase, *Acc. Chem. Res.*, 32, 360–367 (1998).
41. S.J. Allen, K. Giles, T. Gilbert, and M.F. Bush, Ion mobility mass spectrometry of peptide, protein, and protein complex ions using a radio-frequency confining drift cell, *Analyst*, 141, 884–891 (2016).
42. M.F. Bush, Z. Hall, K. Giles, J. Hoyes, C.V. Robinson, and B.T. Ruotolo, Collision cross sections of proteins and their complexes: A calibration framework and database for gas-phase structural biology, *Anal. Chem.*, 82, 9557–9565 (2010).
43. D.H. Holden, J. Schartz, E. Zhuk, and J.S. Brodbelt, Top down protein analysis by ultraviolet photodissociation in an Orbitrap Tribrid Mass Spectrometer, *62nd American Society for Mass Spectrometry Conference*, June 2014, Baltimore, MD.
44. M.W. Duncan, R. Aerbersold, and R.M. Caprioli, The pros and cons of peptide-centric proteomics, *Nat. Biotechnol.*, 28(7), 659–664 (2010).
45. R.A. Bradshaw, A.L. Burlingame, S. Carr, and R. Aerbersold, Protein identification, the good, the bad, and the ugly, *Mol. Cell. Proteomics*, 4, 1221–1222.
46. S.J. Berger, K.M. Millea, I.S. Krull, and S.A. Cohen, Middle-out proteomics: Incorporating multidimensional protein fractionation and intact protein mass analysis as elements of a proteomic analysis workflow, In *Separations in Proteomics*, G. Smejkal and A. Lazarev (Eds.) Taylor & Francis Group, New York, 2005.

5 Is the Number of Peaks in a Chromatogram Always Less Than the Peak Capacity?
A Study in Memory of Eli Grushka

Joe M. Davis and Mark R. Schure

CONTENTS

5.1 INTRODUCTION

Eli Grushka was a unique individual who wrote unique papers. The subjects of these papers were interesting to Eli and in the truest spirit of a scientist he investigated various aspects of separation science that others did not pursue. He was a curious man and sought many answers, especially those that could be expressed mathematically and in compact form. Many of these papers were ahead of their time and those who study the mechanics of chromatographic separations need to return to these papers on occasion.

Many of these interests were spawned as extensions of the studies Eli performed during his postdoctoral time with Professor J. Calvin Giddings at the University of Utah, Salt Lake City, Utah. We list citations of two of them here:

- Chromatographic Peak Capacity and the Factors Influencing It, *Analytical Chemistry* 42 (1970) 1142–1147.
- Effect of High Speed on Peak Capacity in Liquid Chromatography, *Journal of Chromatography* 316 (1984) 81–93.

These two papers and many more were part of the curious mind of Eli Grushka. One of our colleagues at the University of Utah, Professor Joel M. Harris, repeatedly sang the praises of Eli Grushka; he would tell us that Eli's papers were a welcome addition to any literature. Eli touched many people and was part of the Giddings group alumni, a strange fraternity of which we both are members.

The reader will notice that the word *peak capacity* is in both the article titles, which are given earlier. Peak capacity is a term that reflects the number of peaks under various peak width assumptions that fit in a specified separation space (either temporal or length-based separations apply to this concept equally). The term was used explicitly by Giddings [1] who showed simple equations for expressing the peak capacity of gel filtration columns. At a later date the book, *An Introduction to Separation Science*, by Karger, Snyder, and Horvath [2] described on page 158: "... the maximum number of separated bands which can be accommodated between the first and last bands in the chromatogram. This number ... is the peak capacity of the chromatogram."

The peak capacity cannot be directly measured but may be calculated using experimental chromatograms with various levels of accuracy [3]. The peak capacity concept is highly useful when dealing with complex separations in which the effective saturation, which is discussed below and is equal to the number of detectable components divided by the peak capacity, is larger than approximately 0.2.

In this study, which we dedicate to Eli Grushka, we consider a question inspired by Professor Peter Carr at the University of Minnesota:

> Is the number of maxima in a chromatogram always less than the peak capacity at unit resolution?

In the work that follows, we attempt to answer this question by two means. The first is based on an analysis of simulated chromatograms containing single component peaks (SCPs) having different distributions of height and different distributions of time between successive SCPs. An SCP is defined as the peak profile of a single detectable mixture component. It contributes to a maximum in a chromatogram, either by itself as a singlet or by fusing with other SCPs to form a doublet, triplet, n-plet, and so on. The second means is based on statistical overlap theory (SOT), which is reviewed in Section 5.3.1. The theory for this problem already exists and requires only simple manipulation.

5.2   BACKGROUND

In a chromatogram containing SCPs with constant standard deviation σ, as is found in gradient elution liquid chromatography (LC) or temperature-programmed gas chromatography (GC), the peak capacity is [4,5]

$$n_c = \frac{{}^1 D}{4 \sigma R_s} \tag{5.1}$$

where ${}^1 D$ is the duration of the chromatogram, which is equal to the difference between the retention times of the first and last SCPs, and R_s is the resolution specified as prerequisite for separation. The units of σ and ${}^1 D$ can be either spatial or temporal, although we will use temporal ones. In simple terms, n_c is the number of times the effective peak width $4\sigma R_s$ fits into ${}^1 D$, and consequently is the number of contiguous units into which equally spaced SCPs can be placed. The heights of SCPs are not specified in the definition of n_c but are usually assumed to be equal.

The peak capacity is important in SOT. In SOT, one considers a chromatogram to be a member of a large ensemble of similar chromatograms, in which the retention times and heights (amplitudes) of SCPs are modeled with statistical functions. The late Professor Francesco Dondi and coworkers at the University of Ferrara named these functions the *interdistance distribution* [6] or *interdistance model* [7] and the *amplitude model* [8], respectively. Here, we keep the name, amplitude model, but use *interpeak statistics function* [9] instead of interdistance distribution or model, because the latter implies spatial coordinates. Several forms of SOT exist, for example, point process [10–17], Fourier analysis [6,7,18,19], and pulse-point [8], with developments for one-dimensional [6–8,10,11,14,18,20], two-dimensional [12,15,16,19,21,22], and various types of n-dimensional [13,15,17] systems. The theory has been used to estimate the number of components and SCP properties (e.g., standard deviation) in real, synthetic, and computer-simulated mixtures [11,16,18,23–34], estimate the fraction of SCPs resolved experimentally as different peak types (e.g., singlets and doublets) [27,29,34], estimate interpeak statistics functions [31,35], investigate retention–structure relationships [35], determine homologous compound classes [36], compare separations of different dimensionality [15,37,38], evaluate column overload [39], and quantify improvements to separation by multivariate selectivity [40]. Reviews of the theory have been published [9,41–46], including two for this series during the editorship of Grushka [41,42].

Regarding the number of maxima that can fill a separation with peak capacity n_c at unit resolution, we realize that sufficiently large values of n_c must exceed p regardless of R_s, because p can never exceed m. The relationship among p, n_c, and R_s is less clear, when n_c is much less than m. One might think that since a multicomponent maximum must have a width that exceeds the SCP width (unless its constituent SCPs coincide), the number of such maxima—or a mix of multi- and single-component maxima—that can fit in ${}^1 D$ must be less than n_c. However, this argument presumes that the maxima profiles are independent of one another, which they are not: the existence of one affects the breadth, height, and location of the other in its vicinity. The late Professor Georges Guiochon and coworkers at the University of Tennessee showed by simulation that the number of maxima has no simple relation to the underlying SCP structure, when n_c is much less than m [24]. Furthermore, maxima containing three or more SCPs are not easily characterized by R_s, which is defined for SCP pairs. The ambiguity warrants study.

5.3   THEORY

5.3.1   REVIEW OF STATISTICAL OVERLAP THEORY

It is useful to review some aspects of SOT. The theory used herein is based on point processes, in which SCPs are modeled by geometrical figures centered about SCP retention times, and peak overlap is modeled by figure clustering. For any interpeak statistics function, the expected number of peaks p in an ensemble of one-dimensional chromatograms of extent ^1D and containing m SCPs with standard deviation σ can be written as

$$p = mf(\alpha,...) \tag{5.2}$$

where f is a function that differs for every interpeak statistics function but always depends on the saturation α

$$\alpha = \frac{4m\sigma R_s^*}{^1D} \tag{5.3}$$

The ellipsis in Equation 5.2 indicates that p may depend on variables other than α, depending on the interpeak statistics function. The attribute R_s^* in Equation 5.3 is a type of resolution, with the geometrical figure modeling the SCP width equal to a line segment of duration, $4\sigma R_s^*$ [10].

Different interpeak statistics functions are used here, with a different form of Equation 5.2 for each function. It is not important to give them explicitly, as they are referenced. Lest Equation 5.2 appear abstract, however, we give an example, appropriate to a random distribution of SCPs [10]

$$p = m\exp(-\alpha) \tag{5.4}$$

Here, $f(\alpha,...)$ in Equation 5.2 equals $\exp(-\alpha)$.

If one were to identify R_s in Equation 5.1 for peak capacity n_c with R_s^* in Equation 5.3 for saturation α, then α could be interpreted as the number of SCPs, m, divided by an SOT-based peak capacity equal to $^1D/(4\sigma R_s^*)$. Early papers on SOT (including ones by us) proposed that α could be so interpreted; some papers continue to do so. However, the implications of that interpretation were not as clear then as now.

It may be appropriate to identify R_s^* with R_s in certain circumstances. However, research has shown that R_s^* differs from R_s, when p is identified with the average number of maxima in the chromatographic ensemble [20,22,47]. The details are not reviewed here; it is sufficient to recognize that R_s (defined as the difference between the retention times of two SCPs, divided by four times their average temporal standard deviation) is independent of SCP heights, whereas R_s^* depends not only on the amplitude model but the saturation and interpeak statistics function as well [20,47]. In effect, the value of R_s^* is constrained by the ways SCPs overlap to produce maxima. For any amplitude model, Professor Attila Felinger of the University of Pécs developed a theory on the distribution of the minimum resolution required to separate two SCPs into two maxima [48]. The average of this distribution is the value of R_s^* as the saturation approaches its minimum value (usually zero) [48,49]. To date, however,

TABLE 5.1

Results for the Comparison of 16 Interpeak Statistics Functions, When
$m = {}^1D/(4\sigma)$

Interpeak Statistics Function	SOT Prediction of p/m, with $m = n_c$ and $R_s = 1$[a]	Simulation Values of p/m, with $m = {}^1D/(4\sigma)$[b]	SOT Prediction of p/m, with $m = {}^1D/(4\sigma R_s^*)$ and $\alpha = R_s^* = 1$
Gamma, $k = 1$ (Poisson)	0.533 (0.628)	0.537 ± 0.020	0.368
Gamma, $k = 3$	0.678 (0.665)	0.683 ± 0.019	0.423
Gamma, $k = 5$	0.743 (0.681)	0.741 ± 0.019	0.440
Gamma, $k = 7$	0.786 (0.691)	0.774 ± 0.020	0.450
Uniform	0.667 (0.667)	0.670 ± 0.015	0.500
Gaussian, RSD = 0.50	0.742 (0.682)	0.739 ± 0.018	0.489
Gaussian, RSD = 0.25	0.878 (0.709)	0.855 ± 0.018	0.499
Gaussian, RSD = 0.15	0.968 (0.722)	0.926 ± 0.015	0.499
Power, $D = 0.2, \beta = 2$	0.954 (0.720)	0.891 ± 0.019	0.465
Power, $D = 0.2, \beta = 10$	0.587 (0.636)	0.608 ± 0.019	0.388
Power, $D = 0.2, \beta = 100$	0.400 (0.583)	0.414 ± 0.019	0.291
Power, $D = 0.2, \beta = 1000$	0.283 (0.564)	0.285 ± 0.017	0.216
Power, $D = 1, \beta = 2$	0.992 (0.724)	0.897 ± 0.005[c]	0.443
Power, $D = 1, \beta = 10$	0.578 (0.630)	0.596 ± 0.007[c]	0.323
Power, $D = 1, \beta = 100$	0.386 (0.548)	0.419 ± 0.009[c]	0.207
Power, $D = 1, \beta = 1000$	0.291 (0.496)	0.322 ± 0.019[c]	0.144

[a] Parentheses contain the values, $\alpha = R_s^*$.
[b] Average and standard deviation of p/m determined from 100 simulated chromatograms. Amplitude model: exponential. $m = n_c = 250$ unless otherwise noted, with $R_s = 1$.
[c] As in footnote b but for $m = n_c = 2500$ to improve precision [47].

theory for R_s^* at larger saturations has been developed only for the exponential amplitude model; it predicts that $R_s^* \equiv R_s^*(\alpha)$ decreases with increasing α [20,22,47].

The following example shows that R_s^* is not a good attribute with which to calculate peak capacity, at least when p is identified with the number of maxima. Suppose that the number of SCPs, m, and the peak capacity, n_c, at unit resolution ($R_s = 1$), are the same in all chromatograms produced by the 16 interpeak statistics functions listed in Table 5.1 (these functions are described later in the chapter). Equation 5.1 shows this supposition implies that $m = {}^1D/(4\sigma)$, which in turn implies that $\alpha = R_s^*$, in accordance with Equation 5.3. Under this condition, the fractions of SCPs that are predicted by SOT to be resolved as peaks, p/m, are listed in the second column of Table 5.1. They are in good agreement with their counterparts in the third column, which were computed as described later in the chapter from an ensemble of simulated chromatograms. The second column of Table 5.1 also reports the values $\alpha = R_s^*$, appropriate to the SOT calculations. Unlike the constant resolution, $R_s = 1$, they differ for different interpeak statistics functions and range from 0.496 to 0.724. One value (0.496) is less than the smallest possible R_s value, 0.5, for separating two SCPs into two maxima.

Suppose instead that $m = {}^1D/(4\sigma)$ and $R_s = 1$ as before, but that $R_s^* = 1$ as well. This leaves the simulation results unchanged but implies that m equals the SOT-based peak capacity ${}^1D/(4\sigma R_s^*)$ and that $\alpha = 1$, in accordance with Equation 5.3. The last column in Table 5.1 lists SOT predictions of p/m under this condition, which agree poorly with the simulation results *because R_s does not equal R_s^*.*

Fortunately, a simple approach allows one to interpret (but not to calculate!) SOT predictions using n_c and R_s. The effective saturation, α_e, is an SOT metric equal to [37]

$$\alpha_e = \frac{\alpha}{R_s^*} = \frac{4\sigma m}{{}^1D} = \frac{m}{n_c R_s} \tag{5.5}$$

where the third and fourth expressions are obtained by introducing Equations 5.3 and 5.1, respectively. Unlike saturation α, the effective saturation depends only on n_c and R_s. It is a mapping of α into a new variable, but one in which the R_s^* value may be constrained. In our case, with p identified with the number of maxima, α and α_e are not linearly proportional. Table 5.1 shows that the same α_e value (unity) corresponds to different α values for different interpeak statistics functions.

It is important to realize that the natural variable of SOT equations is saturation α. One must *calculate* SOT predictions relative to α, but one should at least consider *interpreting* them relative to α_e. Some advantages to this interpretation are discussed [37,38,40]. The interpretation requires knowledge of any dependence of R_s^* on α.

5.3.2 RELATIVE VALUES OF p AND n_c

It is simple to use SOT to compare the relative values of n_c and p, with the latter equaling the average number of maxima in an ensemble of chromatograms containing SCPs following the exponential amplitude model. The answer applies to all interpeak statistics functions in Table 5.1 (and possibly others). We reexpress Equation 5.2 as

$$\frac{p}{m} = f(\alpha,\ldots) = g\left(\frac{1}{\alpha_e},\ldots\right) \tag{5.6}$$

where $g(1/\alpha_e,\ldots)$ is a new SOT expression obtained on mapping α into $1/\alpha_e$, calculated from Equation 5.5 as

$$\frac{1}{\alpha_e} = \frac{R_s^*}{\alpha} = \frac{{}^1D}{4\sigma m} = \frac{n_c R_s}{m} \tag{5.7}$$

Since p is identified with peak maxima, the value of R_s^* in the second part of Equation 5.7 is constrained as discussed earlier.

It is not necessary to evaluate $g(1/\alpha_e,\ldots)$. Rather, it is sufficient to evaluate $f(\alpha,\ldots)$ from the published SOT equations for α and associate its value with the corresponding coordinate, $1/\alpha_e$. This requires knowledge of the R_s^* value at that α.

Consider a graph of p/m versus $1/\alpha_e = {}^1D/(4\sigma m)$, as expressed by Equations 5.6 and 5.7. In such a graph, one has also a relationship between p and ${}^1D/(4\sigma)$ that is the same for any m, except for scale. This is true, because one can multiply both p/m and $^1D/(4\sigma m)$ by m to obtain p and $^1D/(4\sigma)$; one changes values of the graph coordinates but not the relationship. Equation 5.1 shows that this relationship can be interpreted as being between p and the peak capacity n_c at unit resolution.

Therefore, if the SOT prediction p/m lies *above* the threshold line, $p/m = {}^1D/(4\sigma m)$ of unit slope and zero intercept, then p is predicted to *exceed* $^1D/(4\sigma)$, the value of n_c at unit resolution. However, if the prediction lies *below* the line $p/m = {}^1D/(4\sigma m)$, then p is predicted to be *less* than n_c at unit resolution.

These two interpretations also apply to the estimates of p/m that are determined by simulating the chromatographic ensemble. By considering amplitude models that are not exponential, we can investigate the problem more thoroughly than by SOT alone. Experimental evidence suggests that SCP heights in complex chromatograms often follow a distribution that is similar to the exponential distribution [24,50,51] or log-normal distribution [52,53]. We shall consider both, as well as the case in which all SCP heights are constant, that is, the same. To our knowledge, this is the first use of log-normal heights in simulated chromatograms.

5.3.3 Interpeak Statistics Functions

The interpeak statistics functions used here are based on homogeneous renewal processes [54]. Each differs in the distribution of time between successive SCPs. The most common is the Poisson distribution, which has a theoretical basis in chromatograms of complex mixtures having diverse composition [6,10,11,55–57]. Other functions have been proposed for less complex chromatograms. Dondi and coworkers developed Fourier-analysis SOT for the gamma, normal (or Gaussian), and uniform distributions [6,7], which were applied experimentally [31,35] and subsequently adapted to point-process SOT [14,20]. The gamma distribution depends on shape parameter k; as k increases, SCPs are more likely to be found near the average time between SCPs. The Poisson distribution is a special case, with $k = 1$. For the Gaussian distribution, the distribution of intervals between SCPs is varied on changing the relative standard deviation (RSD), with small RSDs producing more periodic spacing of SCPs. The uniform distribution has the same likelihood of spacing between two adjacent SCPs. Using the truncated Pareto power-law distribution, the authors developed an interpeak statistics function depending on both the dimension D, with small values favoring small spacing between SCPs, and the Boltzmann-weighted incremental free energy β, with large values favoring *wild* behavior [47]. Other interpeak statistics functions exist (e.g., Weibull and log-normal distributions), as well as the ones expressed by weighted linear combinations of different functions [58].

The ellipsis in Equation 5.2 signifies a possible dependence of peak number p on parameters other than saturation α. Examples of such dependencies include k for the gamma distribution, RSD for the Gaussian distribution, and D and β for the power-law distribution.

The 16 interpeak statistics functions listed in Table 5.1, which were investigated in previous SOT studies [20,47], are used here as model systems to produce markedly different chromatograms. The panels of Figure 5.1 are examples of simulated

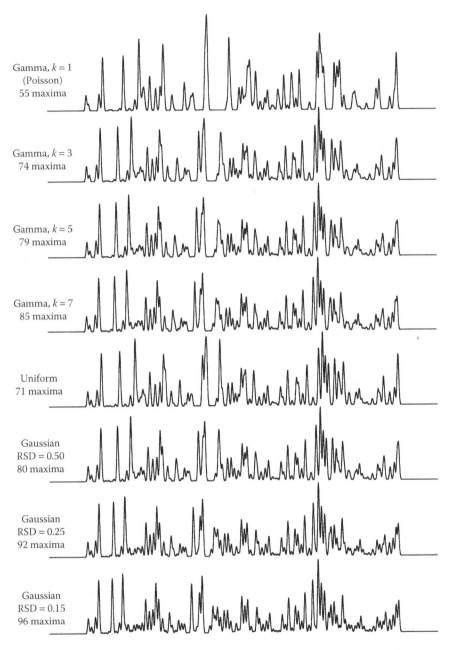

FIGURE 5.1 Simulated chromatograms produced by 16 interpeak statistics functions listed in Table 5.1. Vertical scale adjusted as needed. Amplitude model: exponential. Numbers of maxima are reported. $m = n_c = 100$, $R_s = 1$, $\alpha_e = 1$.   (*Continued*)

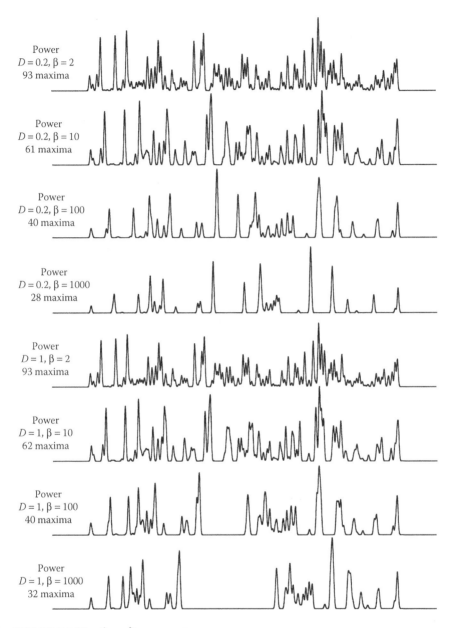

FIGURE 5.1 (Continued)

chromatograms produced by these interpeak functions for an exponential amplitude model ($m = n_c = 100$, $R_s = 1$, $\alpha_e = 1$). The chromatograms are similar, in that the sequential SCP heights are the same and the sequential intervals between successive SCPs are similar (specifically, sequential values of the cumulative distributions of the interpeak statistics functions are the same). The latter attribute preserves small spacings between successive SCPs in all chromatograms, but only in accordance with what small means for a given interpeak statistics function. Similarly, the large spacings are large in all chromatograms, but only in accordance with what large means for a given interpeak statistics function. Various maxima groupings can be seen to nearly align but not exactly align in different chromatograms, and to be resolved to different levels, as the interpeak statistics function and SCP spacing change. The numbers of maxima are reported in the figure, and agree well with the average results in Table 5.1. These chromatograms will be useful to our interpretation.

5.4 PROCEDURES

SOT predictions of p/m versus α and of R_s^* versus α for the exponential amplitude model were recomputed to verify published results using equations for the gamma distribution ($k = 1, 3, 5,$ and 7) [20], Gaussian distribution (RSD = 0.15, 0.25, and 0.50) [20], uniform distribution [20], and power-law distribution ($D = 0.2$ and $\beta = 2, 10, 100,$ and 1000; $D = 1$ and $\beta = 2, 10, 100,$ and 1000) [47]. All previous predictions were verified. The new p/m values were associated with $1/\alpha_e$, Equation 5.7, with R_s^* therein taken from the computations of R_s^* versus α.

New simulation values of p/m were recomputed, as published values of p/m versus α are insufficient to span the desired ranges of $1/\alpha_e$. For each interpeak statistics function, chromatograms containing either $m = 250$ or 2500 SCPs were simulated for various $1/\alpha_e = {}^1D/(4\sigma m)$ values, with σ calculated relative to a ${}^1D$ value spanning the coordinates, 0.1 and 0.9 (i.e., ${}^1D = 0.8$). (All coordinates are relative values unrelated to actual times.) The retention times of SCPs were computed over the range, 0.1–0.9, by inverse transform sampling (i.e., by mapping uniform random numbers into the cumulative distributions of interpeak statistics functions) and then linearly scaling them between the limits, 0.1 and 0.9. The heights of SCPs were constant, exponentially distributed, or log-normally distributed. The exponential and log-normal heights were computed by inverse transform sampling with the scale parameter σ_s of the log-normal distribution varying between 0.1 and 2.5 (the location parameter is immaterial as it only scales the heights). Each chromatogram was simulated by summing Gaussian SCPs over the coordinates, 0–1, with the duration between successive points equal to $\sigma/5$. For each $1/\alpha_e$, 100 chromatograms were simulated, maxima were counted, and mean and standard deviations of p/m were computed. The results for SCPs obeying the exponential amplitude model were verified by a comparison to earlier computations.

5.5 RESULTS AND DISCUSSION

Figures 5.2 through 5.5 are graphs of p/m versus ${}^1D/(4\sigma m) = n_c R_s/m$. Each panel contains results for one of the 16 interpeak statistics functions listed in Table 5.1. The bold line is a graph of the threshold relation, $p/m = {}^1D/(4\sigma m)$. The solid curve

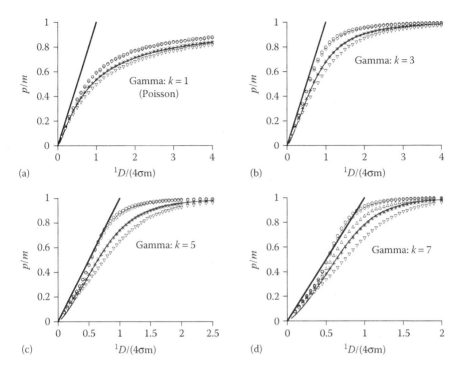

FIGURE 5.2 Graphs of p/m versus $^1D/(4\sigma m)$ for gamma interpeak statistics function, with k equal to (a) 1, (b) 3, (c) 5, and (d) 7. Bold line is the graph of $p/m = {}^1D/(4\sigma m)$. Curve is SOT prediction for exponential amplitude model. $m = 250$. Symbols represent average p/m simulation values for different amplitude models: $\bigcirc$, constant; $\square$, log-normal distribution (LND), $\sigma_s = 0.05$; $\Diamond$, LND, $\sigma_s = 0.1$; $\triangle$, LND, $\sigma_s = 0.5$; $\times$, exponential; $+$, LND, $\sigma_s = 1.19$; $\triangledown$, LND, $\sigma_s = 2.5$.

is the SOT prediction for the exponential amplitude model. The symbols represent simulation results for different amplitude models and are identified in the caption to Figure 5.2. The results for every proposed log-normal amplitude model are not shown in every graph, if p/m varies only slightly with σ_s.

5.5.1   RELATIVE VALUES OF p AND n_c

An examination of these graphs shows that in most cases both SOT predictions and simulation results lie *below* the threshold line, $p/m = {}^1D/(4\sigma m)$. This means that the number of maxima p is usually *less* than the peak capacity n_c at unit resolution for the chosen conditions. This outcome is unsurprising for large $^1D/(4\sigma m)$, because as noted earlier n_c can increase without bound (at least in principle), whereas p cannot exceed m. The SOT predictions for the exponential amplitude model and simulation results for all amplitude models are similar for very small $^1D/(4\sigma m)$, where the underlying SCP structure is lost [24]. Therefore, they are plausible, although one should not infer that the observed trend extends to limiting cases of p and n_c approaching unitary values. Of greater interest are the results in the region, $0.5 \leq {}^1D/(4\sigma m) \leq 1$, where p is sometimes greater and sometimes less than n_c at unit resolution.

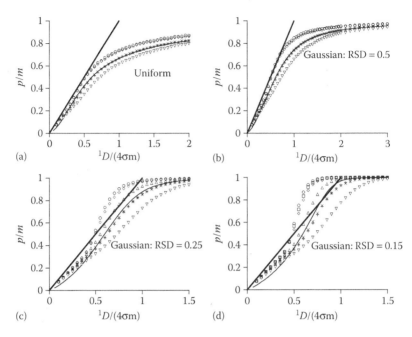

FIGURE 5.3 As in Figure 5.2 but for (a) uniform interpeak statistics function and Gaussian interpeak statistics function, with RSD equal to (b) 0.5, (c) 0.25, and (d) 0.15.

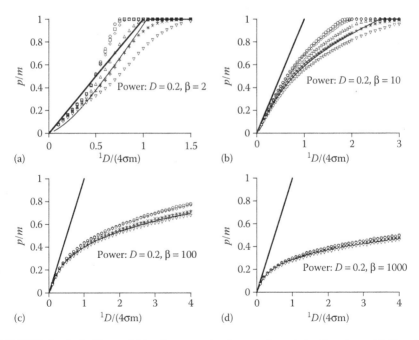

FIGURE 5.4 As in Figure 5.2 but for power-law interpeak statistics function, with $D = 0.2$ and β equal to (a) 2, (b) 10, (c) 100, and (d) 1000.

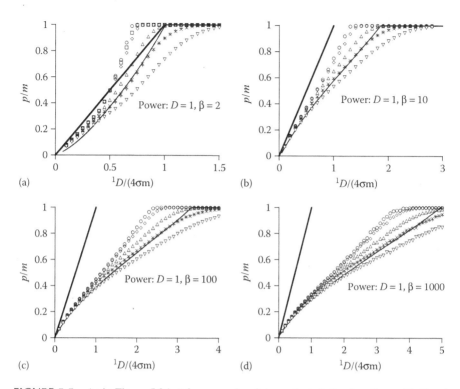

FIGURE 5.5   As in Figure 5.2 but for power-law interpeak statistics function, with $D = 1$ and β equal to (a) 2, (b) 10, (c) 100, and (d) 1000, and with m equal to 2500 to improve precision.

To understand why this occurs, it is important to understand a trend toward the periodicity of SCP spacing. For any amplitude model, p/m increases with increasing k for the gamma interpeak statistics function and with decreasing RSD for the Gaussian interpeak statistics function. Similarly, p/m increases with decreasing β at fixed D and with increasing D at fixed β for the power-law function. As these parameters change as indicated, the intervals between SCPs become more periodic and peak overlap decreases.

This near-periodicity of SCP spacing is associated with a reversal of the general relationship between p and n_c, but only over a narrow range of conditions. Specifically, p can exceed n_c at unit resolution over the range, $0.5 \leq {}^1D/(4\sigma m) \leq 1$, when SCPs have equal or similar heights (e.g., the constant and small-σ_s log-normal amplitude models). The most extreme results are found for these interpeak statistics functions and parameters: Gaussian, RSD = 0.25 (Figure 5.3c); Gaussian, RSD = 0.15 (Figure 5.3d); power-law, $D = 0.2$, $\beta = 2$ (Figure 5.4a); and power-law, $D = 1$, $\beta = 2$ (Figure 5.5a). A tendency toward this result also occurs for the gamma interpeak statistics function with $k = 7$ (Figure 5.2d). The near-periodicity of SCP spacing can be seen in the corresponding simulated chromatograms in Figure 5.1, even though the amplitude model is exponential.

Over the range, $0.5 \leq {}^1D/(4\sigma m) \leq 1$, the abscissa ${}^1D/(4\sigma m)$ in Figures 5.2 through 5.5 describe a chromatogram containing from one to two times as many SCPs, m, as the peak capacity at unit resolution, ${}^1D/(4\sigma)$. However, as shown by Equation 5.1, it can equally describe a chromatogram containing a number of SCPs, m, equal to a peak capacity n_c defined relative to a resolution R_s between 0.5 and 1. This correspondence is more useful to our interpretation. It is well known that two SCPs of equal height are resolved as maxima if R_s equals or exceeds 0.5, and similar arguments can be made for two SCPs of similar height if R_s is increased slightly. With a sequence of nearly equally spaced SCPs of similar height, as in the cases discussed earlier, this tendency toward separation still exists because the SCPs that precede or follow any SCP pair make only peripheral contributions to the pair's overlap. It is unsurprising that p can exceed n_c at unit resolution under such conditions: less than unit resolution is needed.

As supportive evidence, SOT predictions show that p/m lies above the threshold line, $p/m = {}^1D/(4\sigma m)$, over at least part of this ${}^1D/(4\sigma m)$ range for the Gaussian interpeak statistics function and RSD values of 0.1, 0.05, and 0.005 (results not shown). Because SCPs become exactly equally spaced, as RSD values approach zero, we are led to the somewhat ironical finding that exceptions to the general trend, $p < n_c$ at unit resolution, are found in chromatograms whose structure mimics the conditions under which the peak capacity itself is defined: equally spaced SCPs with equal heights.

In contrast, when the SCP spacing is less periodic and SCP heights vary widely, it is more likely that maxima numbers decrease in the range in question. For widely varying heights, even a near-periodic spacing produces the usual behavior, as is seen in the graphs identified previously when the amplitude model is log-normal with large σ_s. Although we could not have predicted that a resolution of unity in Equation 5.1 would satisfy the trend, $p < n_c$, our findings suggest that it generally is true. This result affirms the usual choice of a resolution of unity to evaluate chromatographic performance.

5.5.2 TRENDS

Before concluding, we summarize several trends in Figures 5.2 through 5.5. Given the diversity of the amplitude models and interpeak statistics functions, we perhaps have the largest compilation of SOT predictions and simulation results outside of a review, facilitating insight into several phenomena.

1. Both SOT predictions and simulation results show that p/m increases with increasing ${}^1D/(4\sigma m)$ for all amplitude models and interpeak statistics functions. This is expected, because for fixed m the peak capacity increases with increasing ${}^1D/(4\sigma m)$ as shown by Equation 5.1.
2. The SOT predictions agree closely with simulation results for the exponential amplitude model. This finding is consistent with previous publications [20,47].
3. The minimum resolution required for separating two SCPs into two maxima increases with the ratio of SCP heights. This fact explains why the p/m results are greatest for the constant amplitude model. It also explains why p/m decreases, when the amplitude model is the log-normal distribution

(LND) and its scale parameter σ_s increases. With increasing σ_s, the LND widens and SCP heights differ more frequently, causing peak overlap to increase. As σ_s approaches zero, the width of the LND approaches zero, resulting in the convergence of the p/m results for the constant and LND amplitude models.

4. A fit to average values of the minimum resolution distribution for the LND amplitude model is reported elsewhere [49] and equals 0.723 for $\sigma_s = 1.19$. This also is the value of R_s^* in SOT at large $^1D/(4\sigma)$, explaining why the p/m results on the right-hand sides of graphs for the LND, $\sigma_s = 1.19$ amplitude model are similar to those for the exponential amplitude model ($R_s^* = 0.725$ under similar conditions). It does *not* explain why these models produce almost equal p/m values as $^1D/(4\sigma m)$ decreases. Most probably, it makes little difference whether SCPs with small heights are abundant (as with the exponential distribution) or sparse (as with the LND); such SCPs are obscured by overlapping with larger SCPs in either case. This is a matter of current interest to us.

5. For all interpeak statistics functions, p/m values for different amplitude models differ most when p/m is large. As peak overlap increases, SCPs blend into one another, producing maxima whose underlying height structure ultimately is lost. This is especially evident as $^1D/(4\sigma m)$ approaches zero.

5.6 CONCLUSION

Our results show that the number of maxima in model chromatograms usually is less than the peak capacity at unit resolution, with exceptions occurring when equally or nearly equally spaced SCPs of equal or similar height are greater in number than the peak capacity by a factor of one to two. It would be foolish to deduce an ironclad guideline from these extensive but ultimately limited results. In principle, one could study an infinite number of interpeak statistics functions and amplitude models, as well as SCP spacing and height distributions that follow no statistical law. Other exceptions may occur. However, we suspect that this trend will be more observed than not observed. We would like to think that our results would have interested Eli Grushka, as they show once again how the peak capacity, a concept that he helped to develop, is useful in characterizing chromatograms.

REFERENCES

1. Giddings, J. C. 1967. Maximum number of components resolvable by gel filtration chromatography. *Analytical Chemistry* 39: 1027–1028.
2. Karger, B. L., Synder, L. R., Horvath, C. 1973. *An Introduction to Separation Science.* New York: John Wiley & Sons.
3. Lan, K., Jorgenson, J. W. 1999. Automated measurement of peak widths for the determination of peak capacity in complex chromatograms. *Analytical Chemistry* 71: 709–714.
4. Grushka, E. 1970. Chromatographic peak capacity and the factors influencing it. *Analytical Chemistry* 42: 1142–1147.
5. Giddings, J. C. 1969. Generation of variance, "theoretical plates," resolution, and peak capacity in electrophoresis and sedimentation. *Separation Science* 4: 181–189.

6. Felinger, A., Pasti, L., Dondi, F. 1990. Fourier analysis of multicomponent chromatograms. Theory and models. *Analytical Chemistry* 62: 1846–1853.

7. Felinger, A., Pasti, L., Dondi, F. 1992. Fourier analysis of multicomponent chromatograms. Recognition of retention patterns. *Analytical Chemistry* 64: 2164–2174.

8. Dondi, F., Bassi, A., Cavazzini, A., Pietrogrande, M. C. 1998. A quantitative theory of the statistical degree of peak overlapping in chromatography. *Analytical Chemistry* 70: 766–773.

9. Schure, M. R., Davis, J. M. 2015. The simple use of statistical overlap theory in chromatography. *LC-GC North America* 33(4): 14–18.

10. Davis, J. M., Giddings, J. C. 1983. Statistical theory of component overlap in multicomponent chromatograms. *Analytical Chemistry* 55: 418–424.

11. Martin, M., Herman, D. P., Guiochon, G. 1986. Probability distributions of the number of chromatographically resolved peaks and resolvable components in mixtures. *Analytical Chemistry* 58: 2200–2207.

12. Oros, F. J., Davis, J. M. 1992. Comparison of statistical theories of spot overlap in two-dimensional separations and verification of means for estimating the number of zones. *Journal of Chromatography* 591: 1–18.

13. Davis, J. M. 1993. Statistical theory of spot overlap for n-dimensional separations. *Analytical Chemistry* 65: 2014–2023.

14. Pietrogrande, M. C., Dondi, F., Felinger, A., Davis, J. M. 1995. Statistical study of peak overlapping in multicomponent chromatograms: Importance of the retention pattern. *Chemometrics and Intelligent Laboratory Systems* 5: 239–258.

15. Martin, M. 1995. On the potential of two- and multi-dimensional separation systems. *Fresenius Journal of Analytical Chemistry* 352: 625–632.

16. Samuel, C., Davis, J. M. 2002. Statistical-overlap theory of column switching in gas chromatography: Application to flavor and fragrance compounds. *Analytical Chemistry* 74: 2293–2305.

17. Davis, J. M., Blumberg, L. M. 2005. Probability theory for number of mixture components resolved by n independent columns. *Journal of Chromatography A* 1096: 28–39.

18. Pietrogrande, M. C., Dondi, F., Felinger, A. 1996. HRGC separation performance evaluation by a simplified Fourier analysis approach. *Journal of High Resolution Chromatography* 19: 327–332.

19. Marchetti, N., Felinger, A., Pasti, L., Pietrogrande, M. C., Dondi, F. 2004. Decoding two-dimensional complex multicomponent separations by autocovariance function. *Analytical Chemistry* 76: 3055–3068.

20. Davis, J. M. 1997. Extension of statistical overlap theory to poorly resolved separations. *Analytical Chemistry* 69: 3796–3805.

21. Davis, J. M. 2005. Statistical-overlap theory for elliptical zones of high aspect ratio in comprehensive two-dimensional separations. *Journal of Separation Science* 28: 347–359.

22. Liu, S., Davis, J. M. 2006. Dependence on saturation of average minimum resolution in two-dimensional statistical-overlap theory: Peak overlap in saturated two-dimensional separations. *Journal of Chromatography A* 1126: 244–256.

23. Giddings, J. C., Davis, J. M., Schure, M. R. 1984. Test of the statistical model of component overlap by computer-generated chromatograms. In: Ahuja, S. (Ed.) *ACS Symposium Series 250: Ultrahigh Resolution Chromatography*. pp. 9–26. Washington, DC: American Chemical Society.

24. Herman, D. P., Gonnord, M.-F., Guiochon, G. 1984. Statistical approach for estimating the total number of components in complex mixtures from nontotally resolved chromatograms. *Analytical Chemistry* 56: 995–1003.

25. Dondi, F., Kahie, Y. D., Lodi, G., Remelli, M., Reschiglian, P., Bighi, C. 1986. Evaluation of the number of components in multi-component liquid chromatograms of plant extracts. *Analytica Chimica Acta* 191: 261–273.
26. Coppi, S., Betti, A., Dondi, F. 1988. Analysis of complex mixtures by capillary gas chromatography with statistical estimation of the number of components. *Analytica Chimica Acta* 212: 165–170.
27. Davis, J. M. 1988. Experimental affirmation of the statistical model of overlap. *Journal of Chromatography* 449: 41–52.
28. Dondi, F., Gianferrara, T., Reschiglian, P., Pietrogrande, M. C., Ebert, C., Linda, P. 1990. Effects of different organic modifiers in optimization of reversed-phase high-performance liquid chromatographic gradient elution of a mixture of natural secoiridoid compounds. *Journal of Chromatography* 485: 631–645.
29. Delinger, S. L., Davis, J. M. 1990. Experimental verification of parameters calculated with the statistical model of overlap from chromatograms of a synthetic multicomponent mixture. *Analytical Chemistry* 62: 436–443.
30. Felinger, A., Pasti, L., Reschiglian, P., Dondi, F. 1990. Fourier analysis of multicomponent chromatograms. Numerical evaluation of statistical parameters. *Analytical Chemistry* 62: 1854–1860.
31. Dondi, F., Betti, A., Pasti, L., Pietrogrande, M. C., Felinger, A. 1993. Fourier analysis of multicomponent chromatograms. Application to experimental chromatograms. *Analytical Chemistry* 65: 2209–2222.
32. Pietrogrande, M. C., Marchetti, N., Dondi, F., Righetti, P. G. 2002. Spot overlapping in two-dimensional polyacrylamide gel electrophoresis separations: A statistical study of complex protein maps. *Electrophoresis* 23: 283–291.
33. Pietrogrande, M. C., Tellini, I., Felinger, A., Dondi, F., Szopa, C., Sternberg, R., Cidal-Madjar, C. 2003. Decoding of complex isothermal chromatograms: Application to chromatograms recovered from space missions. *Journal of Separation Science* 26: 569–577.
34. Campostrini, N., Areces, L. B., Rappsilber, J., Pietrogrande, M. C., Dondi, F., Pastorino, F., Ponzoni, M., Righetti, P. G. 2005. Spot overlapping in two-dimensional maps: A serious problem ignored for much too long. *Proteomics* 5: 2385–2395.
35. Pietrogrande, M. C., Pasti, L., Dondi, F., Rodriguez, M. H. B., Diaz, M. A. C. 1994. PCB Separation by HRGC-MS. Fourier analysis for characterizing Aroclor chromatograms. *Journal of High Resolution Chromatography* 17: 839–850.
36. Pietrogrande, M. C., Zampolli, M. G., Dondi, F. 2006. Identification and quantification of homologous series of compound in complex mixtures: Autocovariance study of GC/MS chromatograms. *Analytical Chemistry* 78: 2579–2592.
37. Davis, J. M., Carr, P. W. 2009. Effective saturation: A more informative metric for comparing peak separation in one- and two-dimensional separations. *Analytical Chemistry* 81: 1198–1207.
38. Davis, J. M. 2011. Dependence on effective saturation of numbers of singlet peaks in one- and two-dimensional separations. *Talanta* 83: 1068–1073.
39. Felinger, A., Vigh, E., Gelencsér, A. 1999. Statistical determination of the proper sample size in multicomponent separations. *Journal of Chromatography A* 839: 129–139.
40. Davis, J. M., Rutan, S. C., Carr, P. W. 2011. Relationship between selectivity and average resolution in comprehensive two-dimensional separations with spectroscopic detection. *Journal of Chromatography A* 1218: 5819–5828.
41. Davis, J. M. 1994. Statistical theories of peak overlap in chromatography. In: Brown, P. R., Grushka, E. (Eds.) *Advances in Chromatography* (vol. 34). pp. 109–176. New York: Marcel Dekker.

42. Felinger, A. 1998. Mathematical analysis of multicomponent chromatograms. In: Brown, P. R., Grushka, E. (Eds.) *Advances in Chromatography* (vol. 39). pp. 201–238. New York: Marcel Dekker.

43. Felinger, A. 1998. Chapter 15: Statistical theory of peak overlap; Chapter 16: Fourier analysis of multicomponent chromatograms. In: Vandeginste, B. G. M., Rutan, S. C. (Eds.) *Data Analysis and Signal Processing in Chromatography* (Data Handling in Science and Technology, vol. 21). pp. 331–409. Amsterdam, the Netherlands: Elsevier.

44. Pietrogrande, M. C., Cavazzini, A., Dondi, F. 2000. Quantitative theory of the statistical degree of peak overlapping in chromatography. *Reviews in Analytical Chemistry* 19: 123–155.

45. Felinger, A., Pietrogrande, M. C. 2001. Decoding complex multicomponent chromatograms. *Analytical Chemistry* 73: 619A–626A.

46. Dondi, F., Pietrogrande, M. C., Felinger, A. 2008. Decoding complex 2D separations. In: Cohen, S. A., Schure, M. R. (Eds.) *Multidimensional Liquid Chromatography: Theory and Applications in Industrial Chemistry and the Life Sciences.* pp. 59–90. Hoboken, NJ: John Wiley & Sons.

47. Schure, M. R., Davis, J. M. 2011. The statistical overlap theory of chromatography using power law (fractal) statistics. *Journal of Chromatography A* 1218: 9297–9306.

48. Felinger, A. 1997. Critical peak resolution in multicomponent chromatograms. *Analytical Chemistry* 69: 2976–2979.

49. Davis, J. M. 2011. Computation of distribution of minimum resolution for log-normal distribution of chromatographic peak heights. *Journal of Chromatography A* 1218: 7841–7849.

50. Nagels, L. J., Creten, W. L., Vanpeperstraete, P. M. 1983. Determination limits and distribution function of ultraviolet absorbing substances in liquid chromatographic analysis of plant extracts. *Analytical Chemistry* 55: 216–220.

51. Nagels, L. J., Creten, W. L. 1985. Evaluation of the glassy carbon electrochemical detector selectivity in high-performance liquid chromatographic analysis of plant material. *Analytical Chemistry* 57: 2706–2711.

52. Enke, C. G., Nagels, L. J. 2011. Undetected components in natural mixtures: How many? What concentrations? Do they account for chemical noise? What is needed to detect them? *Analytical Chemistry* 83: 2539–2546.

53. Gundlach-Graham, A., Enke, C. G. 2015. Effect of response factor variations on the response distribution of complex mixtures. *European Journal of Mass Spectrometry* 21: 471–479.

54. Cox, D. R. 1967. *Renewal Theory,* Science Paperback edition. New York: Chapman and Hall.

55. Felinger, A. 1995. Superposition of chromatographic retention patterns. *Analytical Chemistry* 67: 2078–2087.

56. Dondi, F., Pietrogrande, M. C., Felinger, A. 1997. Decoding complex multicomponent chromatograms by Fourier analysis. *Chromatographia* 45: 435–440.

57. Davis, J. M., Pompe, M., Samuel, C. 2000. Justification of statistical overlap theory in programmed temperature gas chromatography: Thermodynamic origin of random distribution of retention times. *Analytical Chemistry* 72: 5700–5713.

58. Zou, M. 1996. *Studies of Statistical Theories of Overlap in One and Two Dimensional Separations.* Masters Thesis, Southern Illinois University at Carbondale, IL.

6 Advances in Organic Polymer-Based Monolithic Columns for Liquid Phase Separation Techniques

Ziad El Rassi

CONTENTS

NONSTANDARD ABBREVIATIONS

(AAPMIm)Br, 1-acetylamino-propyl-3-methylimidazolium bromide; [BMIM]BF$_4$, 1-butyl-3-methylimidazolium tetrafluoroborate; 1,12-DoDDMA, 1,12-dodecanediol dimethacrylate; 1,6 HEDA, 1,6-hexanediol ethoxylate diacrylate; ACN, aceto-nitrile; acryl-POSS, acrylopropyl polyhedral oligomeric silsesquioxane; AHA, 6-azidohexanoic acid; AIBN, 2,2′-azobis(isobutyronitrile); AM, acrylamide; AMIM$^+$Cl$^-$, 1-allyl-3-methylimidazolium chloride; AMPS, 2-acrylamido-2-methyl-1-propane-sulfonic acid; APTMS, amino-propyl-trimethoxysilane; ATRP, atom transfer radical polymerization; AZT, 3′-azido-3′-deoxythymidine; β-CD, β-cyclodextrin; BCMBP, 4,4′-bis(chloromethyl)-1,1′-biphenyl; BIDMA, bisphe-nol A dimethacrylate; BIGDMA, bisphenol A glycerolate dimethacrylate; BMA, butyl methacrylate; BMIM$^+$Cl$^-$, 1-butyl-3-methylimidazolium chloride; BTEE, ethyl-2-methyl-2-butyltellanyl propionate; BUDMA, tetramethylene dimethacry-late; BVPE, 1,2-bis(p-vinylphenyl) ethane; CuAAC, Cu (I) catalyzed 1,3-dipolar azide–alkyne cycloaddition; DCX, α,α′-dichloro-p-xylene; DEGDE, diethylene glycol diethyl ether; DiEDMA, dioxyethylene dimethacrylate; DMDA, N,N-methyl-N-dodecylamine; DMF, dimethylformamide; DPEPA, dipentaerythritol penta-/hexa-acrylate; DVB, 1,4-divinylbenzene; EDMA, ethylene glycol dimethacrylate; FSNPs, fumed silica nanoparticles; GCMA, glycerol carbonate methacrylate; GMA, glycidyl methacrylate; GMM, glyceryl monomethacrylate; GNPs, gold nanopar-ticles; GPTMS, γ-glycidoxypropyltrimethoxysilane; GSH, glutathione; HDDMA, 1,6-hexanediol dimethacrylate; HDT, 1,6-hexanedithiol; HEDMA, hexamethylene dimethacylate; HEMA, 2-hydroxyethylmethacrylate; HMA, hexyl methacrylate; HPMA, 3-chloro-2-hydroxypropyl methacrylate; MAA, methacrylic acid; LMA, lau-ryl methacrylate; MA-L-Phe-OMe, N-methacryloyl-L-phenylalanine methyl ester; MBA, N,N′-methylenebisacrylamide; MEDSA, N,N-dimethyl-N-methacryloxyethyl-N-(3-sulfopropyl)ammonium betaine; META, methacrylatoethyl trimethyl ammo-nium chloride; MFSNP, methacryloyl fumed silica nanoparticles; MNPs, magnetic nanoparticles; MOFs, metal-organic frameworks; MPC, 2-methacryloyloxyethyl phosphorycholine; MPS, sodium 2-methylpropene-1-sulfonate; MPTMS, 3-mercap-topropyltrimethoxysilane; NAHAM, N-acryloyltris(hydroxymethyl)aminomethane; NAPMH, N-(3-aminopropyl) methacrylamide hydrochloride; NAPM, 2-naphthyl methacrylate; NAS, N-acryloxysuccinimide; NVC, N-vinylcarbazole; NVP, N-vinyl-2-pyrrolidinone; ODT, 1-octadecanethiol; ODY, 1,7-octadiyne; PAHs, polycyclic aromatic hydrocarbons; PDA, 1,4-bis(acryloyl)piperazine; PEDAS, pentaerythritol diacrylate monostearate; PEG, polyethylene glycol; PEGDA, poly(ethylene glycol) diacrylate; PETA, pentaerythritol triacrylate; PETAC, pentaerythritol tetraacrylate; PhDA, 1,4-phenylene diacrylate; PMA, propargyl methacrylate; POSS, polyhedral oligomeric silsesquioxanes; PTM, pentaerythritol tetrakis(3-mercaptopropionate); PTMS, phenyltrimethoxysilane; SPDA, N,N-dimethyl-N-acryloyloxyethyl-N-(3-sulfopropyl)ammonium betaine; SPE, N,N-dimethyl-N-methacryloxyethyl-N-(3-sul-fopropyl)ammonium betaine; SPMA, 3-sulfopropyl methacrylate potassium; SPP, N,N-dimethyl-N-(3-methacryl-amidopropyl)-N-(3-sulfopropyl)ammonium betaine; TAIC, triallyl isocyanurate; TeEDMA, tetraethyleneglycol dimethacrylate; TFA, trifluoroacetic acid; TMOS, tetramethyloxysilane; TMPTA, trimethylolpropane triacrylate; TPTM, trimethylolpropane tris(3-mercaptopropionate); TRIM, trimethylolpropane trimethacrylate; TVCH, 1,2,4-trivinylcyclohexane; VBC-ILs, 1-vinyl-3-butylimidazolium chloride; VBSIm, 1-vinyl-3-(butyl-4-sulfonate) imidazolium; VBTA, vinylbenzyl trimethylammonium chloride; $VC_{18}HIm^+Br^-$, 1-vinyl-3-octadecylimidazolium bromide; VDC, vinyl decanoate; VMNPs, vinylized iron oxide nanoparticles; VPV, vinyl pivalate.

6.1 INTRODUCTION

Monolithic stationary phases are increasingly employed in high-performance liquid chromatography (HPLC) and capillary electrochromatography (CEC) for the separation of a wide range of species ranging in size from small ions to large molecules [1–5]. Monoliths can be divided into two main categories, namely inorganic silica-based monoliths made from alkoxysilane and organic polymer-based monoliths made from acrylamide, styrene, or acrylate/methacrylate monomers. In both categories, monoliths with the desired functionality and porosity can be prepared to meet the need of a particular application. This is facilitated by the number of available choices among a variety of possible parameters, including the nature of monomers, cross-linkers and porogens, postpolymerization functionalization, temperature, and time of polymerization.

This chapter, which is concerned only with organic polymer-based monoliths, summarizes the recent progress made in this field over the past 5 years, a period that witnessed rapid growth in the area of polymer-based monoliths. The first part of this chapter is about polar organic polymer monoliths that have been introduced for hydrophilic interaction liquid chromatography (HILIC) and hydrophilic CEC (HICEC), including mixed-mode monoliths, polar nanoparticle incorporated monoliths, and polar ionic liquid monoliths. The second part of this contribution is concerned with nonpolar organic polymer monoliths whereby novel trends in their preparation as well as mixed-mode monoliths, nonpolar nanoparticle incorporated monoliths, and nonpolar ionic liquid monoliths will be described. In addition, the progress made in organic-silica hybrid monoliths for polar and nonpolar compounds will also be discussed.

6.2 POLAR ORGANIC MONOLITHS

HILIC has gained wide acceptance in the past decade due to its unique selectivity in separating polar analytes and its complementary role to other chromatographic techniques, and in particular reversed-phase chromatography (RPC) because most of the polar species are poorly retained on reversed-phase columns eluting at the column dead time or exhibiting little or no resolution. In addition, despite the fact that normal-phase chromatography (NPC), which uses polar stationary phases, for example, silica microparticles with nonpolar organic mobile phases such as hexane, heptane, ethyl acetate, chloroform, or mixtures of them, provides a variety of chromatographic solutions for the separation and purification of a wide range of solutes; it has its own limitations in (1) separating highly polar analytes, (2) dissolving polar compounds in nonaqueous mobile phases, (3) suffering from poor reproducibility, and in (4) exhibiting poor ionization efficiency in mass spectrometry. All of these facts triggered Alpert about 25 years ago to introduce a novel chromatographic technique called *hydrophilic interaction liquid chromatography* (HILIC) for the separation of peptides, nucleic acids, and other polar compounds [6]. In HILIC, a polar-bonded stationary phase is used with an organic rich hydro-organic mobile phase (e.g., ACN/H$_2$O) in order to separate polar analytes. The organic-rich mobile phases used in HILIC provide some additional advantages, including low-column back pressure, which allows fast separation of analytes with shorter analysis time and permits the direct coupling of HILIC with MS detection.

Similarly to other modes of interactive chromatography, the mobile-phase composition has strong influence on polar solutes' retention in HILIC and HICEC. Usually, with an organic-rich hydro-organic mobile phase, a water-rich layer is formed on the polar stationary phase surface. The separation is achieved by the partitioning of polar solutes in between this adsorbed water layer on the stationary phase surface and the organic-rich hydro-organic mobile phase. However, a simple retention mechanism is not possible for most compounds because HILIC encompasses all kind of polar interactions (e.g., electrostatic, hydrogen bonding, and dipole–dipole) [7].

6.2.1 DIRECT COPOLYMERIZATION

The majority of the reported polar organic polymer monoliths for HILIC separations are based on acrylate/methacrylate-derived monoliths. This may be due to the unique properties of acrylate/methacrylate monoliths, including, among other things, high chemical stability over a wide pH range and excellent mechanical properties. In addition, they are much less nonpolar than styrene-based monoliths.

The most convenient approach for the preparation of a polar or nonpolar type of monolithic stationary phase having the desired surface interactive ligands has been the direct *in situ* copolymerization of the suitable functional monomer and cross-linker in the presence of an appropriate porogen. There are a variety of monolithic HILIC columns described in the recent literature, and consequently the author of this chapter adopted a practical classification based on the nature of the surface charge of the different monoliths to conveniently discuss the monoliths reported thus far. In addition, whether polar or nonpolar monoliths are the subjects of the ensuing discussion, the reader is advised to consult the listing provided in Tables 6.1 and 6.2 for the names, abbreviations, and particularly the structures of the monomers, cross-linkers, and functionalization chemicals used in the preparation of a given monolithic stationary phase.

6.2.1.1 Nature of the Surface Charge

As the polarity and charge of the functional monomers are critical to the design and chromatographic behaviors of the resulting monoliths, the ensuing discussion provides a comprehensive review organized according to the following classification: (1) anionic monoliths (2) cationic monoliths, (3) neutral monoliths, and (4) zwitterionic monoliths. See Tables 6.1 and 6.2 for the structures of functional monomers and cross-linkers.

TABLE 6.1
Names, Abbreviations, and Structures of Anionic, Cationic, Zwitterionic, and Neutral Functional Monomers. The Species Are Arranged Alphabetically in the Ascending Order According to Their Abbreviations

Name	Abbreviation	Structure
Anionic Functional Monomers		
6-Azido-hexanoic acid	AHA	
2-Acrylamido-2-methyl-1-propanesulfonic acid	AMPS	
Methacrylic acid	MAA	
Sodium 2-methylprop-2-ene-1-sulfonate	MPS	
3-Sulfopropyl methacrylate potassium	SPMA	
4-Vinylphenylboronic acid	VPBA	
Cationic Functional Monomers		
2-(Methacryloyloxy) ethyltrimethylammonium methyl sulfate	META	
N-(3-Aminopropyl) methacrylamide hydrochloride	NAPMH	
Vinylbenzyl trimethylammonium chloride	VBTA	

(*Continued*)

TABLE 6.1 (*Continued*)
Names, Abbreviations, and Structures of Anionic, Cationic, Zwitterionic, and Neutral Functional Monomers. The Species Are Arranged Alphabetically in the Ascending Order According to Their Abbreviations

Name	Abbreviation	Structure
Zwitterionic Functional Monomers		
N,N-Dimethyl-N-methacryloxyethyl-N-(3-sulfopropyl) ammonium betaine	MEDSA	
2-Methacryloxyethyl phosphorylcholine	MPC	
N,N-Dimethyl-N-acryloyloxyethyl-N-(3-sulfopropyl)ammonium betaine	SPDA	
N,N-Dimethyl-N-(3-methacrylamidopropyl)-N-(3-sulfopropyl)-ammonium betaine	SPP	
Neutral Functional Monomers		
3-Methylolacrylol-3-oxapropyl-3-(NN'-dioctadecylcarbomyl)-propionate	AOD	
Butyl methacrylate	BMA	
Cyclohexyl methacrylate	CHM	
2-Ethylhexyl methacrylate	EHM	
Glycerol carbonate methacrylate	GCMA	

(Continued)

TABLE 6.1 (*Continued*)
Names, Abbreviations, and Structures of Anionic, Cationic, Zwitterionic, and Neutral Functional Monomers. The Species Are Arranged Alphabetically in the Ascending Order According to Their Abbreviations

Name	Abbreviation	Structure
Glyceryl methacrylate	GMM	
Glycidyl methacrylate	GMA	
2-Hydroxyethyl methacrylate	HEMA	
Hexyl methacrylate	HMA	
3-Chloro-2-hydroxylpropyl methacrylate	HPMA	
Lauryl methacrylate	LMA	
N-Acryloyl tris(hydroxymethyl) aminomethane	NAHAM	
2-Naphthyl methacrylate	NAPM	
N-Acryloxysuccinimide	NAS	
N-Vinyl Carbazole	NVC	

(*Continued*)

TABLE 6.1 (*Continued*)
Names, Abbreviations, and Structures of Anionic, Cationic, Zwitterionic, and Neutral Functional Monomers. The Species Are Arranged Alphabetically in the Ascending Order According to Their Abbreviations

Name	Abbreviation	Structure
N-Vinyl pyrrolidone	NVP	
Octadecylacrylate	ODA	
Propargyl methacrylate	PMA	
Stearyl methacrylate	SMA	
Vinyl decanoate	VDC	
Vinyl pivalate	VPV	

TABLE 6.2
Names, Abbreviations, and Structures of Cross-Linkers. The Species Are Arranged Alphabetically in the Ascending Order According to Their Abbreviations

Alkanediol dimethacrylate n = 4–12	ADDMA	
4,4′-Bis(chloromethyl)-1,1′-biphenyl	BCMBP	
1,3-Butanediol dimethacrylate	BDDMA	

(*Continued*)

TABLE 6.2 (*Continued*)
Names, Abbreviations, and Structures of Cross-Linkers. The Species Are Arranged Alphabetically in the Ascending Order According to Their Abbreviations

Bisphenol A dimethacrylate	BIDMA	
Bisphenol A ethoxylate diacrylate	BIEDA	
Bisphenol A glycerolate dimethacrylate	BIGDMA	
Tetramethylene dimethacrylate or 1,4-butanediol dimethacrylate	BUDMA	
1,2-Bis(*p*-vinyl phenyl) ethane	BVPE	
1,4-Cyclohexanediol dimethacrylate	CHDMA	1,4-Cyclohexanediol dimethacrylate
α,α′-Dichloro-*p*-xylene	DCX	
Dioxyethylene dimethacrylate	DiEDMA	
1,12-dodecanediol dimethacrylate	1,12-DoDDMA	
Dipentaerythritol penta/hexa-acrylate	DPEPA	R = H or
Divinylbenzene	DVB	

(*Continued*)

TABLE 6.2 (*Continued*)
Names, Abbreviations, and Structures of Cross-Linkers. The Species Are Arranged Alphabetically in the Ascending Order According to Their Abbreviations

Ethylene glycol dimethacrylate	EDMA	
1,6-Hexanediol dimethacrylate	HDDMA	
1,6-Hexanediol ethoxylate diacrylate	1,6 HEDA	
Hexamethylene dimethacrylate	HEDMA	
N,N'-Methylene bisacrylamide	MBA	
Methylene dimethacrylate n = 1, 2, 3	MEDMA	
Neopentyl glycol dimethacrylate	NPGDMA	
1,7-Octadiyne	ODY	
Oxymethylene dimethacrylate n = 1, 2, 3	OMDMA	
1,4-Bis(acryloyl) piperazine or piperazine diacrylamide	PDA	
Pentaerythritol diacrylate monostearate	PEDAS	

(*Continued*)

TABLE 6.2 (*Continued*)
Names, Abbreviations, and Structures of Cross-Linkers. The Species Are Arranged Alphabetically in the Ascending Order According to Their Abbreviations

Poly(ethylene glycol) diacrylate	PEGDA	
Pentaerythritol triacrylate	PETA	
Pentaerythritol tetraacrylate	PETAC	
1,4-Phenylene diacrylate	PhDA	
Pentaerythritol tetrakis(3-mercaptopropionate)	PTM	
Triallyl isocyanurate	TAIC	
Tetraethylene glycol dimethacrylate	TeEDMA	
Trimethylolpropane triacrylate	TMPTA	

(Continued)

TABLE 6.2 (*Continued*)
Names, Abbreviations, and Structures of Cross-Linkers. The Species Are Arranged Alphabetically in the Ascending Order According to Their Abbreviations

Trimethylolpropane tris(3-mercaptopropionate)	TPTM	
Trimethylolpropane trimethacrylate	TRIM	
1,2,4-Trivinyl cyclohexane	TVCH	
Bisphenol A epoxy vinyl ester resin	VER	

6.2.1.1.1 Anionic Monoliths

An anionic poly(methacrylic acid-*co*-EDMA) [poly(MAA-*co*-EDMA)] monolith has been prepared and tested in capillary HILIC (cHILIC) [8]. On deprotonation of the carboxylic acid group at pH > 5.0, the monolithic surface becomes an anionic surface. As reported, this monolithic column showed homogeneous and continuous column bed, good permeability, and narrow pore size distribution. In order to investigate the effect of the amount of the functional monomer MAA on the porosity of the poly(MAA-*co*-EDMA) anionic monolith, the concentrations of MAA in the polymerization mixture were varied. The results revealed that the permeability increased as well as the ion-exchange capacity of the column increased at elevated ratio of MAA to EDMA, but at the expenses of yielding a heterogeneous column bed [8]. Furthermore, the effects of hydrophilic and hydrophobic porogenic systems on the porosity of the poly(MAA-*co*-EDMA) monolith and its separation efficiencies in HILIC have been reported. Two different sets of poly(MAA-*co*-EDMA) monoliths were prepared using hydrophilic porogen system consisting of PEG in dimethyl sulfoxide and hydrophobic porogen system containing dodecanol and toluene. The scanning electron microscopic examinations showed that the homogeneity of the hydrophobic porogen monolith was not as good as that in the hydrophilic porogen one. The chromatographic evaluation revealed that the separation efficiency of the monolithic column prepared with the hydrophilic porogen was much higher than the column prepared with the hydrophobic porogen. In addition, the larger ion-exchange capacity of the former column provided stronger hydrophilic interaction for polar compounds as well as more pronounced electrostatic interactions for charged compounds. Based on these results, it was believed that the hydrophilic porogen might have promoted the exposure of the carboxylic acid groups of the MAA on the monolithic surface, and as a consequence a larger ion-exchange capacity resulted. In addition, increasing the molecular weight of PEG increased the permeability of the monolith at the expense of decreasing its specific surface area. This was explained by the fact that increasing the molecular weight of PEG produced a solvated system with higher steric hindrance, and consequently larger through pores were obtained. On another front, permeability of the prepared monoliths increased with increasing the amount of PEG in the polymerization mixture. Five nucleosides were baseline separated using the poly(MAA-co-EDMA) monolithic under HILIC mobile phase composition. This monolithic column was also used to separate aniline and benzoic acid derivatives, and the retention behaviors of these charged solutes were typical of HILIC behaviors on this anionic monolithic column [8].

Another hydrophilic polymethacrylate-based monolithic column for HI–CEC applications has been reported [9]. It was prepared via the copolymerization of 2-acrylamido-2-methyl-1-propanesulfonic acid (AMPS) functional monomer and PETA cross-linker. A decrease in the column permeability and pore size was observed with increasing wt% of AMPS, whereas an increase in the average pore size and column permeability was realized with increasing the methanol wt% in the polymerization solution. On account of the low solubility of AMPS in organic solvents, water was added as one of the porogens for dissolving AMPS. The anionic stationary phase surface was fully ionized at pH 4.5, a pH that is much higher than 1.67, which is the pK_a of the sulfonate groups of the AMPS monomer. The polar AMPS monomer together with the polar PETA cross-linker imparted a hydrophilic character to the poly(AMPS-co-PETA) monolithic column. A mixed-mode type of interaction consisting of HI–strong cation exchange (SCX) was observed in the analysis of charged peptides with high column separation efficiencies without peak tailing (Figure 6.1). The favorable hydrophilicity and satisfying stability of the poly(AMPS-co-PETA) monolithic column were further exploited in the separation of polar analytes such as neutral polar amides and phenols.

A trimodal polymer monolith with three modes of interactions, including reversed phase, hydrophilic, and cation-exchange interactions has been introduced [10]. This monolith was prepared by the in situ copolymerization of glycidyl methacrylate (GMA) and 4-vinylphenylboronic acid (VPBA) functional monomers with the EDMA cross-linker [10]. The boronic acid groups on the monolithic surface become ionized at high pH values yielding an anionic monolithic surface. VPBA was chosen as one of the monomers to prepare the mixed-mode monolith because it was believed that monoliths containing hydrophilic/ionizable $B(OH)_2$ groups would afford more flexible adjustment of selectivity in terms of hydrophobic, hydrophilic, as well as cation-exchange interactions. Higher content of VPBA in the polymerization mixture led to the formation of smaller microglobules inside the capillary during the copolymerization process, which caused a higher back pressure. The polar functionalities of

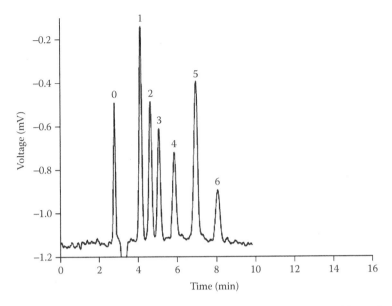

FIGURE 6.1 Electrochromatogram of peptides. Experimental conditions: capillary column B, 30 cm effective length, 50 cm total length, 100 μm i.d.; mobile phase, 80 mmol/L ammonium formate pH 4.5 in ACN/H$_2$O (50/50 (v/v)); applied voltage: −10 kV; supplement pressure: 3.4 MPa; detection wavelength, 214 nm. Solutes: 0. Toluene, 1. Tyr–Gly–Gly–Phe–Met, 2. Tyr–Gly–Gly–Phe–Leu, 3. Gly–Gly–Phe–Met, 4. Gly–Gly–Phe–Leu, 5. Tyr–Gly, 6. Tyr–Gly–Gly. (Reproduced from Lin, J. et al., *J. Chromatogr. A*, 1218, 4671, 2011. With permission.)

boronic acid-bonded monolith could contribute to the separation of polar analytes in the HILIC mode. As boronic acid functionalities are weakly acidic (pK_a = 8.86), the monolithic surface becomes more negatively charged as the pH of the mobile phase becomes highly alkaline. Among several binary porogenic solvents investigated when preparing the poly(GMA-*co*-VPBA-*co*-EDMA) monolith, BDO/diethylene glycol (DEG) was best suited for the solubilization of the hydrophobic GMA and EDMA, and the hydrophilic VPBA in order to prepare the polymerization solution. As the polymerization is temperature sensitive, the effect of temperature on the morphology and permeability of the resultant monolithic column was also studied. These results suggested that a high reaction temperature leads to a fast polymerization and tends to form denser and smaller flow-through pores. The column performance was assessed by the separation of series of amides and anilines. Poly(GMA-*co*-VPBA-*co*-EDMA) monolith was used to separate alkaloids and proteins successfully due to its RPC and HILIC mixed-mode retention behavior.

Yet, another hydrophilic methacrylate-based monolith has been reported in Reference 11 *via* the *in situ* copolymerization of the functional monomer 2-hydroxyethylmethacrylate (HEMA) and the polar cross-linker *N*,*N'*-methylenebisacrylamide (MBAA). The authors have successfully used this column to separate amines, nucleosides, and narcotics with good reproducibility in the

HI–CEC mode [11]. AMPS was incorporated into the poly(HEMA-*co*-MBAA) monolithic surface in order to generate an EOF whereby the zeta potential was mainly provided by the sulfonic acid groups of AMPS, which are totally dissociated when the pH value of the mobile phase exceeds 3.0, thus ensuring a permanent negative charge density on the monolithic surface. The polar sites on the surface of the monolith responsible for hydrophilic interactions are the hydroxyl, amide, and sulfonic acid groups. The sulfonated monolith provided typical HI–CEC retention behavior for polar solutes. An increase in the content of AMPS increased the charge density of the monolith, and in turn, increased the EOF velocity and reduced the analysis time.

A convenient procedure has been developed by combining free radical polymerization and azide–alkyne cycloaddition *click* reaction for the one-pot synthesis of a polymer monolithic column carrying 6-azidohexanoic acid (AHA) without using any postmodification [12]. The preparation of poly(AHA-*co*-propargyl methacrylate-*co*-ethylene dimethacrylate) [poly(AHA-*co*-PMA-*co*-EDMA)] involved two simultaneous processes, namely the polymerization of PMA and EDMA to yield the poly(PMA-*co*-EDMA) monolith skeleton coupled with the Cu (I) catalyzed 1,3-dipolar azide-alkyne cycloaddition (CuAAC) click reaction between AHA and alkyne-containing monolith skeleton to produce the poly(AHA-*co*-PMA-*co*-EDMA) monolith. This monolith showed HILIC/RPC mixed-mode retention behavior depending on the %ACN content in the mobile phase. Mixtures of nucleobases/nucleosides, polar alkaloids, and phenolic compounds were separated on the poly(AHA-*co*-PMA-*co*-EDMA) monolith under HILIC conditions, whereas a mixture of polycyclic aromatic hydrocarbons was separated on the same column under RPC conditions. In addition, the poly(AHA-*co*-PMA-*co*-EDMA) monolith showed electrostatic interactions with charged analytes due to the existence of carboxyl functionalities in the monolith.

Derivatized β-cyclodextrin (β-CD) functionalized monolithic columns have been prepared by a *one-step* strategy using click chemistry [13]. The intended derivatized β-CD functional monomers were first synthesized by a click reaction between PMA and mono-6-azido-β-CD, which was followed by sulfonation or methylation of the resulting derivatized β-CD functional monomers. Thereafter, monolithic columns were prepared through a one-step *in situ* copolymerization of the derivatized β-CD functional monomers and EDMA cross-linker. Three monolithic columns were prepared using underivatized β-CD, sulfated β-CD, and methylated β-CD. The hydroxyl functionalities in unmodified β-CD monolith and sulfonic acid groups in sulfated β-CD monolith (anionic monolith) imparted hydrophilic character to those monoliths and polar solutes, including nucleosides, small peptides, and some other polar solutes were well separated under HILIC conditions, whereas methylated β-CD monolith was useful for the separation of nonpolar compounds and drug enantiomers in capillary RPC. Furthermore, the retention of highly hydrophilic analytes on sulfated β-CD monolith (i.e., the anionic monolith) was significantly higher than that of unmodified β-CD monolith due to the substitution of hydroxyl groups with the more polar sulfonic acid groups.

6.2.1.1.2 Cationic Monoliths

These positively charged monoliths have been used in both HILIC and HI–CEC, and they can be viewed as hydrophilic interaction/strong anion-exchange (HI–SAX) monoliths. Gunasena and El Rassi [14] reported a cationic HI–CEC monolith column designated as AP monolith that resulted from the *in situ* copolymerization of the cross-linker EDMA and the functional monomer *N*-(3-aminopropyl) methacrylamide hydrochloride (NAPMH). This cationic AP monolith possesses amine/amide functionalities on its surface. Noticeable variations in the chromatographic properties, including retention factor (k) and the selectivity factor (α) were observed on subtle changes in the monomer composition in the polymerization mixture. The k values for polar solutes decreased as the percentage of the cross-linker EDMA in the polymerization solution was decreased. Over the pH range studied, the AP monolith showed a strong anodal EOF. The EOF velocity was increased with increasing the pH of the mobile phase up to pH 6 and thereafter the EOF velocity remained unchanged. This behavior was explained by the ionization of surface amino groups and the interplay of the ionization and the shielding effects of the adsorbed mobile-phase ions to the surface of the monolithic stationary phase. With increasing the mobile phase pH, the ionization of the amino groups of the AP monolith decreased, which would lower the electrostatic attractions and in turn the neutralization/shielding effect exercised by the negatively charged acetate ions of the mobile phase. This led to an apparent increase in the EOF velocity with increasing pH in the range 4–6. The AP monolith was characterized using various polar solutes, including phenols, substituted phenols, and amides, see Figure 6.2 for a typical electrochromatogram of some polar compounds.

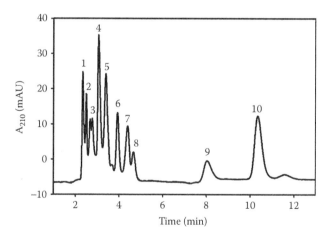

FIGURE 6.2 Separation of 10 polar compounds obtained on the NAPM/EDMA monolith (called, AP-monolith). Conditions: AP–monolithic capillary column, 33.5 cm (effective length 25 cm) × 100 μm i.d.; running voltage, −20 kV; column temperature, 20°C; hydro-organic mobile phase, 5 mM NH$_4$Ac at 95% ACN (v/v); EOF tracer, toluene. Solutes: 1, cresol; 2, 2-amino-4-cresol; 3, phenol; 4, 4-amino-2-nitrophenol; 5, 4-cyanophenol; 6, resorcinol; 7, catechol; 8, hydroquinone; 9, 4-nitrophenol; 10, pyrogallol. (Reproduced from Gunasena, D.N. and El Rassi, Z., *J. Chromatogr. A*, 1317, 77, 2013. With permission.)

A HILIC monolith with surface fixed positively charged groups has been introduced [15]. It was prepared by copolymerizing the functional monomer vinylbenzyl
trimethylammonium chloride (VBTA) with the cross-linker bisphenol A glycerolate
dimethacrylate (BIGDMA). It is believed that the multiple polar hydroxyl and ester
groups of BIGDMA imparted a hydrophilic backbone to the polymeric monolith
[15], whereas the ammonium groups of the VBTA functional monomer afforded
strong electrostatic interactions with anionic solutes, and its vinylbenzyl groups provided rigid phenyl framework for the fabrication of a stable polymeric structure. This
reported poly(VBTA-co-BIGDMA) monolith afforded hydrophilic and electrostatic
interactions in HI–CEC. In addition, this monolith showed RP–HILIC mixed-mode
retention behavior depending on the organic content of the mobile phase. Finally,
this poly(VBTA-co-BIGDMA) monolith was successfully used to separate benzoic
acid derivatives, phenols, nucleosides, and nucleobases.

The same research group reported another HI–SAX monolith for CEC [16],
which was prepared by the copolymerization of 2-(methacryloyloxy)ethyltrimeth-
ylammonium methyl sulfate (META) as the functional monomer and PETA as the
cross-linker. As the percentage of META increased while keeping the monomer to
porogen ratio constant, a noticeable increase in the permeability of the monolith
was observed. This was attributed to the reduced polymeric aggregation caused by
the decrease in the amount of cross-linker. In addition, increasing the percentage of
META led obviously to increase the hydrophilicity of the monolith as well as the
monolith charge density, and in turn, to increase its zeta potential. As expected, the
ammonium groups on the monolithic surface generated an anodic EOF. The retention patterns of polar analytes, including phenolic compounds and various amides
clearly showed hydrophilic interaction/anion exchange mixed-mode retention behavior. Polar phenols and charged basic nucleic acid bases and nucleosides were also
used for the evaluation of the retention mechanism of the column. This HI–SAX
monolith was successfully used in the separation of carboxylic phytohormones.

Similarly, another research article reported the preparation of HI-weak anion
exchange (WAX)-RP trimodal polymer monoliths by one-step copolymerization
of amino acid-based monomers and MBA cross-linker [17]. First, amino acid containing monomer, namely N-methacryloyl-L-phenylalanine methyl ester (MA–L–
Phe–OMe) was synthesized through the acylation reaction between L-phenylalanine
methyl ester hydrochloride and methacryloyl chloride. Then, the functional monomer
(MA–L–Phe–OMe) and the cross-linker MBA were in situ copolymerized in a short
stainless steel column of 4.6 mm i.d. yielding the poly(MA–L–Phe–OMe-co-MBA)
monolith. In this trimodal monolith, MA–L–Phe–OMe has a hydrophobic aromatic
ring and MBA has a more hydrophilic amide group. As one would expect, the monolithic column showed mixed-mode retention behavior and allowed the separation
of hydrophobic polycyclic aromatic hydrocarbons and hydrophilic nucleobases and
nucleosides by just altering the ACN content of the mobile phase.

A one-pot approach was reported for the preparation of poly(3'-azido-3'-
deoxythymidine-co-propargyl methacrylate-co-pentaerythritol triacrylate) [poly(AZT-
co-PMA-co-PETA)] monolith by combining free radical polymerization with CuAAC
click chemistry [18]. The antiretroviral drug azidothymidine (AZT) used for the
treatment of HIV/AIDS infection has a unique chemical structure possessing both

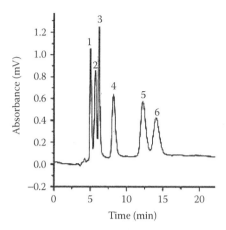

FIGURE 6.3 (a) Separation of anesthetics. Conditions: mobile phase: 20 mM PB containing 90.5% (v/v) CAN at pH 7.0; flow rate (actual flow rate after splitting): 0.035 mL min^{-1} (140 nL min^{-1}); pump pressure: 3.8 MPa; detection wavelength: 214 nm; the analytes are (1) papaverine (100 ppm); (2) triazolam (100 ppm); (3) pemoline (100 ppm); (4) codeine (100 ppm); (5) morphine (100 ppm); (6) adanon (100 ppm). (Reproduced from Lin, Z. et al., *Analyst*, 140, 4626, 2015. With permission.)

the hydrophilic/ionizable nucleoside group and the azido group. The chromatographic property of the poly(AZT-*co*-PMA-*co*-PETA) monolith in terms of hydrophobic, hydrophilic, and cation exchange retention mechanisms were evaluated with alkylbenzenes, amides, anilines, and benzoic acids. In addition, the potentials of the monolithic columns were demonstrated in the chromatography of sulfonamides, nucleobases, nucleosides, anesthetics (Figure 6.3), and proteins. The reported results proved the presence of electrostatic interactions between poly(AZT-*co*-PMA-*co*-PETA) monolith and charged analytes due to the existence of multiple ionizable moieties.

6.2.1.1.3 Neutral Monoliths

A neutral hydroxyl monolith for HI–CEC that resulted from the copolymerization of the polar cross-linker, pentaerythritol triacrylate (PETA), and the functional monomer, glyceryl methacrylate (GMM) was introduced by Gunasena and El Rassi [14]. Both GMM and PETA have hydroxyl groups in their structures, which impart the poly(GMM-*co*-PETA) monolith with hydrophilic interaction sites. By considering the fact that the polymer backbone of methacrylate-based monolith is somewhat hydrophobic [19], a polar cross-linker PETA has been used instead of the traditional ethylene glycol dimethacrylate (EDMA) cross-linker in order to realize a more hydrophilic polymer backbone. Although without fixed charges on the surface, the hydroxyl monolith showed cathodal EOF. It is believed that the EOF is due to the adsorption of mobile-phase ions (e.g., acetate ions) onto the monolithic surface, which impart the monolith with the zeta potential necessary to generate an adequate EOF. Most interestingly, the magnitude of the EOF increased with increasing the %ACN in the mobile phase (Figure 6.4), thus confirming the

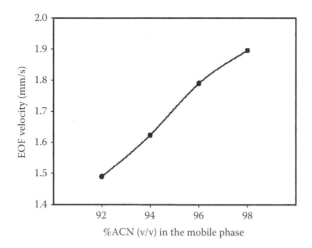

FIGURE 6.4   Plot of EOF velocity (mm/s) vs. %ACN (v/v) in the mobile phase. Conditions: hydroxy monolithic capillary column, 33.5 cm (effective length 25 cm) × 100 μm i.d.; running voltage, 20 kV; column temperature, 20°C; hydro-organic mobile phase, 5 mM NH$_4$Ac (pH 7.0) at various %ACN; EOF tracer, toluene; sample injection, pressure at 5 bar for 10 s. (Reproduced from Gunasena, D.N. and El Rassi, Z., *J. Chromatogr. A*, 1317, 77, 2013. With permission.)

mobile-phase ion adsorption onto the neutral polar monolithic surface. The generation of EOF by neutral monoliths has been reported first by Okanda and El Rassi and later confirmed by other contributions from the laboratories of the author of this chapter. Returning to the hydroxyl monolith, it was successfully used to separate polar solutes, including phenols and phenolic derivative, nucleic acid bases, and nucleosides [14].

A novel poly(*N*-acryloyltris(hydroxymethyl)aminomethane-*co*-pentaerythritol triacrylate), poly(NAHAM-*co*-PETA) monolith has been prepared [19], and its usefulness investigated in capillary liquid chromatography (cLC). It was synthesized by the *in situ* copolymerization of the functional monomer NAHAM and the cross-linker PETA. The poly(NAHAM-*co*-PETA) monolith can be viewed as a neutral but polar monolith with hydroxyl and amide functionalities. Separations of polar analytes on this column are mainly due to hydrophilic interactions between the separated analytes and the surface hydroxyl groups of the monolith. The performance of poly(NAHAM-*co*-PETA) monolith column has been evaluated using five standard nucleosides, seven standard benzoic acids, and five standard anilines.

An amide functionalized hydrophilic monolith was synthesized by the *in situ* photopolymerization of *N*-vinyl-2-pyrrolidinone (NVP), acrylamide (AM), and *N,N'*-methylenebisacrylamide (MBA) in fused silica capillaries, and applied for HILIC-based enrichment of *N*-linked glycopeptides [20]. Although AM and MBA were selected as the functional monomer and the cross-linker, respectively, the NVP functioned as the stabilizing monomer to impart a good mechanical stability to the monolith at high mobile-phase flow velocities. *N*-Linked glycopeptides were retained on the polar amide column with 80%–85% ACN mobile phase and then eluted from the column by lowering the %ACN content to 70%. According to the report, nonglycosylated peptides were hardly seen in the elution fraction, demonstrating the high specificity of the monolith for glycopeptide enrichment. In addition, the amide functionalized hydrophilic monolith was shown useful in the N-glycosylation profiling of HeLa cells and human serum samples.

6.2.1.1.4 Zwitterionic Monoliths

Zwitterionic monoliths combining on their surfaces both positive and negative fixed charges are known for their weak electrostatic interactions with charged analytes, which are advantageous when combined with the selectivity of hydrophilic interactions [21]. Polar sulfoalkylbetaine monomers, possessing both positively charged quaternary ammonium and negatively charged sulfonic groups, have been used in the preparation of zwitterionic monolithic columns for HILIC and HI–CEC of polar solutes [7,21–23].

As the chemistry of the cross-linking monomer is an important parameter affecting the properties of monolithic columns prepared by using single-step polymerization procedures [9,11], a brief discussion of reported effects in recent literature is in order here. A study aiming at evaluating the effects of the nature of the cross-linker on the performance of polar monolithic columns intended for use in HILIC applications was reported recently in Reference 24. Seven different polymethacrylate monolithic capillary columns have been synthesized using N,N-dimethyl-N-methacryloxyethyl-N-(3-sulfopropyl)ammonium betaine (MEDSA) functional monomer and various cross-linking monomers differing in polarity and size, including EDMA, tetramethylene dimethacrylate (BUDMA), hexamethylene dimethacrylate (HEDMA), dioxyethylene dimethacrylate (DiEDMA), PETA, bisphenol A dimethacrylate (BIDMA), and BIGDMA. The authors have reported the following findings based on their experimental results: (1) the type of the cross-linker affects the pore size and its distribution, (2) the efficiency of monolithic columns for polar, low molecular weight compounds in the HILIC mode depends rather on the polarity than on the size of the cross-linker molecules and no clear correlation is apparent between the cross-linker methylene chain length and the column separation efficiency, (3) the proportion of mesopores in the polar zwitterionic polymethacrylate monolithic columns depends on the length of the cross-linking dimethacrylate, whereas the mesopore size and distribution depend on the polarity of the cross-linkers, and (4) the zwitterionic monolithic columns prepared with polar cross-linkers show dual RPC–HILIC separation mechanisms, depending on the sample properties, and on the mobile-phase composition [24].

Another study involved the preparation of monolithic microcolumns by the *in situ* copolymerization of zwitterionic sulfobetaine functional monomer MEDSA with BIGDMA and DiEDMA cross-linkers in the presence of 1-propanol, BDO, and water as the porogens [25]. The separation efficiency and selectivity of the poly(MEDSA-*co*-BIGDMA) column did not change significantly even after more than 3 weeks of continuous use, which is much better than the poly(MEDSA-*co*-DiEDMA) column, which lost approximately half of its separation efficiency after 30 h of use, probably because the pore morphology in the monolith prepared with the DiEDMA cross-linker is less stable. Some significant differences between the separation selectivity on the poly(MEDSA-*co*-DiEDMA) and poly(MEDSA-*co*-BIGDMA) monolithic columns were observed, including the reverse order of elution for some solutes. The poly(MEDSA-*co*-BIGDMA) column provided a remarkable baseline separation of diastereomers (+) catechin and (−) epicatechin, containing two chiral carbon atoms, whereas those two compounds completely coeluted on the poly(MEDSA-*co*-DiEDMA) column. Both poly(MEDSA-*co*-BIGDMA) and poly(MEDSA-*co*-DiEDMA) columns were used for the separation of phenolic acids and flavonoid compounds in one-dimensional chromatography (1D) in the HILIC mode. The HILIC analysis of a mixture of antioxidants containing 32 phenolic acids and flavonoids on the monolithic columns under gradient conditions showed an incomplete separation. To achieve a better separation, the authors have coupled these zwitterionic HILIC columns with a short nonpolar core shell column in the second dimension, for comprehensive 2D LC separations of phenolic and flavonoid compounds [25].

Another monolithic HILIC column was prepared through the *in situ* copolymerization of the zwitterionic monomer *N,N*-dimethyl-*N*-(3-methacryl-amidopropyl)-*N*-(3-sulfopropyl)ammonium betaine (SPP) and the cross-linker EDMA [21]. The influence of the EDMA content of the monolith was investigated. Increasing the EDMA weight fraction in the polymerization mixture resulted in a decrease of theoretical plate height. It was also observed that the porosity of monolithic column increased significantly with increasing the weight proportion of EDMA. SEM measurements showed that increasing the weight fraction of EDMA yielded a polymer with bigger microglobules. This observation was not consistent with the previously reported effect of EDMA, which necessitated further investigation to clarify the reason for this behavior [21]. Methanol has been used as the porogen because it provides sufficient solubility to both the functional monomer and the cross-linker when preparing poly(SPP-*co*-EDMA) monolith [21]. It was also observed that the porosity of the monolithic column slightly increased with increasing the proportion of methanol in the polymerization mixture. The zwitterionic surface of the poly(SPP-*co*-EDMA) monolithic column offered the possibility of weak electrostatic interactions with charged analytes, making it possible to develop HILIC separations by manipulating the pH value, ionic strength, and the organic solvent content of the mobile phase. Typical HILIC retention mechanism was exhibited by the poly(SPP-*co*-EDMA) monolithic column at high ACN content (>60% v/v). This poly(SPP-*co*-EDMA) monolith was successfully used to separate benzoic acids, phenols, and a series of basic and neutral compounds (Figure 6.5). In addition, highly polar ascorbic acid and dehydroascorbic acid were simultaneously separated using the poly(SPP-*co*-EDMA) monolithic column.

The highly polar monomer *N,N*-dimethyl-*N*-acryloyloxyethyl-*N*-(3-sulfopropyl) ammonium betaine (SPDA) and the hydrophilic cross-linker MBA were used for the preparation of a HILIC monolith to be used in micro-HPLC [23]. For the purpose of comparison, another novel monolithic column containing SPDA as the functional monomer and EDMA as the cross-linker has been prepared. The polarity and separation efficiency of poly(SPDA-*co*-MBA) monolith column was compared with that of the poly(SPDA-*co*-EDMA) monolithic column. The functional monomer SPDA showed poor solubility in commonly used porogenic solvents, including methanol.

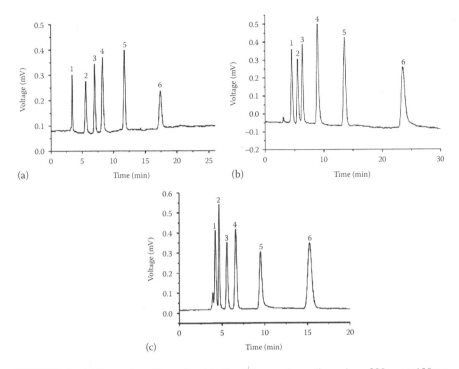

FIGURE 6.5 (a) Separation of benzoic acids. Conditions: column dimensions, 280 mm × 100 μm id; mobile phase, 50 mM ammonium formate pH 5.3 in ACN/H$_2$O (83:17, v/v); samples: (1) benzoic acid; (2) 2-hydroxybenzoic acid; (3) 4-hydroxybenzoic acid; (4) 2,4-dihydroxybenzoic acid; (5) 3,5-dihydroxybenzoic acid; (6) 3,4,5-trihydroxybenzoic acid; detection wavelength, 234 nm; flow rate 500 nL/min. (b) Separation of basic and neutral samples. Conditions: mobile phase, ACN/H$_2$O (95:5, v/v); detection wavelength, 254 nm; samples: (1) toluene; (2) thymine; (3) uracil; (4) adenine; (5) thiourea; (6) cytosine; others as in (A). (c) Separation of phenols. Conditions: mobile phase, ACN/H$_2$O (90:10, v/v); detection wavelength, 214 nm; samples: (1) 3-hydroxybenzotrifluoride; (2) phenol; (3) p-aminophenol; (4) resorcinol; (5) pyrogallol; (6) phloroglucinol; others as in (a). (Reproduced from Liu, Z. et al., *J. Sep. Sci.*, 36, 262, 2013. With permission.)

To overcome this problem water was used as the cosolvent. But, the cross-linker EDMA showed a low solubility in water. The contrasting solubility between SPDA and EDMA provided a very narrow space for the optimization of the polymerization mixture composition of poly(SPDA-*co*-EDMA) monolith. Therefore, a relatively high content of MeOH in the porogen was required to dissolve both SPDA and EDMA. Compared to EDMA, the hydrophilic cross-linker MBA showed good solubility in water and it allowed the preparation of monoliths with higher proportion of the functional monomer SPDA in poly(SPDA-*co*-MBA) monolith. The authors have observed a decrease of column permeability when the weight content of the cross-linker MBA was increased in the polymerization mixture, whereas all other conditions remained the same. The porosity of the poly(SPDA-*co*-MBA) monolith column clearly increased as the weight content of the porogen increased. The authors

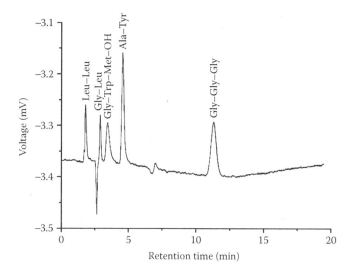

FIGURE 6.6 Separation of peptides on poly(SPDA-*co*-MBA) monolithic column. Experimental conditions: column dimension, 222 mm × 100 μm I.D.; mobile phase, 20 mM ammonium formate (pH 3.0) in ACN/H$_2$O (75/25, v/v); detection wavelength, 214 nm; flow rate, 800 nL/min. (Reproduced from Yuan, G. et al., *J. Chromatogr. A*, 1301, 88, 2013. With permission.)

have reported that the poly(SPDA-*co*-MBA) monolith yields better overall separation efficiencies, and greater polarity than the poly(SPDA-*co*-EDMA) monolith. In addition, a significantly enhanced hydrophilicity was observed for the poly(SPDA-*co*-MBA) monolith compared to the previously reported zwitterionic sulfobetaine monolithic columns, which could be evidenced by the lowered critical composition of the mobile-phase corresponding to the transition from the HILIC to the RPC retention mode. It was suggested that the increased hydrophilicity could be the result of the incorporation of the more hydrophilic functional monomer SPDA over that of the highly polar cross-linker MBA. The retention mechanism studies showed that electrostatic interactions could also contribute to the overall retention of charged analytes. The final optimized poly(SPDA-*co*-MBA) monolith was successfully applied to the separation of a series of polar compounds, such as phenols, bases, benzoic acid derivatives, and peptides (Figure 6.6). As shown in Figure 6.6, the elution order of the peptides investigated was in the order of increasing peptide polarity as Gly–Gly–Gly the most polar tripeptide among the tested solutes eluted last, whereas the most nonpolar Leu–Leu dipeptide eluted first under isocratic HILIC conditions. It was also suitable for the separation of highly polar compounds, such as allantoin and urea and their determination in one cosmetic product.

In another report, a porous monolith for HILIC applications has been described using *N,N*-dimethyl-*N*-methacryloxyethyl-*N*-(3-sulfopropyl)ammonium betaine (SPE) as the functional monomer and poly(ethylene glycol) diacrylate (PEGDA) as the cross-linker in a binary porogenic system comprising isopropanol and decanol. PEGDA, which has an acrylate group at each end of the molecule and a three unit ethylene glycol connecting chain, is a good cross-linker that has been shown to be more hydrophilic than EDMA [7]. SPE has positively and negatively charged functionalities that can participate in both hydrophilic and ion-exchange interaction mechanism. Methanol was used as the porogenic solvent for preparing poly(SPE-co-PEGDA) monolith and no monolith was observed during the first attempt. This indicates that the methanol was a poor solvent, leading to large pore formation. Therefore, isopropanol was substituted for methanol as the porogen, and a monolith was obtained. Finally, a long chain alcohol, decanol was selected to complement the short-chain alcohol isopropanol, as a secondary porogenic solvent. The combination of SPE with the relatively hydrophilic cross-linker PEGDA, yielded poly(SPE-co-PEGDA) monolith, which showed typical HILIC retention mechanism when the content of ACN in the mobile phase was higher than 60%. As SPE contains both positive and negative charges, the poly(SPE-co-PEGDA) monolith may exhibit ionic interactions with charged compounds in addition to hydrophilic interactions. As expected, varying the organic solvent concentration, pH, and salt concentration of the mobile phase largely affected the selectivity, resolution, and peak shapes. The poly(SPE-co-PEGDA) monolith was successfully used to separate amides, phenols, and benzoic acid derivatives.

Yet another zwitterionic monolith has been prepared [22] by the copolymerization of the zwitterionic functional monomer SPE and the cross-linker 1,2-bis(p-vinylphenyl) ethane (BVPE). In a set of experiments aiming at investigating the effect of the polymerization reaction time on the porosity of the poly(SPE-co-BVPE) monolith revealed via SEM images that the formed polymer clusters became bigger as the polymerization time was increased. As the mesoporosity can be regarded as a function of the size of microglobules, then it follows that the smaller the microglobules, the higher the fraction of mesopores. SEM images showed that the fraction of mesopores and the specific surface area were optimal when the polymerization time was reduced, whereas the column back pressure increased with increasing the polymerization time. It was reported that 2 h polymerization time yielded the best column efficiencies. The chromatographic properties of the optimized poly(SPE-co-BVPE) monolithic column were evaluated with test mixtures containing both basic and neutral compounds in the HILIC mode using a gradient elution. This poly(SPE-co-BVPE) monolith was demonstrated in the rapid and high-resolution separation of low molecular weight compounds such as pyrimidines and purines producing sharp and symmetrical peaks using HILIC gradient separation mode.

Still another zwitterionic HILIC monolith for CEC has been prepared by the thermal copolymerization of zwitterionic monomer 2-methacryloyloxyethyl phosphorycholine (MPC) with a relatively polar cross-linker PETA [26]. In order to generate an EOF in CEC, either a positively charged quaternary amine monomer such as META or a negatively charged sodium 2-methylpropene-1-sulfonate (MPS) (1% with respect to monomers) was used. As zwitterionic functionalities possess zero net charge, no EOF was observed without the inclusion of charged monomers into the zwitterionic monoliths. Salt concentration in the mobile phase showed no effect on the retention of negatively charged nucleotides implying that electrostatic interaction between charged solutes and the monolith is too small and retention is dominantly controlled by hydrophilic interactions. Satisfactory separation selectivity

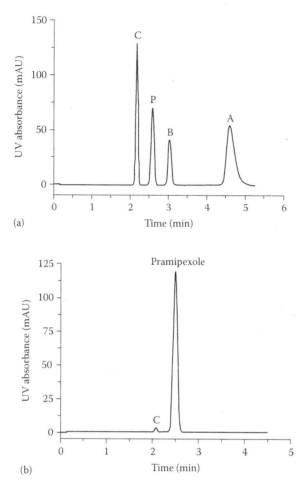

FIGURE 6.7 Chromatograms for the separation of pramipexole and its related impurities. Conditions: poly(MPC-*co*-PETA-*co*-MPS), 20.5 cm effective length, 28.5 cm total length Å~ 100 μm; mobile phase, ACN/20 mM aqueous ammonium formate solution (pH 3.0; 80/20, v/v); applied voltage, 12 kV; supplement pressure, 0.3 MPa (inlet); injection, 10 kV, 5 s; detection wavelength, 262 nm; (a) standard mixture of pramipexole and its related impurities and (b) a crude product of pramipexole, sample concentration: 1 mg/mL. (Reproduced from Qiu, D. et al., *Electrophoresis*, 37, 1725, 2016. With permission.)

and column efficiency were observed for the separation of neutral polar amides, nucleotides, and nucleosides under HILIC conditions. In addition, the poly(MPC-*co*-PETA-*co*-META) monolith was demonstrated in the quantitative analysis of the impurities of drug pramipexole (Figure 6.7).

Although in the preparation of polar monolithic column for HILIC, the selection of the nature of the hydrophilic functional monomer is very critical for the polarity of the monolith in question, this selection may not be sufficient to reach the desired hydrophilicity for achieving the best separation performance of a given monolithic stationary phase. Therefore, in addition to the functional monomer, the selection of nature of the cross-linker may also be important in order to realize the desired hydrophilicity of the polymer backbone, which together with the polar functional monomer would affect the chromatographic properties of the monolithic column thus impacting the separation selectivity. In order to investigate the effect of the cross-linker on the separation performance of polar zwitterionic sulfoalkylbetaine-type monolithic columns, three cross-linkers, including 1,4-bis(acryloyl)piperazine (PDA), EDMA, and MBA, were copolymerized with the hydrophilic monomer SPDA [27]. The reported results showed that the polarity of sulfoalkylbetaine type monolithic column is related to the polarity of the cross-linker, which further affected the column selectivity and efficiency. Although EDMA is the popular methacrylate cross-linker, its low solubility in water limited its utilization with polar functional monomers. In contrast to EDMA, MBA seems to be a suitable cross-linker for poly-acrylamide gels because it possesses a higher solubility in water, which makes MBA more compatible with highly polar functional monomers. PDA is another polar cross-linker that has been successfully employed for preparing organic polymers because again PDA could also be a better match for the solubility of highly hydro-philic functional monomers in water. The poly(SPDA-*co*-PDA) and poly(SPDA-*co*-MBA) that contains polar cross-linkers have more uniform structure, better column efficiencies, and better column permeability compared to those of the poly(SPDA-*co*-EDMA) monolith. The reason for the abovementioned observations might be related to the degree of compatibility between the functional monomer and the cross-linker regarding the solubility in the porogenic mixture. The critical composi-tion, which corresponds to the %ACN composition at which the change of retention mechanism from HILIC to RPC occurred, was used as the parameter to assess the hydrophilicity of the prepared monoliths. The results showed that the hydrophilicity of poly(SPDA-*co*-MBA) and poly(SPDA-*co*-PDA) was significantly higher than that of poly(SPDA-*co*-EDMA) and this hydrophilicity was in accordance with the polar-ity order of the cross-linkers. This was manifested by the chromatograms obtained on the three monolithic columns under identical conditions (Figure 6.8) for the separa-tion of nucleobases and nucleosides, which clearly show some major differences in the selectivity with the type of the cross-linker used. Furthermore, the separation of four benzoic acid derivatives indicated that electrostatic interactions depend solely on the properties of the functional monomer but not on the type of the cross-linker.

6.2.2 Postpolymerization Functionalization

Although the direct copolymerization of functional monomers with cross-linkers to yield monolithic stationary phases with the desired functionalities may look at a first glance simple and can be readily achieved, the choice of porogens, polymerization temperature, and other conditions still involve many trial and error to reach an opti-mum monolith. Therefore, time-consuming and labor-intensive reoptimization of the polymerization conditions from one monolith to another is required to get a monolith with satisfactory chromatographic performances. In addition, the HILIC functional monomers are not always directly and/or easily available and some monomers are incompatible with a certain copolymerization process. As an alternative route of

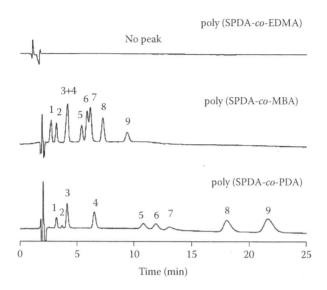

FIGURE 6.8 Separation of benzoic acid derivatives. Experimental conditions: mobile phase, 50 mM ammonium formate (pH 3.0) in ACN/H$_2$O (3/97, v/v); samples: (1) benzoic acid; (2) 2-hydroxybenzoic acid; (3) 4-hydroxybenzoic acid; (4) 2,4-dihydroxybenzoic acid; (5) 3,5-dihydroxybenzoic acid; (6) 3,4,5-trihydroxybenzoic acid. Detection wavelength, 214 nm; flow rate, 400 nL/min. (Reproduced from Liu, C. et al., *J. Chromatogr. A*, 1373, 73, 2014. With permission.)

monolithic column preparation, postpolymerization functionalization of precursor monoliths allows the introduction of various polar ligands onto the monolithic surfaces starting from one generic template monolith [14,28–32]. Glycidyl methacrylate (GMA) containing precursor monoliths are the most popular for this purpose because the reactive epoxy group on GMA allows addition of various functionalities through simple nucleophilic substitution reactions. During recent years, there was a strong interest among chromatographers to use *click reactions* for the surface modification of template monolithic columns in order to obtain HILIC stationary phases. The concept of performing organic reactions by *click reactions* was introduced in 2001 [33], and involves the reaction of thiols with enes, whether proceeding it by a radical (termed thiol-ene reaction) or by an anionic chain (termed thiol Michael addition). These processes carry many of the attributes of click reactions, including achieving quantitative yields, rapid reaction rates, and requiring essentially no clean up. Accordingly, both thiol-ene radical and thiol Michael addition reactions are now routinely referred to in the literature as thiol click reactions [34]. For a detailed review of the thiol click reactions and their wide use in materials chemistry, the reader is directed to a recent review article on this topic by Nair et al. [35].

6.2.2.1 Nature of the Surface Charge

As in Section 6.2.1, a classification in terms of the nature of surface charge provides a convenient way to assess the difference between the various monoliths. In addition, one can conveniently justify the rationale behind the various chemistries in terms of the ensuing use of such monoliths. Although the author tried his best to group items under specific topics, however that does not exclude some minor overlap.

6.2.2.1.1 Anionic Monoliths

A novel approach based on thiol-mediated Michael addition click reaction for the postpolymerization functionalization (PPF) of porous monoliths has been reported recently [36]. The approach involved the synthesis inside a capillary of a generic matrix containing *N*-hydroxysuccinimide esters (NHS) as surface reactive functionalities through photo-induced polymerization of NAS and EDMA. This reactive monolith was further surface functionalized through nucleophilic substitution of the NHS residues with 2-aminoethanethiol. This step yielded free thiols for reacting with any functional maleimides *via* C–S bond formation under mild conditions through Michael addition reaction. The thiol-containing monolith thus obtained was allowed to react with hydrophilic *N*-maleoyl-β-alanine yielding a hydrophilic monolith with surface-grafted COOH groups, which were evaluated under HILIC conditions in the separation of mixtures of amides and phenolic compounds. Furthermore, the authors anticipated that this strategy might provide a convenient approach for functionalization of polymer monoliths with maleimide-labeled biomacromolecules.

6.2.2.1.2 Cationic Monoliths

Following the abovementioned click reaction chemistries, a poly(NAS-*co*-EDMA) monolith has been used for developing amine-based monolith *via* thiol-yne click surface grafting [31]. During the two-step grafting procedure, the monolith was first reacted with propargylamine, which leads to an amide bond attachment of alkyne groups. In the second step, click grafting of cysteamine was performed under UV irradiation in the presence of 2-hydroxy-2-methyl-1-phenyl-propane as a photoinitiator. These surface grafting yielded NH$_2$ groups on the monolithic pore surface. A stable anodic EOF was observed up to pH 5, above which its magnitude decreased. The reversal of the EOF direction was observed above a pH value of 7 and further increase in the buffer pH values led to a stable cathodal EOF. Though thiol-yne reaction is mechanistically comparable to the thiol-ene one, it may allow higher grafting density because the surface-grafted alkyne can react with two thiols. This monolith was successfully used to separate hydroxyl-substituted aromatic solutes under HILIC conditions.

In a different approach, Gunasena and El Rassi have exploited the previously discussed AP-monolith in HI–CEC by modifying its surface with neutral, mono- and oligosaccharides to produce a series of the so-called sugar modified AP-monoliths (SMAP-monolith), which are considered as stratified hydrophilic monoliths possessing a sublayer of polar amine/amide groups and a top layer of sugar (i.e., a polyhydroxy top layer) [14]. In this regards, *N*-acetylglucosamine (GlcNAc), chitotriose, and chitopentose were covalently attached to the amine-bearing monolithic surface *via* reductive amination, which converted the primary amine to the much less reactive secondary amines. The SMAP-monolith yielded an anodal EOF, due to the fact that the reductive amination does not remove the amine groups; it rather transforms the primary amines to secondary amines, and therefore the positive charges are still present on the surface of the monoliths to give the positive zeta potential required to maintain the anodal EOF. By modifying the AP-monolith with sugars, different HILIC retention properties were observed. Furthermore, the polarity of the stationary phases and the k values of polar solutes increased with increasing the number of repeating sugar units in the immobilized saccharides.

Another cationic stationary phase was obtained by functionalizing poly(HPMA-Cl-*co*-EDMA) monolith with triethanolamine (TEA-OH), and the resulting stationary phase was evaluated in nano-HILIC [37]. TEA-OH attached poly(HPMA-Cl-*co*-EDMA) monolith showed a stable and reproducible retention behavior in the separation of nucleosides and benzoic acid derivatives under HILIC conditions.

6.2.2.1.3 Neutral Monoliths

A HILIC monolithic column containing hydroxysuccinimide reactive sites for PPF has been designed by Tijunelyte et al. [30]. The functional monomer *N*-acryloxysuccinimide (NAS) and the cross-linker EDMA were copolymerized in the presence of toluene as the porogenic solvent to yield poly(NAS-*co*-EDMA) monolith with surface succinimide groups. This macroporous monolith was used as a starting material for the double step surface thiol-ene click PPF. During the first step, allylamine was used for the grafting of alkene units on the surface of poly(NAS-*co*-EDMA) monolith, *via* hydroxysuccinimide/alkene nucleophilic substitution. The obtained polymeric network with the surface-grafted allyl units serves as a *clickable* monolith that allows modification of its interfacial properties *via* the thiol-ene click reaction. During the second step, the pore surface with pendant allyl moieties was further functionalized *via* two-step thiol-ene click reaction with thiol containing oligo(ethylene glycol) and mercaptoethanol as shown in Figure 6.9. This surface grafting yielded a neutral monolith for HI–CEC. Oligo(ethylene glycol) (OEG) was selected as a good candidate because it is well established that ethylene oxide-based molecular fragments tethered on a solid surface are prone to hydration. After the OEG grafting, the remaining and accessible double bonds were scavenged with mercaptoethanol in a second step thiol-ene click coupling.

Very recently, a hydroxy monolith referred to as the OHM capillary column was used in PPF whereby the surface hydroxyl groups were first reacted with glycerol diglycidyl ether in the presence of boron trifluoride (BF_3) to convert the surface into epoxy activated OHM [38]. The column thus activated with epoxy functions was further functionalized with polar compounds such as glycerol and polyamines to yield polyol OHM and polyamine OHM columns, respectively, for use in HI–CEC (Figure 6.10). In another PPF of the epoxy-activated OHM, hydroxypropyl-β-cyclodextrin (HP-β-CD) was grafted onto the epoxy OHM surface to produce HP-β-CD OHM for enantioseparation by CEC. The polyol OHM capillary column and the polyamine OHM capillary columns exhibited the typical hydrophilic interaction columns vis-à-vis polar solutes such as phenolic compounds and nucleobases. Figure 6.11a shows typical electrochromatograms of some phenols obtained on the precursor OHM capillary and on the polyol monolithic column. It can be clearly seen

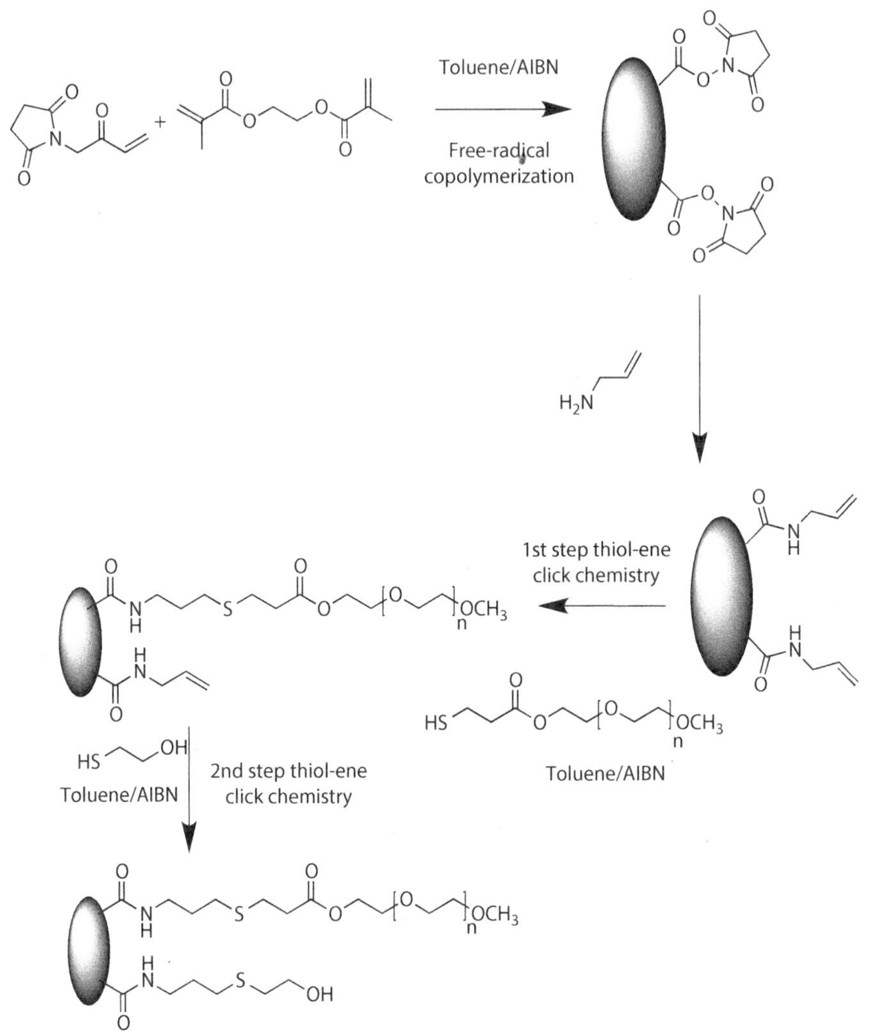

FIGURE 6.9 Schematic illustration of the synthetic pathway for the preparation of poly(NAS-*co*-EDMA) with surface bound oligo(ethylene glycol) chains. (Adapted from Tijunelyte, I. et al., *Polymer*, 53, 29, 2012.)

the significant enhancement in the hydrophilic property of the polyol OHM capillary column when compared to the precursor OHM capillary column. The HP-β-CD OHM column was able to resolve some racemic compounds as well as positional isomers as illustrated in Figure 6.11b. All the columns exhibited good reproducibility from run-to-run, day-to-day, and column-to-column.

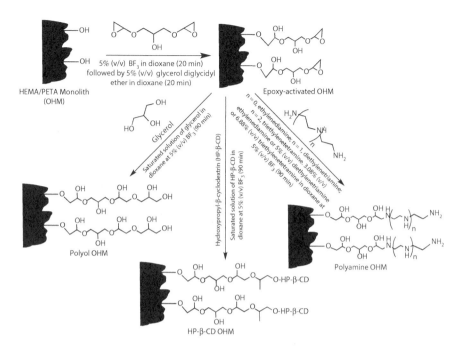

FIGURE 6.10 Reaction schemes of postpolymerization modification of poly(HEMA-co-PETA) precursor monolith for making polyol, polyamine, and HP-β-CD monoliths. (Reproduced from Khadka, S. and El Rassi, Z., *Electrophoresis*, 37, 2016. With permission.)

6.2.2.1.4 Zwitterionic Monoliths

A hydrophilic monolithic column has been prepared by PPF of a GMA-based monolith [39]. The template monolithic column was first prepared *via* the *in situ* polymerization of AM, GMA, MBA, and AMPS in a ternary porogenic solvent system consisting of cyclohexanol, dodecanol, and toluene. The epoxide groups in the obtained monolith were subsequently modified with ammonia by simply passing 0.1 M solution of ammonia through the column. The hydrophilic monolith column was tested in CEC with five polar compounds, including thiourea, aniline, naphthylamine, diphenylamine, and dimethyl acetamide. This hydrophilic monolithic column was further used for the separation of five alkaloids. As the surface of the monolith has (1) sulfonic acid groups supplied by AMPS, which were intentionally introduced to produce a stable cathodal EOF in the pH range 4.0 to 10.0 and (2) primary amine groups introduced by PPF, which slightly affected the EOF at low pH due to their partial protonation to positively charged sites, the monolith can be regarded as *zwitterionic* or more accurately amphoteric because of the presence of both positively and negatively charged sites. In fact, the monolith exhibited hydrophilic interactions as well as ion-exchange interactions due to the multiplicity of interactive sites, including hydroxyl, amine, and sulfonic acid groups.

A porous monolith with zwitterionic functionalities has been prepared *via* the *thiol-ene* click chemistry [32]. This involved the reaction of the epoxy groups of the

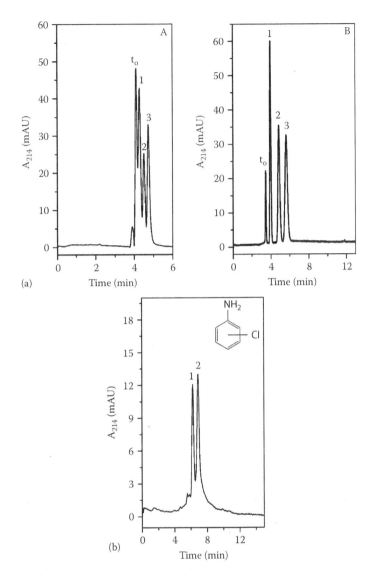

FIGURE 6.11 (a) Electrochromatograms of some phenols obtained on the precursor OHM capillary column in (A) and on the polyol monolith capillary column modified in dioxane (B). Conditions: capillary column, 33.5 cm (effective length 25 cm) × 100 μm i.d.; running voltage, 20 kV; sample injection at a pressure of 10 bars for 9 s; column temperature, 25°C; hydro-organic mobile phase, 5% (v/v) of 3.5 mM NH₄Ac, pH 7.0, 95% (v/v) ACN. Solutes: 1, phenol; 2, resorcinol; 3, pyrogallol; EOF tracer or t_0, toluene. (b) Separation of chloroaniline positional isomers obtained on the HP-β-CD OHM capillary column. Conditions: capillary column, 33.5 cm (effective length 25 cm) × 100 μm i.d.; running voltage, 20 kV; sample injection at a pressure of 10 bars 9 s; column temperature, 25°C; hydro-organic mobile phase, 5% (v/v) of 10 mM NH₄AC (pH 7.0), and 95% (v/v) ACN. Solutes: 1) 4-chloroaniline and 2) 3-chloroaniline. (Reproduced from Khadka, S. and El Rassi, Z., *Electrophoresis*, 37, 2016. With permission.)

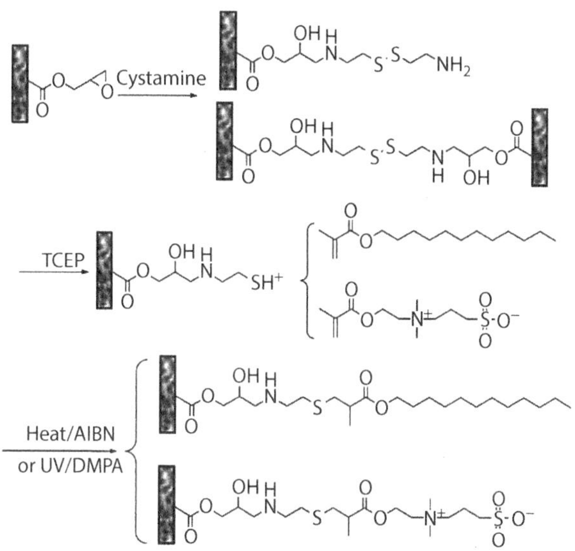

FIGURE 6.12 Reaction scheme for the preparation of monoliths and their functionalization *via thiol-ene* click reaction. (Reproduced from Lv, Y. et al., *Analyst*, 137, 4114, 2012. With permission.)

traditional poly(GMA-*co*-EDMA) monolith with cystamine followed by cutting the disulfide bond using tris(2-carboxylethyl)phosphine (TCEP) that liberates the desired thiol groups. Most of the cystamine reacts through both amine functionalities to yield a high degree of functionalization. Then, the thiol-containing monolith was clicked with the zwitterionic SPE functional monomer (Figure 6.12). This reaction yielded a monolith with a hydrophilic surface chemistry suitable for HILIC separations as it was demonstrated in the separation of six peptides by gradient elution with a decreasing percentage of acetonitrile in the mobile phase. Four nucleosides were also selectively separated on this HILIC monolith with relatively high separation efficiencies.

A two-step surface modification of poly(styrene-*co*-vinylbenzyl chloride-co-divinylbenzene) monolithic stationary phase has been described in Reference 28. It involved a hypercrosslinking step followed by a second step that consisted of thermally initiated surface grafting of [2-(methacryloyloxy)ethyl]dimethyl(3-sulfo-propyl)ammonium hydroxide. The prepared monolith column provided a dual RPC and HILIC retention mechanism. The surface grafting reaction was optimized using the so-called response surface methodology. During the model evaluation, three experimental factors, including activation time, time of grafting, and temperature of grafting were investigated, whereas all other variables were kept constant. The main factor that affected the surface modification of hypercrosslinked monoliths was the synergistic effect of grafting time.

A novel approach has been reported for the preparation of porous polymeric hypercrosslinked monoliths of large surface area bearing surface zwitterionic functionalities through the attachment of gold nanoparticles (GNPs) in a layered architecture [29]. The poly(4-methylstyrene-*co*-vinylbenzyl chloride-*co*-divinylbenzene) precursor monolith was prepared in a capillary and hypercrosslinked *via* a Friedel–Crafts alkylation forming a large amount of mesopores within the monolith. Then, the free radical bromination was used to modify the surface chemistry of the monolith. The brominated monolith thus obtained was reacted with cystamine in a microwave oven. Then, GNPs were attached to this thiol containing monolithic surface. In order to prepare a HILIC monolithic column, these GNPs were further modified with cysteine and polyethyleneimine (PEI). Cysteine is a zwitterionic compound that can be utilized for the preparation of monoliths suitable for HILIC separations. Similarly, PEI is known to be an excellent anion-exchange ligand, which also exhibit good hydrophilicity. When comparing the two monoliths, the attachment of PEI produced a column exhibiting better performance than those modified with mainly cysteine alone because of the formation of a dense hydrophilic polymer layer on the surface of the GNPs. Finally, the authors have prepared monolithic columns with layered architecture by embedding a second layer of nanoparticles after modification with PEI, which served as a spacer enabling construction of a dual layer of GNPs that were then functionalized with cysteine. This approach formed a hydrophilic layer consisting of both PEI and cysteine functionalities held together *via* interactions with GNPs. This dual layered monolith was demonstrated in the separation of nucleosides and peptides.

Another zwitterionic stationary phase was prepared *via* the poly(HPMA-Cl-*co*-EDMA) monolith, which is considered as a novel reactive starting material for the preparation of various chromatographic stationary phases through PPF. The 3-chloro-2-hydroxypropyl groups on this monolith can covalently attach various different ligands *via* simple and one-stage reactions. In addition, the poly(HPMA-Cl-*co*-EDMA) monolith is a hydrophilic support due to its hydroxyl functionality and it is therefore ideal as a base material for the synthesis of HILIC stationary phases. In fact, this novel reactive monolith has been postfunctionalized with taurine (2-aminoethane sulfonic acid) to obtain a zwitterionic stationary phase for CEC, which contained charged groups such as secondary amine and sulfonic acid [40]. The magnitude and direction of the EOF varied with the pH of the mobile phase. At acidic pH, an anodal EOF was observed due to the protonated positively charged amine groups, whereas at basic pH a cathodal EOF was obtained because of the negatively charged deprotonated sulfonic acid groups. This zwitterionic monolith was demonstrated in the separation of nucleosides under HILIC conditions. In addition, a mixture of alkyl benzenes was also separated using the same monolith column under RPC conditions. It was thought that the source of the nonpolar character of the monolith originates from methylene groups of both EDMA and HPMA–Cl.

Recently, hydrophilic monoliths suitable for the separation of small polar molecules were prepared by adjusting the prepolymerization mixture compositions, early termination of the polymerization reaction, and hypercrosslinking postpolymerization modification [41]. The investigators of this work demonstrated an application of nucleophilic substitution reaction in the preparation of hypercrosslinked monolithic stationary phases suitable for the analysis of low-molecular weight polar compounds. Generic monoliths prepared a priori in capillaries by the *in situ* radical copolymerization of styrene, vinylbenzyl chloride, and divinylbenzene were perfused for 2 h with solutions of diaminoalkanes (e.g., 1,2-diaminoethane,

1,4-diaminobutane, 1,6-diaminohexane, and 1,8-diamonooctane in tetrahydrofuran) and the hypercrosslinking reaction was carried out at 80°C–120°C for 1–4 h. In order to perform a further modification of the surface residual chloromethyl groups of the hypercrosslinked columns, the columns were filled with 2-aminoethanesulfonic acid (i.e., taurine) and the nucleophilic substitution reaction was carried out at elevated temperatures. The column separation efficiency was found to increase with the use of longer diaminoalkane (e.g., 1,8-diaminoocatane) possibly due to the fact that shorter diaminoalkanes are not able to cross-link the polymer enough to provide an efficient separation of small molecules. In addition, it was demonstrated that the decrease in the polymerization reaction time followed by a hypercrosslinking modification led to a significant increase in column performance. The retentive property of columns depended largely on the composition of the used mobile phase. With a mobile phase containing 60% (v/v) ACN, the hypercrosslinked zwitterionic columns retained both alkylbenzenes and polar compounds *via* the RPC mode. Increasing the ACN concentration in the mobile phase up to 95% (v/v), led to the loss of retention of the alkylbenzenes, and the elution order of low-molecular weight polar compounds such as phenolic compounds followed by HILIC behavior when using a mobile phase at 80% (v/v) ACN in the sense that the retention of phenolic acids increased with an increase in the number of substituted hydroxyl groups.

A novel two-step polymerization approach has been proposed as an effective method to introduce small pores into the monolithic structure to realize an increased surface area for a better separation of small polar molecules [42]. A poly(lauryl methacrylate-*co*-tetraethyleneglycol dimethacrylate) [poly(LMA-*co*-TeEDMA)] polymer scaffold monolithic column was first prepared by copolymerizing LMA and TeEDMA, and then used without further modification as a support for the second monolithic layer of poly(MEDSA-*co*-BIGDMA) with zwitterionic functionalities. In comparison to the single-step polymerized zwitterionic monolithic columns, the formation of a secondary monolithic layer provides additional cross-linking within the pores, producing columns with an increased proportion of mesopores. The separation of nucleobases and nucleosides, phenolic acids, and flavones were achieved on the two-steps zwitterionic polymer monolith under HILIC conditions. Figure 6.13 shows the separation of nucleobases and nucleosides obtained on the two-step zwitterionic monolithic column using a shallow gradient elution at decreasing the %ACN in the mobile phase.

6.2.3 Monoliths with Incorporated Nanoparticles

The unique chemical and physical properties of nanoparticles (NPs), including primarily their well-developed surfaces, which offer different adsorption selectivity toward a variety of substances have triggered the use of NPs in separation sciences. On account of their nano size, the NPs must be supported on a solid medium with favorable flow characteristics that can be readily prepared and confined in columns or channels of all sizes. Polymeric monoliths are the preferred support media for NPs because they can be (1) readily formed by *in situ* polymerization in columns and channels of all sizes, (2) tailor-made from many available monomers, (3) used conveniently to trap or incorporate NPs covalently, and (4) very permeable with favorable

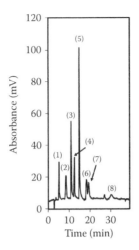

FIGURE 6.13 Gradient separation of nucleosides and nucleobases obtained on two-step zwitterionic monolithic columns. Peaks; thymine (1), 2-deoxyadenosine (2), adenine (3), adenosine (4), cytosine (5), 2-deoxyguanosine (6), cytidine (7), and guanosine (8). Mobile-phase A:10 mM ammonium acetate pH 3; B: ACN 10 mM ammonium acetate. Gradient conditions; 0–3 min 95% B, 3–12 min 95% B to 85% B, 13 min 90% B. Flow rate 4 μL/min. Detection at 214 nm. (Reproduced from Currivan, S. et al., *J. Chromatogr. A*, 1402, 82, 2015. With permission.)

flow characteristics *via* their large flow through pores. On these bases, monoliths with incorporated NPs have recently emerged as a new class of stationary phases in chromatography. Two main strategies have been used for the immobilization of NPs onto monolithic surfaces: (1) immobilization of premodified or unmodified NPs by physical entrapment in the monolithic structure *via* first dispersing the NPs in the prepolymerization solution containing the monomer, cross-linker, and porogen followed by *in situ* polymerization and (2) immobilization of functionalized NPs *via* chemical bonding and/or electrostatic attractions by either dispersion of NPs and *in situ* polymerization as in (1) or by simply passing a solution containing dispersed NPs through a given monolithic column with surface reactive groups capable of either reacting chemically or attracting electrostatically the prefunctionalized NPs.

Very recently, Aydoğan and El Rassi reported a monolithic HILIC column with incorporated fumed silica nanoparticles (FSNPs) into a polymethacrylate monolithic column obtained by copolymerizing glyceryl monomethacrylate (GMM) and EDMA [poly(GMM-*co*-EDMA)] [43]. When compared to the hydrophilic parent monolithic column without incorporated FNSPs, the same poly(GMM-*co*-EDMA) monolithic column with incorporated FSNPs yielded enhanced HILIC behavior. As one would expect, the incorporation of FSNPs yielded a monolithic surface with a dual retentive property consisting of hydrophilic interactions with polar solutes and hydrophobic interactions with nonpolar solutes due to the surface silanol groups and the surface siloxane bridges, respectively, as well as to the hydroxylated part (conferred by the functional monomer GMM) and hydrophobic part (made up of the cross-linker EDMA) of the monolith, respectively. This dual surface retentive property was revealed when studying solute retention behavior over a range of mobile phase composition in terms of %ACN in the hydro-organic mobile phase. The poly(GMM-co-EDMA) with incorporated FNSPs was evaluated over a wide range of mobile-phase compositions with polar acidic, weakly basic, and neutral analytes, including hydroxybenzoic acids, nucleotides, nucleosides, and polar amides. The retention of these analytes was mainly controlled by hydrophilic interactions with the FSNPs and electrostatic repulsion from the negatively charged silica surface in the case of hydroxybenzoic acids and nucleotides. As shown in Figure 6.14a, five different nucleosides were separated in about 10 min using a 5 min linear gradient at decreasing ACN concentration from 95% to 85% (v/v) in water–ACN mobile phase. The solute retention was largely affected by the amount of incorporated FSNPs in the monolith. This is shown in Figure 6.14b whereby the retention factor k of thiourea and some nucleosides increased linearly with the amount of fumed silica per unit column length in the range studied. This means that increasing the amount of incorporated FNSPs corresponds to increasing the phase ratio of polar sites in the column, whereas the partition coefficient K seems to be not significantly affected by the amount of added FNSPs to the column. In all cases, the solutes tested exhibited weak HILIC with the monolith without incorporated FNSP as manifested by the y-intercept of the lines in Figure 6.14b.

A poly(GMA-co-PEGDA) monolith was modified with GNPs to yield a monolithic surface with enhanced reactive sites that facilitate further modification in the aim of achieving a good hydrophilicity [44]. Variable functional modification of GNPs with cysteine and N-glycosidase F (PNGase F) was carried out for developing stationary phases capable of HILIC-based enrichment and online

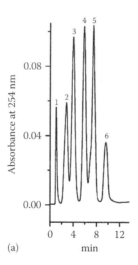

(a)

FIGURE 6.14 (a) Chromatogram of some nucleosides on the monolithic column with incorporated FSNPs at 3.1 mg fumed silica per cm of column length. Column, 10 cm × 4.6 mm i.d.; linear gradient elution in 5 min at decreasing ACN content in the mobile phase from 95% ACN to 85% ACN (v/v); flow rate, 1 mL/min; detection wavelength, 254 nm. Peaks order, (1) toluene, (2) adenosine, (3) uridine, (4) cytidine, (5) inosine, (6) guanosine.   (*Continued*)

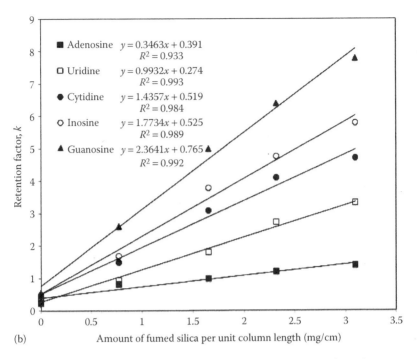

(b)

$y = 0.3463x + 0.391$
$R^2 = 0.933$

$y = 0.9932x + 0.274$
$R^2 = 0.993$

$y = 1.4357x + 0.519$
$R^2 = 0.984$

$y = 1.7734x + 0.525$
$R^2 = 0.989$

$y = 2.3641x + 0.765$
$R^2 = 0.992$

Amount of fumed silica per unit column length (mg/cm)

FIGURE 6.14 (Continued) (b) Dependence of solute retention factor (k) on the amount of incorporated fumed silica per unit column length. Column, 10 cm × 4.6 mm i.d.; mobile phase, 95:5 ACN/water (v/v); detection wavelength, 254 nm; flow rate, 1 mL/min. (Reproduced from Aydoğan, C. and El Rassi, Z., *J. Chromatogr. A*, 1445, 55, 2016. With permission.)

deglycosylation of glycopeptides, respectively. The polymer monolithic matrix was first prepared by the *in situ* copolymerization of GMA and PEGDA in a fused capillary column of 250 μm i.d. The poly(GMA-*co*-PEGDA) was further reacted with cystamine dihydrochloride followed by reducing the disulfide groups of cystamine with TCEP to generate the desired thiol functionalities required for the GNPs attachment. Finally, a solution containing GNPs was pumped through the monolith to prepare the GNP functionalized poly(GMA-*co*-PEGDA) monolith. For cysteine functionalization of the attached GNPs, 0.5 M cysteine solution was continuously pumped through the column. Similarly, for PNGase F immobilization, 200 μL of 25000 UmL⁻¹ PNGase F were continuously pumped through the column. The cysteine-modified monolith matrix was shown to allow the enrichment of glycopeptides by HILIC with high selectivity, and the subsequent elution under weak alkaline conditions. It was also reported that during glycopeptide elution the same monolith with immobilized PNGase F could be directly coupled with the HILIC column to achieve simultaneously the online deglycosylation without buffer exchange and pH adjustment. The developed columns were used in the analysis of human plasma glycoproteome.

6.2.4 IONIC LIQUID MONOLITHS

The development and applications of ionic liquids (ILs) systems for CE and CEC [45] and liquid chromatography [46] have been recently reviewed. Obviously, in most instances the monomers used in making the IL immobilized monoliths have polar characters. As reviewed and documented in Reference 45, ILs are generally defined as salts in the liquid state at ambient temperatures, which are composed of bulky, nonsymmetrical, organic cations (e.g., imidazolium, pyrrolidinium, pyridinium, tetraalkyl ammonium, or tetraalkyl phosphonium), and numerous different organic or inorganic anions (e.g., tetrafluoroborate, hexafluorophosphate, and bromide). See Table 6.3 for typical ILs used so far in monolithic column preparations.

TABLE 6.3

Names, Abbreviations, and Structures of Ionic Liquids. The Species Are Arranged Alphabetically in the Ascending Order According to Their Abbreviations

Name	Abbreviation	Structure
1-Acetylamino-propyl-3-methylimidazolium bromide	(AAPMIm)Br	
1-Allyl-3-methylimidazolium chloride	AMIM+Cl−	
1-Butyl-3-methylimidazolium tetrafluoroborate or chloride	[BMIM]BF$_4$ or BMIM+Cl−	
1-Vinyl-3-butylimidazolium chloride	VBC-ILs	
1-Vinyl-3-(butyl-4-sulfonate) imidazolium bromide	VBSIm	
1-Vinyl-3-octadecylimidazolium bromide	VC18HIm+Br−	

When compared to common organic solvents, ILs stand out for their unique and outstanding characteristics, such as very low volatility, high thermal stability, tunable viscosity, and the ability to dissolve easily in a wide range of inorganic and organic solvents. These characteristics have drawn considerable interest and triggered their use in separation sciences. As documented in a recent review article [46], a few different approaches have been developed to covalently attach ILs onto a given monolithic surface. In general, it is thought that IL stationary phases exhibit multimode mechanism, including hydrophobic, hydrophilic, hydrogen bonding, anion exchange, π–π interactions, and dipole–dipole interactions, and consequently they could be used in different chromatographic modes, including ion exchange, RPC, NPC, and HILIC to separate various classes of compounds.

A novel IL-based zwitterionic organic polymer monolithic column has been developed by a single-step copolymerization of 1-vinyl-3-(butyl-4-sulfonate) imidazolium (VBSIm), acrylamide, and N,N'-methylenebisacrylamide in a quaternary porogenic solvent consisting of formamide, dimethyl sulfoxide, PEG 8000, and PEG 10000 for cHILIC [47]. The optimized monolith exhibited favorable selectivity and retention for nucleosides and benzoic acid derivatives. The retention of nucleosides on the zwitterionic poly(VBSIm-AM-MBA) monolithic column was reported to be based on hydrophilic interaction and repulsive electrostatic interaction mechanisms, whereas the benzoic acid derivatives were separated based on hydrophilic interaction, electrostatic interaction, and hydrogen bonding mechanisms.

Another IL-bonded multifunctional monolithic stationary phases has been developed by an *in situ* polycondensation of urea–formaldehyde (UF) and acrylamino-functionalized IL such as 1-acetylamino-propyl-3-methylimidazolium bromide (AAPMIm)Br [48]. In this process, the commercially available 3-aminopropyltrimethoxysilane (APTMS) and the IL reagent 1-(3-aminopropyl)-3-methylimidazolium bromide were functionalized with acetic anhydride. This acrylamino-functionalized APTMS was used for the modification of the capillary wall with a layer of acetylamino groups for anchoring the monolithic stationary phase. Then, the polymerization solution containing urea, formaldehyde, (AAPMIm)Br, and HCl was filled into the capillary and the polycondensation reaction was carried out at 65°C for 10 min (Figure 6.15). Due to the presence of positively charged imidazole groups in the (AAPMIm)Br-bonded monolith, a stable anodal EOF over a wide pH range was observed in CEC runs. The hydrophilicity of (AAPMIm)Br-bonded monolithic matrix and the positively charged imidazole rings on the surface of the monolithic column allowed hydrophilic interaction (HI) and anion-exchange retention mechanism for polar and charged solutes. Various polar compounds, including phenols, benzoic acid, and its homologues and enkephalins (Figure 6.15) have been well separated, which demonstrated a satisfactory separation performance of the obtained monolith.

6.2.5 POLAR ORGANIC-SILICA HYBRID MONOLITHS

Organic-silica hybrid monolithic stationary phases have been proposed to combine the best features of the two kinds of monoliths: (1) organic polymer monoliths and (2) inorganic silica-based monoliths. Polymer-based monoliths, which are known

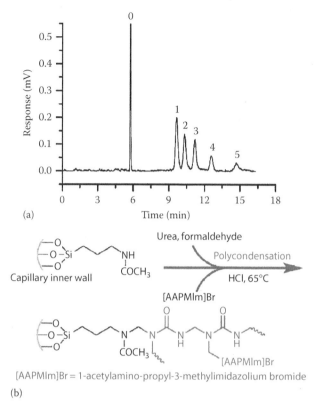

FIGURE 6.15 (a) Electrochromatogram of enkephalins. Mobile phase: ACN: Ammonium formate (40 mmol/L pH 3.0) = 60:40 (v/v); Effective column length: 35 cm (total length: 55 cm) × 100 μm i.d. running voltage: +10 kV, supplementary pressure: 1000 psi, pump flow rate: 0.05 mL/min. Solutes: 0,1,3,5-tri-*tert*-butylbenzene; 1, Gly–Gly–Phe–Met (GM4); 2, Tyr–Gly–Gly–Phe–Met (YM5); 3, Tyr–Gly–Gly–Phe–Leu (YL5); 4, Gly–Gly–Phe–Leu (GL4); 5, Gly–Phe–Leu (GL3). (b) The scheme of UF polycondensation for (AAPMIm) Br-bonded monolith. (Reproduced from Wang, J. et al., *J. Chromatogr. A*, 1449, 100, 2016. With permission.)

for their excellent pH stability and ease of preparation, undergo swelling in organic solvents leading to mechanical unstability, whereas the silica-based monoliths offer high mechanical stability, high separation efficiencies, and good solvent resistance, but their preparation are more tedious involving bare silica matrix formation by a sol-gel process, followed by a chemical modification of the monolithic surface, which is not only time-consuming but also prone to poor reproducibility [49]. Therefore, organic-silica hybrid monolithic stationary phases should in principle combine the ease of preparation and pH stability of the organic polymer monoliths to the good mechanical stability and high separation efficiencies for small molecules provided by inorganic silica-based monoliths, thus achieving the best of the two kinds of monolithic stationary phases.

The organic-silica hybrid stationary phases are synthesized by *one-pot* sol-gel process of monomers containing both organic and alkoxysilane functionalities in the presence of suitable porogenic solvents. In these monomers, at least one organic moiety is covalently linked *via* a nonhydrolyzable Si–C bond to a siloxane species, which will hydrolyze and polycondense to form the silica network. Inorganic silica network in the hybrid monoliths provides the mechanical and structural stability, whereas the organic moiety provides the desired chromatographic ligand.

A facile one-pot approach in combination with *thiol-ene* click reaction has been reported for the synthesis of the glutathione (GSH)–silica hybrid monolithic columns by using hydrolyzed tetramethyloxysilane (TMOS) and γ-methacryloxypropyltrimethoxysilane (γ-MAPS) as coprecursors and GSH as the functionalized organic monomer [50]. The proposed one-pot approach involves two main processes: (1) the hydrolysis and condensation of TMOS and γ-MAPS and (2) the *thiol-ene* click reaction between GSH and vinyl-end silica monolithic matrix. A stepwise reaction temperature profile was used to control the polymerization process in which the initial polycondensation reaction was carried out at 40°C for the preparation of silica monolithic matrix and then the temperature was increased to 60°C–65°C to decompose the free radical initiator AIBN in order to start the *thiol-ene* click reaction. Glutathione is a tripeptide comprised of three amino acids, namely cysteine, glutamic acid, and glycine with a pendant thiol group, which can be reacted with vinyl containing monomers *via thiol-ene* click reaction. Furthermore, GSH contains two free carboxylic acid groups and one amine group exhibiting hydrophilic and ion-exchange characteristics. This hybrid monolithic column showed HILIC/RPC mixed-mode retention behavior depending on the %ACN content of the mobile phase. Mixtures of phenolic compounds, amides, and nucleotides were baseline separated using GSH–silica hybrid monolith under HILIC conditions. In addition, a hydrophilic partitioning/cation-exchange mixed-mode type mechanism was observed in the separation of aniline compounds.

A similar one-pot approach has been used for the preparation of organic-silica hybrid monolithic columns by using simultaneous polycondensation and *thiol-ene* click reaction [51]. In this process, the thiol containing 3-mercaptopropyltrimethoxysilane (MPTMS) was polycondensed with TMOS in the presence of PEG 6000 as the porogen to form the silica skeleton with simultaneous *thiol-ene* click reaction between MPTMS and vinyl end of an organic monomer (Figure 6.16). The temperature for the preparation of monolith was kept at 40°C, which was sufficient for both *thiol-ene* click reaction and polycondensation. The relatively lower temperature was chosen to minimize the self-polymerization of vinyl-end organic monomers on the monolithic surface, so that the porous structure of the prepared organic-silica hybrid monolith would not be affected. Two types of organic-silica hybrid monoliths were prepared using META ([2-(methacryloyloxy)ethyl]trimethylammonium) and acrylamide (AM) with the proposed method. The two hybrid monoliths possess polar groups, namely the quaternary ammonium groups on the META–silica hybrid monolith surface and the amide groups on the AM–silica monolith surface, which allowed typical HILIC retention behavior when tested with neutral polar compounds, for

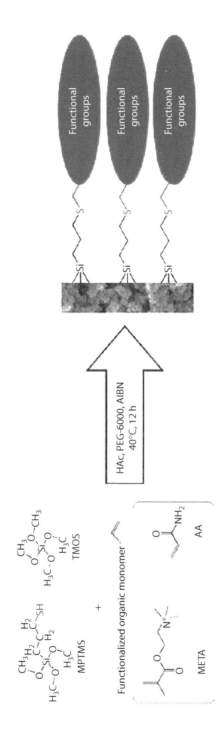

FIGURE 6.16 Scheme of the preparation of organic–silica hybrid monolith. (Reproduced from Chen, M.-L. et al., *J. Chromatogr. A*, 1284, 118, 2013. With permission.)

example, thiourea, formamide, and DMF. In fact, as the ACN content of the mobile phase was increased from 70% to 95% (v/v), the retention of the three polar analytes also increased on META– and AM–silica monolithic columns. Moreover, these three analytes were baseline separated on META– and AM–silica monolithic columns, but not well separated on the SH–silica hybrid column (i.e., without META or AM functionalization) demonstrating the successful grafting of organic monomers on the silica matrix. In addition, the META–silica and AM–silica hybrid monolithic columns showed good resolution in the separation of nucleotides, nucleosides, and benzoic acids in the HILIC mode.

Another hybrid organic-silica monolithic stationary phase has been prepared by a one-pot approach using TMOS and γ-glycidoxypropyltrimethoxysilane (GPTMS) as coprecursors and PEG 10,000 as the porogen [52] in the presence of peptides such as GSH and/or somatostatin (ST) or a protein such as ovomucoid (OV). The peptides or protein were simultaneously attached to the silica monolithic skeleton *via* the epoxy ring opening reaction between the epoxy groups of the GPTMS and the amine group in the side chain of the peptide and/or protein alongside the polycondensation of TMOS and GPTMS (Figure 6.17a). As peptides and proteins are naturally zwitterionic and chiral compounds with hydrophilic properties, these hybrid monolithic stationary phases demonstrated pH-dependent EOF direction and magnitude, HILIC behavior, and chiral separation ability. In addition, gold nanoparticles (AuNPs) were used to fabricate an AuNP-mediated GSH–AuNP–GSH monolithic stationary phase *via* strong interaction between sulfhydryl groups of GSH and AuNPs, see Figure 6.17b to improve its chiral separation ability using a postmodification method. The prepared GSH–silica, ST–silica, and OV–silica showed typical HILIC behaviors when separating neutral amides, nucleotides, and peptide standards (Figure 6.18).

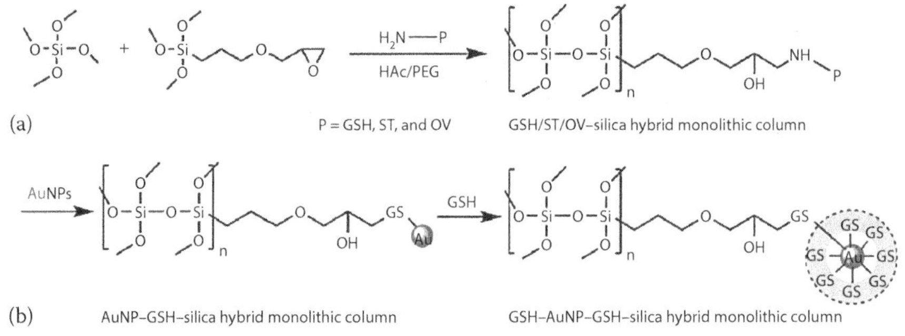

FIGURE 6.17 Schematic one-pot process for the preparation of peptide– and protein–silica hybrid monolithic columns (a); and AuNP-modified AuNP–GSH–silica and AuNP-mediated GSH–AuNP–GSH–silica hybrid monolithic columns (b). (Reproduced from Zhao, L. et al., *J. Chromatogr. A*, 1446, 125, 2016. With permission.)

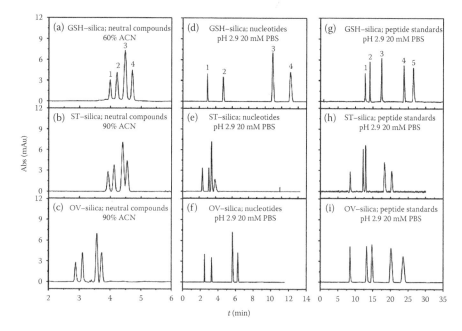

FIGURE 6.18 Electrochromatograms of neutral compounds (a–c); nucleotides (d–f) and the HPLC peptide standard mixture (g–i) on the GSH–, ST–, and OV–silica hybrid monolithic columns. Experimental conditions: 20 mM ammonium formate buffer containing 60% ACN at pH 2.6 for GSH–silica hybrid monolithic column (a), 90% ACN at pH 2.6 for ST– and OV–silica hybrid monolithic columns (b) and (c); applied voltage, 20 kV; detection wavelength: 190 nm (a–c); mobile phase, 20 mM PBS at pH 2.9; applied voltage, 20 kV; detection wavelength: 260 nm (d–f); mobile phase, 20 mM PBS at pH 2.9; applied voltage: 10 kV; detection wavelength: 190 nm (g–i). Solutes: (1) toluene, (2) DMF, (3) formamide, and (4) thiourea (a–c); (1) UMP, (2) AMP, (3) CMP, and (4) GMP (d–f); (1) Leu enkephalin, (2) Met enkephalin, (3) Val–Try–Val, (4) Gly–Try, and (5) angiotensin II (g–i). Sample injection, 8 psi for 5 s. The concentration of neutral compound each: 5 mM; nucleotide each: 0.3 mM and HPLC peptide standard mixture: 0.05 mg/mL. (Reproduced from Zhao, L. et al., *J. Chromatogr. A*, 1446, 125, 2016. With permission.)

6.3 NONPOLAR ORGANIC MONOLITHS

Nonpolar monolithic columns have found increased use in RPC and RP–CEC. Both closely related liquid-phase separation techniques are used for the separations of both small and large molecules. Methacrylate and acrylate-based monoliths as well as styrene-based monoliths are by far the most widely used nonpolar organic polymer stationary phases. They are usually prepared by direct *in situ* copolymerization of nonpolar functional monomers and cross-linkers as well as by functionalization of a precursor or a parent monolith using PPF methodologies.

6.3.1 DIRECT COPOLYMERIZATION

As aforementioned, the most commonly practiced method for polymer monolith fabrication is by copolymerizing the monomers and cross-linkers together in the presence of the proper porogens and an initiator. As most of the nonpolar monoliths are essentially neutral or are not charged under the mobile-phase conditions used, there will be no classification according to the charges of the monomers as in the abovementioned part pertaining to polar monoliths. In this part of the chapter, the monoliths are treated according to the chemical nature of the monomers making the monoliths, for example, acrylate/methacrylate or styrene-based monoliths.

6.3.1.1 Poly Acrylate/Methacrylate-Based Monoliths

As in part 2 of this chapter, and for the sake of clarity and convenience for the readers, the structures of the functional and cross-linking monomers used for nonpolar monoliths are shown in Tables 6.1 and 6.2.

In order to evaluate the effect of the alkyl chain length and structure of the functional monomer on the structural features of the monolith, different alkyl methacrylate monomers with butyl, cyclohexyl, 2-ethyl hexyl, lauryl, and stearyl functional groups were used in the polymerization mixture with EDMA as the cross-linker [53]. Alkylbenzenes and standard proteins were used as test solutes, and the standard parameters such as column permeability, methylene and phenyl selectivity were evaluated. The highest permeability was observed for butyl methacrylate (BMA) monolith and the least for the lauryl methacrylate (LMA), whereas the other three monoliths showed similar permeability. For the methylene group and phenyl group selectivity there was no direct correlation with the alkyl chain length of the monomer molecule. The lowest selectivity was observed with the LMA-based monolith, whereas BMA-based monolith showed the highest phenyl selectivity and the stearyl methacrylate (SMA)-based monolith possessed the highest methylene group selectivity. LMA and cyclohexyl methacrylate provided marginally better separations for the tested standard proteins by gradient elution in HPLC (Figure 6.19). This finding corroborates with the one reported subsequently [54] in the sense that a shorter alkyl chain monolith (e.g., C8) performed better in the separation of proteins by CEC.

The effect of the alkyl chain length of the functional monomers on solute retention was also investigated by Puangpila et al. in CEC [54]. The authors reported neutral monoliths (void of fixed charges) to completely eliminate electrostatic interactions nuisance created between charged solutes and the otherwise surface-attached charged moieties, an approach that was originally described by Okanda and El Rassi [55] and later by Karenga and El Rassi [5]. Two different series of monolithic columns were prepared with C8, C12, and C16 surface-bound chains using the same cross-linking monomer PETA. One series of monoliths (called series A) was produced by adjusting the composition of functional monomers and cross-linker to obtain comparable solute retention regardless of the alkyl chain length and another series of monoliths (referred to as series B) was prepared by maintaining the same composition of functional monomers and cross-linker yielding chromatographic retention, which increased as expected in the order of increasing the n-alkyl chain

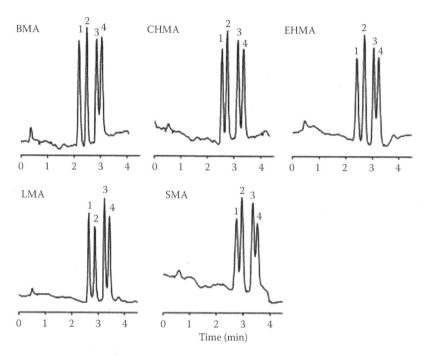

FIGURE 6.19 Separation of proteins on the monolithic polymethacrylate columns: (1) insulin, (2) cytochrome c, (3) BSA, and (4) β-lactoglobulin. Gradient: 20–80% ACN10.1% TFA in 4 min, flow rate 20 mL/min, UV detection at 214 nm. (Reproduced from Urban, J. et al., *J. Sep. Sci.*, 34, 2054, 2011. With permission.)

length. The C16-monolith of the A series yielded the highest separation efficiency toward small solutes, but the A columns series was inadequate for protein separation. The C8-monolith of the B series provided the best separation efficiency for proteins (Figure 6.20a). For tryptic peptide mapping, the C16-monolith of the A series seems to provide the best separation (Figure 6.20b). For large protein molecules, the energetically *softer* C8 surface allowed faster sorption kinetics and in turn improved the separation efficiency, whereas an energetically *harder* C16 surface favored better separation of the smaller-size peptide solutes. In brief, neutral, polar, and charged solutes (e.g., proteins and peptides) were separated efficiently and the results were in agreement with the previously reported work on neutral monoliths by El Rassi and coworkers [55,56].

A 1,6-hexanediol ethoxylate diacrylate cross-linker was reported for making nonpolar monoliths for RPC separations in capillaries [57]. The suitability of this cross-linker was studied with three alkyl methacrylate functional monomers, namely BMA, LMA, and SMA. For comparing the retentive abilities of these monoliths, the authors also made alkyl methacrylate monolithic columns with EDMA and 1,6-hexanediol dimethacrylate as the cross-linker in fused silica capillaries. As reported by the authors, due to the presence of the two ether linkages, the 1,6-hexanediol ethoxylate diacrylate monolithic columns showed an added advantage

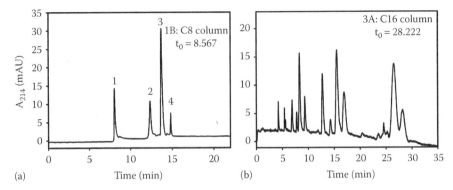

FIGURE 6.20 (a) Electrochromatogram of some standard proteins obtained on monolith 1B: C8 column, 20 cm effective length, 27 cm total length × 100 μm ID; mobile phase, 10 mM sodium phosphate, pH 7.0, at 45 % v/v ACN, running voltage 10 kV, electrokinetic injection for 3 s at 10 kV. Solutes: 1, lysozyme; 2, cytochrome C; 3, ribonuclease A; 4, ovalbumin. (b) Electrochromatogram of the tryptic digest of cytochrome C obtained on 3A: C16 capillary column, 20 cm effective length, 27 cm total length × 100 μm ID; mobile phase, 10 mM sodium phosphate, pH 6.0, at 35% ACN (v/v); running voltage 10 kV, electrokinetic injection for 5 s at 10 kV. (Reproduced from Puangpila, C. et al., *Electrophoresis*, 33, 1431, 2012. With permission.)

of dipole–dipole interactions in the separation of polar analytes. A 30% (w/w) alkyl methacrylate concentration was maintained in all the cases, and the conversion of the monomer was maximal with 1,6-hexanediol ethoxylate diacrylate than for EDMA and 1,6-hexanediol dimethacrylate. Among 18 different monoliths, the 1,6-hexanediol ethoxylate diacrylate-based alkyl methacrylate monolith was found to have the highest column efficiencies of up to 14,800 plates/m for uracil. The poor performance of the EDMA and the 1,6-hexanediol dimethacrylate columns was attributed by the authors to the poor conversion rate of these two cross-linkers in the formation of the polymer.

Vinyl ester-based monoliths have been reported as stationary phases for CEC [58]. They were prepared by using two vinyl ester monomers with different alkyl chain lengths namely vinyl pivalate (VPV) and vinyl decanoate (VDC) as the main functional monomers, AMPS as the charge-bearing monomer, and EDMA as the cross-linker. The electrochromatograms obtained on both poly(VPV-*co*-EDMA) and poly(VDC-*co*-EDMA) for the separation of alkylbenzenes demonstrated typical RPC behavior. It was reported that increasing the microporogen (i.e., isoamyl alcohol) content caused an increase in the volume fraction of small pores, which led to an increase in the total porosity. In addition, it was found that the vinyl ester-based monolith with a shorter alkyl chain length, that is, poly(VPV-*co*-EDMA), is more hydrophobic and exhibited a better separation efficiency possibly due to the branched nature of the VPV.

Eight different dimethacrylate cross-linkers were copolymerized with the LMA functional monomer in order to prepare a series of reversed-phase monolithic capillary columns for use in nano-LC [59]. The used cross-linkers have repeated methylene and oxymethylene units, including EDMA, tetramethylene dimethacrylate, hexamethylene dimethacrylate, dioxyethylene dimethacrylate, trioxyethylene dimethacrylate, tetraoxyethylene dimethacrylate, pentaerythritol dimethacrylate, and bisphenol A dimethacrylate. As reported by the authors, the separation efficiency for small molecules obtained on the monolithic columns significantly improved with increasing the number of repeat nonpolar methylene groups in the cross-linking monomers, which correlated with increasing the number of small pores with size less than 50 nm. The thinnest plate height for alkylbenzenes of 25 μm was obtained on the columns prepared with hexamethylene dimethacrylate. In addition, the authors reported that the columns prepared with polar (poly) oxyethylene dimethacrylate cross-linkers showed enhanced separation efficiency with increasing chain length and in particular better performance than that of the (poly)methylene dimethacrylate cross-linkers of comparable size with less apparent effect of the chain lengths on pore size distribution. In fact, a thinner plate height of 15 μm was obtained on the columns made with tetraoxyethylene dimethacrylate cross-linker with ca. 70,000 plates/m. For the separation of proteins, baseline resolution was achieved for column prepared with hexamethylene dimethacrylate as the cross-linker, and the separation of four standard proteins (insulin from bovine serum pancreas, cytochrome C, bovine serum albumin, and β-lactoglobulin) was achieved in less than 4 min. Conversely, the authors reported that the monoliths made with the cross-linkers of repetitive oxymethylene groups showed very poor efficiency for the separation of proteins.

The effects of the cross-linker length on the separation efficiencies of small molecules have been evaluated by utilizing hexyl methacrylate (HMA) as the functional monomer with two different cross-linkers, namely 1,6-hexanediol ethoxylate diacrylate (1,6 HEDA) and EDMA [60]. It was found that the efficiency and the selectivity of the separation were improved ten-fold when using the longer chain cross-linker (i.e., 1,6 HEDA). This improvement was attributed by the authors to increasing the number of methylene groups, which resulted in increasing the number of mesopores. The separation of nonpolar analytes on the 1,6-HEDA-based monolith showed typical RPC behavior in which the k values for the nonpolar solutes decreased as the %ACN in the mobile phase increased. Furthermore, the 1,6 HEDA-based monolith demonstrated baseline separation of neutral nonpolar molecules, weak acid, and basic solutes under RPC conditions.

The influence of multifunctional cross-linker on the separation of small molecules has been reported in Reference 61. A novel polymeric monolith was prepared with LMA as the functional monomer and dipentaerythritol penta-/hexa-acrylate (DPEPA) as the cross-linker poly(LMA-co-DPEPA), which have five or six acrylate groups to participate in the polymerization for the formation of highly cross-linked polymeric monoliths (Figure 6.21) for small molecule separation with high separation efficiencies reaching 111,000–165,000 plates/m for alkylbenzenes. It was reported that the use of multifunctional cross-linker possibly prevents the formation of gel-like micropores, which in turn reduces the mass transfer resistance thereby increasing the efficiency of separation. Furthermore, ethylene glycol was chosen as the macroporogen (i.e., poor solvent), with hexyl alcohol as the microporogen (i.e., good solvent). The authors found that the proportion of porogenic solvents had a little influence on the permeability of highly cross-linked polymeric monoliths

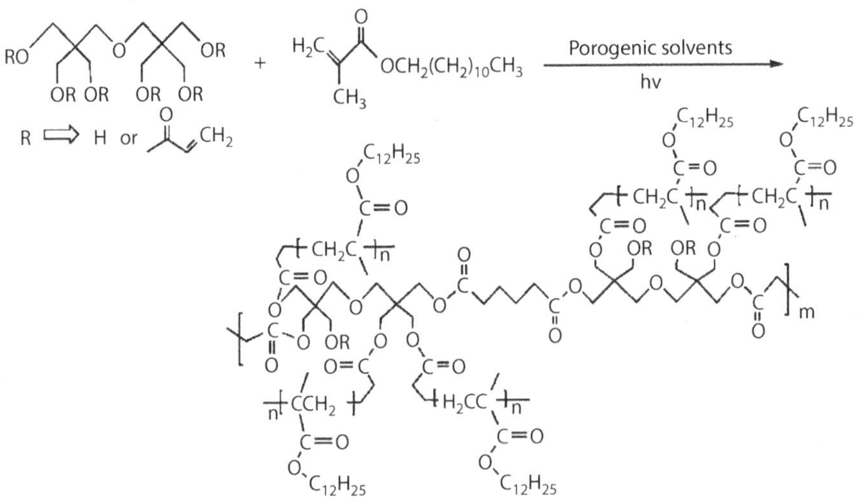

FIGURE 6.21 Schematic preparation of polymeric monoliths [poly(LMA-*co*-DPEPA)] *via* photoinitiated polymerization. (Reproduced from Zhang, H. et al., *Anal. Chim. Acta*, 883, 90, 2015. With permission.)

fabricated with a high content of the cross-linker investigated. The effect of ACN content on the *k* values of five alkylbenzenes followed the principles of RPC in which the *k* of alkylbenzenes decreased with increasing the ACN content of the mobile phase. The columns were also used to separate mixtures of phenolic compounds, basic compounds, and standard protein mixtures under RPC conditions.

A C60 fullerene-containing methacrylate monomer has been reported for the preparation of porous monoliths for the separations of small molecules [62]. The porous monoliths consisted of poly(glycidyl methacrylate (GMA)-*co*-EDMA) and poly(BMA-*co*-EDMA) capillary columns, which incorporated the new monomer [6,6]-phenyl-C_{61}-butyric acid 2-hydroxyethyl methacrylate (PCB-HEM) (Figure 6.22a). Both monoliths showed added advantages in the separation of alkylbenzenes. A very poor separation of alkylbenzenes was observed for the parent poly(GMA-*co*-EDMA) monolith with a number of plates of around 4400 plates/m for benzene, which is the first retained compound exhibiting the highest separation efficiency. On the addition of PCB–HEM, the plate number for benzene increased to around 72,000 plates/m. As shown in Figure 6.22b, the parent monolith poly(GMA-*co*-EDMA) yielded virtually no resolution for the alkylbenzene homologous series as compared to the poly(GMA-*co*-EDMA) with 1 wt% PCB–HEM. The peak asymmetry and separation efficiency were further improved on the poly(GMA-*co*-EDMA) with 1 wt% PCB–HEM when adding 2.5% THF to the mobile phase. In fact, the peak asymmetry factor A_s, which measures the peak tailing, was reduced from A_s <2.0 to A_s <1.7, and the column efficiency for benzene increased from 72,000 plates/m to 85,000 plates/m. On the other hand, the parent poly(BMA-*co*-EDMA) monolith, which is a more hydrophobic monolith than the poly(GMA-*co*-EDMA) monolith exhibited enhanced retention for alkylbenzene solutes than the

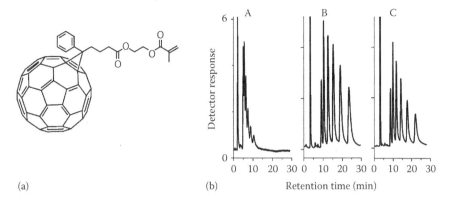

(a) (b) Retention time (min)

FIGURE 6.22 (a) Structure of [6,6]-phenyl-C_{61}-butyric acid 2-hydroxyethylmeth-acrylate ester. (b) Separation of uracil and alkylbenzenes using a parent monolithic poly(glycidylmethacrylate-*co*-ethylene dimethacrylate) capillary column (A) and using a column containing 1 wt % PCB–HEM (B, C), both prepared at a temperature of 70°C. Conditions: column 53 mm × 100 μm i.d., flow rate 0.15 μL/min, UV detection at 254 nm; (A) mobile phase 50:50 vol % acetonitrile—water, back pressure 15 MPa; (B) mobile phase 50:50 vol % acetonitrile—water, back pressure 25 MPa; (C) mobile phase 47.5:2.5:50 vol % acetonitrile–tetrahydrofuran–water, back pressure 27 MPa; peaks in order of elution: uracil, benzene, toluene, ethylbenzene, propylbenzene, butylbenzene, and amylbenzene. (Reproduced from Chambers, S.D. et al., *Anal. Chem.*, 83, 9478, 2011. With permission.)

less hydrophobic poly(GMA-*co*-EDMA) monolith, and yielded columns with plate counts for benzene exceeding 110,000 plates/m.

Jonnada and El Rassi have introduced an organic monolithic column by *in situ* copolymerization of the functional monomer 2-naphthyl methacrylate (NAPM) and the cross-linker trimethylolpropane trimethacrylate (TRIM) in stainless steel columns of 4.6 mm i.d. for RPC [63]. The column showed high mechanical stability and good hydrodynamic characteristics with virtually *zero* compressibility over a long period of daily use with a wide range of mobile-phase compositions. Compared to an octadecyl silica (ODS) column, the π-electron–rich naphthyl ligands in the monolith offer additional π–π interaction with aromatic molecules in addition to hydrophobic interactions under RPC conditions. In fact, and due to the additional π–π interactions, the NMM column was able to fully resolve 3,4,5-trichlorophe-nol from 2,3,4,5,6-pentachlorophenol, whereas a silica-based C18 column did not exhibit enough selectivity to resolve the same solute pair (Figure 6.23a). In addition, when compared to the same C18 silica-based column, the NMM column showed higher retention toward *m*-nitrotoluene than toluene, and the elution order of these two toluene derivatives on the NMM column was the inverse of that obtained on the C18 silica-based column (Figure 6.23b), again indicating the presence of π–π interactions on the former column and its absence on the latter column. Overall, the NMM column was useful in the separation of aromatic compounds such as benzene, toluene, phenol, and aniline derivatives. In addition, mixtures of standard proteins were well separated in less than 1 min using an ultra-steep linear gradient elution at increasing ACN concentration in the mobile phase.

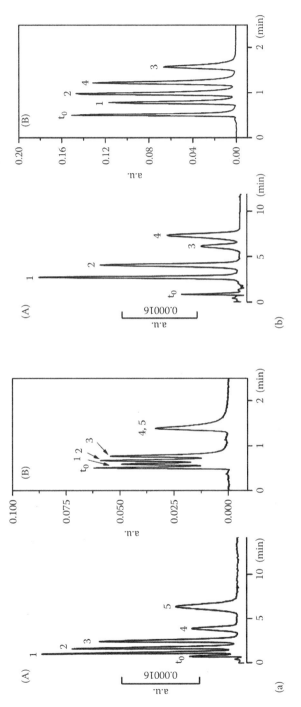

FIGURE 6.23 (a) Chromatograms of phenol and phenol derivatives obtained on NMM column (10 cm × 4.6 mm I.D.) in (A) and ODS column (4.5 cm × 4.6 mm I.D.) in (B). Conditions: mobile phase, ACN:H$_2$O (55:45 v/v); flow rate, 1 mL/min; UV detection at 214 nm. Peaks: t$_0$, thiourea; 1, catechol; 2, phenol; 3, 2-chlorophenol; 4, 3,4,5-trichlorophenol; 5, 2,3,4,5,6-pentachlorophenol. (b) Chromatograms of toluene and toluene derivatives obtained on NMM column (10 cm × 4.6 mm I.D.) in (A) and ODS column (4.5 cm × 4.6 mm I.D.) in (B). Conditions as in (A). Peaks: t$_0$, thiourea; 1, m-toluidine; 2, m-tolualdehyde; 3, toluene; 4, m-nitrotoluene. (Reproduced from Jonnada, M. and El Rassi, Z., *J. Chromatogr. A*, 1409, 166, 2015. With permission.)

A novel monolithic column with double C18 chains was recently introduced [64] for RPC whereby the functional monomer 3-methylacrylol-3-oxapropyl-3-(N,N-dioctadecylcarbomyl)-propionate (AOD) was copolymerized with the cross-linker EDMA yielding the poly (AOD-*co*-EDMA) monolith. Fused silica capillary columns (100 µm i.d.) were used in the preparation of the monolithic column for µHPLC. For comparison studies, a C18 monolith namely poly (SMA-*co*-EDMA) was also prepared according to the author's previous work [65]. Poly (AOD-*co*-EDMA) monolith showed a smaller theoretical plate height of 19.2 µm than poly (SMA-*co*-EDMA) monolith, which was 32.1 µm at a linear flow velocity of 0.85 mm/s using similar conditions. Van Deemter plots and the methylene selectivity studies were also carried out to evaluate the column efficiency and the reversed-phase retention behavior, respectively. Alkylphenones were used as test solutes and the methylene selectivity was observed to be as 1.68, which was similar to 1.70 obtained with the poly (SMA-*co*-EDMA) monolithic column. Sixteen polyaromatic hydrocarbons (PAH) were separated on the poly (AOD-*co*-EDMA) monolithic column by using a gradient elution (Figure 6.24). Again, TOH was used to test this column and a complete separation of α-, β-, γ-, and δ-TOH isomers was obtained in less than 30 min. Baseline separation was achieved for three different standard proteins in less than 8 min in reversed-phase gradient elution.

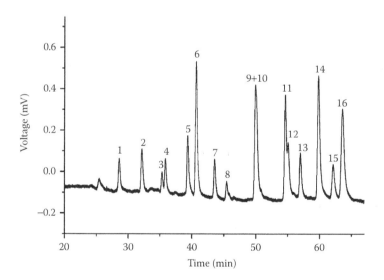

FIGURE 6.24 Separation of a standard mixture of 16 PAHs. Conditions: Poly(AOD-*co*-EDMA) monolithic column: 180 mm × 100 µm I.D.; mobile phase: H_2O (A) and ACN (B), linear gradient, 20% B to 80% B in 40 min; detection wavelength: 254 nm; flow rate: 500 nL/min; injection volume: 20 nL; samples: (1) naphthalene; (2) acenaphthylene; (3) acenaphthene; (4) fluorene; (5) phenanthrene; (6) anthracene; (7) fluoranthene; (8) pyrene; (9)+(10) benz[a]anthracene; chrysene; (11) benzo[b]fluoranthene; (12) benzo[k]fluoranthene; (13) benzo[a]pyrene; (14) dibenz[a,h]anthracene; (15) indeno[1,2,3-cd]pyrene; (16) benzo[g,h,i]perylene. (Reproduced from Duan, Q. et al., *J. Chromatogr. A*, 1345, 174, 2014. With permission.)

In a novel approach for achieving a unique selectivity, mixed ligand monolithic (MLM) columns were prepared for CEC [66]. These columns were made by the copolymerization of different compositions of octadecyl acrylate (ODA) and 2-naphthyl methacrylate (NAPM) functional monomers in the presence of TRIM cross-linker. The combined retentive property of the ODA ligand, which is solely hydrophobic, with that of the NAPM ligand, which is both hydrophobic and π–π interaction provider was exploited in the CEC of neutral, polar, and charged solutes. As expected the magnitude of the EOF changed with the composition of the MLM. As the percent of the monomer ODA in the polymerization mixture was increased, the EOF increased to a maximum at 50 mol% ODA and then leveled off at 75 mol% and 100 mol% ODA, an indication that the ODA ligand, in general, exhibited a higher binding for the mobile-phase ions than the NAPM monomer. This is due to the fact that the ODA is an acrylate-based monomer, whereas the NAPM is a methacrylate-based monomer. It was found that columns with a given composition of both ligands yielded a unique selectivity for a given set of solutes that was not matched by columns made by either ODA or NAPM alone. Several test mixtures were used in the evaluation of the MLM columns, including polycyclic aromatic hydrocarbons, alkyl phenyl ketones, nitroalkanes, alkylbenzenes, toluene derivatives, peptides, and proteins. Peptide mapping of the tryptic digest of the standard lysozyme protein was also studied.

Furthermore, Karenga and El Rassi also developed a series of segmented monolithic columns (SMC) [67] and demonstrated their applications in CEC. These SMCs were made up of two adjoining segments each filled with a different monolith in the aim of controlling and manipulating the EOF, retention, and selectivity in RP–CEC. In these SMCs one capillary segment was filled with a naphthyl methacrylate monolith (NMM) to provide hydrophobic and π–π interactions, whereas the other capillary segment was filled with an octadecyl monolith (ODM) to provide solely hydrophobic interaction. The ODM segment not only provided hydrophobic interactions but also functioned as the EOF accelerator segment. The average EOF of the SMC increased linearly with increasing the fractional length of the ODM segment. The neutral SMC provided a convenient way for tuning EOF, selectivity, and retention in the absence of annoying electrostatic interactions and irreversible solute adsorption. The SMCs allowed the separation of a wide range of neutral solutes, including PAHs that are difficult to separate using conventional alkyl-bonded stationary phases. In all cases, the k of a given solute was a linear function of the fractional length of the ODM or NMM segment in the SMCs, thus facilitating the tailoring of a given SMC to solve a given separation problem. At some ODM fractional length, the fabricated SMC allowed the separation of charged solutes such as peptides and proteins that could not otherwise be achieved on a monolithic column made from NMM as an isotropic stationary phase due to the lower EOF exhibited by this monolith.

A monolithic column for HPLC was prepared by the free-radical copolymerization of a mixture of two functional monomers, namely TMPTA and N-isopropylacrylamide (NIPAAm), with the cross-linker EDMA [68]. The polymerization process yielded highly cross-linked poly (TMPTA-co-NIPAAm-co-EDMA) monoliths in short stainless steel columns of 4.6 mm i.d. Flow rates of up to 7 mL/min were achieved indicating good mechanical stability and permeability of the columns. As expected, the analysis time for some small aromatic solutes was shortened from ~20 min to ~10 min by increasing the column temperature from 25°C to 70°C, which reflects the usual temperature effect in LC. In another work from the same laboratory [69], TMPTA was used along with EDMA to produce a dense network of nonpolar monolith in a 50 mm × 4.6 mm i.d. stainless steel column and used it for the RPC separations of small aromatic molecules.

A recent research article reported a monolith made of bisphenol A epoxy vinyl ester resin (VER) as the functional monomer along with EDMA as the cross-linker [70]. The *in situ* polymerization was carried out in 150 μm i.d. fused capillaries and the ratio of VER to EDMA was carefully adjusted along with the porogens to obtain relatively high permeability monoliths. The mechanical stability of the column was also evaluated by studying the column back pressures *versus* flow rate. Furthermore, the separation of benzene derivatives was demonstrated by HPLC, with a column efficiency of around 200,000 plates/m. The separations seemed to be superior on this column when compared to that obtained on a poly (BMA-*co*-EDMA) monolith.

In order to evaluate the repeatability of in-column preparation of a reversed-phase C18 monolith, the monolith poly(SMA-*co*-EDMA) was prepared by varying the polymerization temperature and performing the polymerization in three different polymerizing devices [71], namely water bath, hot air, and GC oven. The monolithic columns were evaluated by measuring the retention times of neutral aromatic compounds in CEC. Regardless of the device used in the preparation of the monolithic columns, increasing the polymerization time yielded increased solute retention time, which might be due to the formation of smaller pores at high temperatures as a result of fast polymerization. In addition, the formation of free radical initiators at elevated temperatures might be another reason for the small pores. The temperatures 55°C and 60°C were selected for water bath and ovens, respectively, for conducting the polymerization and the % RSD for the results obtained for the three prepared columns were 9.0%, 6.5%, and 12.5% for GC oven, hot air oven, and water bath, respectively. These results indicate that the hot air oven produced more reproducible results than the rest. Furthermore, the interday and intraday repeatability studies showed good precision for data obtained for retention times, peak area, and efficiency for the hot air oven. The solute retention selectivity of the monolithic columns using tocopherols (TOH) or vitamin E homologues was evaluated and compared with other C8 and C18 monolithic columns and silica-based columns that were previously reported. The difficult to separate components, mainly the β-TOH, γ-TOH isomers were baseline resolved with R_s of 1.1. The authors claim that this monolith offer a better selectivity due to presence of the residual polar ester groups on the surface, which would provide polar interactions with the hydroxyl groups of the tocopherols. The previously reported studies performed on C8 and C18 silica-based columns showed no resolution for the β and the γ isomers.

A methacrylate ester-based monolithic column has been developed for cLC applications using butyl methacrylate (BMA) as the functional monomer and EDMA as the cross-linker [72]. This investigation reported the influence of the porogen composition of the prepolymerization mixture on monolithic morphology. As one would expect, the reduction of the porogenic solvent content yielded monoliths with small macropores and globule sizes, leading to a high resistance to flow with the given mobile phase. The optimized monolith was used for the ultra-low determination of parabens, which are a class of antimicrobial agents in human urine and serum.

6.3.1.2 Styrene Divinyl Benzene-Based Monoliths

A poly(styrene-co-DVB-co-MAA) monolithic column was prepared and applied to the separation of some aromatic compounds by RPC in the cLC format [73]. The results were compared to those obtained on the monolith prepared without adding MAA. Baseline separation was achieved with a column separation efficiency of around 28,000 plates/m. These results indicate that the contribution of the MAA is not only merely imparting charges to the monolithic surface for the generation of EOF in CEC but also affected the morphological features of the separation monolith. In fact, although no separation was achieved using the column polymerized without MAA and all the components coeluted in a single large peak, the monolith with added MAA allowed the baseline separation of all the eight analytes, with a separation efficiency of 28,000 plates/m.

Organotellurium-mediated living radical copolymerization has been reported as an effective approach to control the porosity of poly(styrene-co-DVB) monoliths [74] by permitting the generation of well-defined macropores and less-heterogeneous cross-linked networks. This was reported to facilitate the separation of small molecules such as alkylbenzenes at a low-pressure drop in RPC using isocratic elution.

The copolymerization of N-vinylcarbazole (NVC) and DVB in both 200 μm and 100 μm i.d. fused silica capillary tubing by a thermally initiated free-radical copolymerization yielded nonpolar monolithic capillary columns for RPC and ion-pair RPC [75]. The retentive properties of these columns were evaluated by nano-LC using peptides, proteins, and oligonucleotides as model solutes. The surface area of the monolith was found to be in the range of 120–160 m^2g^{-1} as determined by Brunauer–Emmett–Teller procedure (BET). In addition, the monolithic column exhibited high permeability for the common RPC mobile phases. A mixture of seven oligonucleotides $d(PT)_{12}$-$d(pT)_{18}$ was baseline separated by ion-pair RPC in less than 7 min by gradient elution and in less than 8 min by isocratic elution as shown in Figure 6.25a and b, respectively. Furthermore, baseline separations of 5 standard proteins and 9 standard peptides were readily achieved on these capillary columns under RPC conditions by gradient elution.

6.3.2 Cross-Linkers–Based Monoliths

In an attempt to obtain highly cross-linked monolithic columns, the polymerization of various single cross-linking monomers (See Table 6.2 for structures of cross-linkers) was carried out to yield cross-linkers–based monoliths. These single cross-linkers possessing specific functional groups provide both the monolithic rigidity and the chromatographic retentive ligands. Polymer monolithic columns synthesized from single cross-linkers provide satisfactory mechanical stability and high surface area due to the highly cross-linked surface and in addition the optimization of the polymerization mixture is easier because only the monomer–porogens ratio needs to be adjusted. El Rassi and coworkers were the first to report a polymer

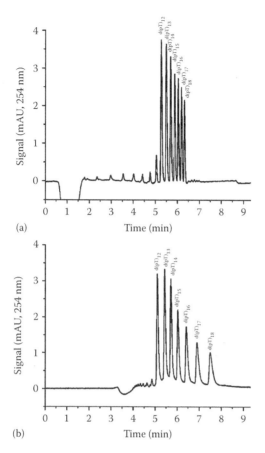

FIGURE 6.25 Ion-pair reversed-phase (IP-RP) separation of seven oligonucleotides $d(pT)_{12}-d(pT)_{18}$: (a) 70 mm × 0.2 mm capillary column. Separation conditions: monolith 2; 15%–40% B in 7 min, 4 µL/min, 50°C, UV 254 nm, inj.: 1 µL, 5 ng total. (b) 70 mm 0.1 mm capillary column. Separation conditions: monolith 5; isocratic elution 0.1 M TEAA pH 7 in 10% acetonitrile, 700 nL/min, 50°C, UV 254 nm, inj.: 300 nL, 3 ng total. (Reproduced from Koeck, R. et al., *Analyst*, 138, 5089, 2013. With permission.)

monolith synthesized from a pentaerythritol diacrylate monostearate (PEDAS) as the single monomer/cross-linker [55,76,77]. PEDAS provided a macroporous nonpolar monolith with C17 surface chains and these monoliths exhibited high separation efficiencies for proteins and peptides.

Along the abovementioned lines, bisphenol A dimethacrylate, bisphenol A ethoxylate diacrylate, and pentaerythritol diacrylate monostearate were used to produce nonpolar monolithic capillary columns for RPC of small molecules [78]. The selection of the porogens was made by the authors on the basis of the dipole moment and the solubility parameter. Tetrahydrofuran and decanol were reported to be useful porogens for the formation of bisphenol A ethoxylate diacrylate (where EO/phenol = 4) monoliths, whereas DMF along with decanol were reported adequate for the formation of bisphenol A dimethacrylate monoliths. For the poly(pentaerythritol diacrylate monostearate) monolith the selection of the porogens was found by the authors to be more challenging. The rigid structure of the pentaerythritol diacrylate monostearate led to either a hard monolith or a soft gel-like substance when using most of the common porogenic solvents. As the structure of the pentaerythritol diacrylate monostearate is similar to a surfactant, the authors predicted that a nonionic surfactant would be a potential candidate for the preparation of the monolith. Tri block poly(ethylene oxide)–poly(propylene oxide)–poly(ethyleneoxide) (PEO–PPO–PEO) or PPO–PEO–PPO, was used as the porogenic solvent, and the *in situ* copolymerization was photoinitiated for a duration of 4 min. The monolithic columns were evaluated using alkylbenzenes and alkylparabens as test solutes in both gradient and isocratic nano-LC. In all cases, baseline separations were achieved for alkylbenzenes and alkylparabens for the three columns in gradient elution. The poly (bisphenol A ethoxylate diacrylate) yielded an average plate number of 30,000 plates/m, whereas the poly (bisphenol A dimethacrylate) and the poly (pentaerythritol diacrylate monostearate) exhibited a maximum of around 61,000 plates/m and 21,000 plates/m, respectively.

The concept of using single cross-linking monomers was further exploited in the same laboratory to produce highly cross-linked polymeric monoliths for RPC in 75 μm i.d. fused silica capillary columns [79]. Alkanediol methacrylates monomers, namely, 1,4-butanediol dimethacrylate, 1,5-pentanediol dimethacrylate, 1,6-hexanediol dimethacrylate, 1,10-decanediol dimethacrylate, 1,12-dodecanediol dimethacrylate, 1,3-butanediol dimethacrylate, and neopentyl glycol dimethacrylate were used in this study. Dodecanol and methanol were found to be adequate porogens for the synthesis of the monoliths. Again, the test solutes alkylbenzenes and parabens were baseline separated using gradient elution. The plate numbers for all the monolithic columns were estimated between 14,000 and 35,000 plates/m.

The same research group as aforementioned introduced a series of highly cross-linked monoliths over the next couple of years (2012–2014) using the concept of single cross-linking monomers. In the first article of the series [80], they demonstrated the use of highly cross-linking polymers made from various dimethacrylate of C6 functional groups namely, 1,6-hexanediol dimethacrylate, cyclohexanediol dimethacrylate and 1,4-phenylene diacrylate (PhDA) in the separation of small molecules (e.g., alkylbenzenes) by RPC using the nano-LC format. The 1,6-hexanediol dimethacrylate column exhibited the highest separation efficiency (86,000 plates/m) with the least permeability. The run-to-run and column-to-column reproducibilities were evaluated with the test analytes alkylbenzenes, which were separated with baseline resolution in less than 10 min on all the columns. The single monomer approach might be the reason for the repeatability of the monolith fabrication [80]. Isocratic elution of alkylbenzenes was performed at an optimum flow rate of 300 nL/min on the capillary columns made with monoliths from these monomers. As reported by the authors, whereas the 1,6-hexanediol dimethacrylate and the 1,4-cyclohexanediol dimethacrylate-based monoliths offered baseline resolution for the alkylbenzenes investigated, the PhDA-based monolith showed very poor resolution.

Along the same lines, a second contribution from the same laboratory as abovementioned reported other monolithic RPC stationary phases based on single multiacrylate/methacrylate, which have been synthesized with monomers such as 1,12-dodecanediol dimethacrylate (1,12-DoDDMA), TRIM, and pentaerythritol tetraacrylate (PETAC) using organotellurium-mediated living radical polymerization (TERP), which was found by the authors of this work to produce more efficient monolithic columns than conventional free-radical polymerization [81]. All prepolymerization solutions were prepared by admixing initiator, monomer, and porogenic solvents with ethyl-2-methyl-2-butyltellanyl propionate (BTEE) as the reaction promoter. Rigid structural monoliths were obtained using all of the monomers, and all monolithic columns could be used to separate alkylbenzenes. As poly(PETA) has a C5 functional group it showed less retention and selectivity than poly(1,12-DoDDMA) and poly(TRIM), which contain more hydrophobic C12 and C6 functionalities, respectively. Plots of logarithmic retention factor of alkylbenzenes versus carbon number indicate that columns with longer hydrocarbon chain length between methacrylate/acrylate end groups showed higher methylene group selectivity. Furthermore, a control column was synthesized to compare columns prepared from the same monomer (TRIM) using TERP or conventional free radical polymerization. Comparison of van Deemter curves for the two columns showed that the column prepared from TERP exhibited much lower HETP.

The last contribution in the series from the same research group reported a highly efficient monolith by a single cross-linker polymerization of poly(ethyleneglycol) diacrylate (PEGDA) monomers for the RPC of small molecules [82]. The authors of this investigation reported a statistical model for optimizing the reagents and conditions in order to simplify the fabrication and the development process. It was reported that the model could predict the probability of obtaining a monolith with 74% accuracy, and the column performance could be correlated to prepolymer solubility and viscosity values. This monolithic column was successfully used to separate low-molecular weight compounds such as hydroxybenzoic acids, phenols, alkyl parabens, nonsteroidal anti-inflammatory drugs (Figure 6.26a), and phenyl urea herbicides (Figure 6.26b). The retention mechanism was found to follow typical reversed-phase behavior with additional hydrogen bonding interactions, making the column suitable for the separation of polar compounds, which are otherwise poorly retained on C18 columns.

6.3.3 Nonclassical Polymerization Approaches and Novel Chemistries

The classical *in situ* free-radical polymerization *via* thermal- or photoinitiation using AIBN as the initiator is the most common and popular approach in preparing monolithic stationary phases. However, *in situ* free-radical polymerization has the disadvantages of slow initiation, fast progression, easy chain transfer, and quick chain termination, which lead to a nonuniform structure and in turn low column separation efficiency and low permeability. To alleviate these disadvantages, atom transfer radical polymerization (ATRP) technique was recently explored as an alternative route for the preparation of monolithic stationary phases [83]. ATRP is a commonly used technique for controlled radical polymerization, which employs transition metal complex as the catalyst with an alkyl halide as the initiator. The ATRP process involves the successive transfer of the halide from the dormant polymer chain to the ligated

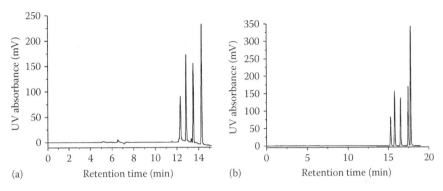

FIGURE 6.26 (a) RPC separation of nonsteroidal anti-inflammatory drugs (NSAIDs) on a PEGDA-700 monolithic column. Conditions: 15 cm × 150 μm i.d. monolithic column; mobile phase component A was acetonitrile with 1% formic acid (pH = 2.5), and B was water with 1% formic acid (pH = 2.5); linear gradient from 10% A to 100% A in 5 min, and then isocratic elution with 100% B; 0.04 cm/s linear velocity; on-column UV detection at 214 nm. Peak identifications in order of elution: paracetamol, aspirin, ibuprofen, and indomethacin. (b) RPC separation of urea herbicides on a PEGDA-700 monolithic column. Conditions: 15 cm × 150 μm i.d. monolithic column; mobile-phase component A was acetonitrile, and B was water; linear gradient from 10% A to 100% A in 15 min, and then isocratic elution with 100% B; 0.04 cm/s linear velocity; on-column UV detection at 214 nm. Peak identifications in order of elution: isoproturon, monuron, monolinuron, diuron, and linuron. (Reproduced from Aggarwal, P. et al., *J. Chromatogr. A*, 1364, 96, 2014. With permission.)

metal complex, thus establishing a dynamic equilibrium between the active species and the dormant ones. In addition, in the ATRP process, the concentration of free radicals is low such that the chain transfer and chain termination could be ignored, which could lead to a narrow molecular weight distribution. In this work [83], the polymer monolith was prepared using triallyl isocyanurate (TAIC) as the functional monomer, TMPTA as the cross-linker, PEG 200 and 1,2-propanediol as coporogens, carbon tetrachloride as the initiator, and ferrous chloride as the catalyst. The monolith was prepared in stainless steel columns (5 cm × 4.6 mm i.d.) and in fused silica capillaries (100 mm × 150 μm i.d.) for their applications in HPLC and cLC, respectively. The characterization indicated that the prepared RPC monolith possessed uniform structure, good permeability, and good mechanical stability. Furthermore, the porous properties of the optimized poly(TAIC-*co*-TMPTA) indicated that the large surface area led to high column efficiency in the separation of small molecules. The separation of some aromatic compounds was demonstrated in both columns of analytical and capillary format.

The use of γ radiation in the preparation of monolithic columns has been described in a recent review article by Svec [84], which revealed that only a few monoliths have been prepared so far *via* this high-energy radiation due to the limited accessibility for scientists to γ-ray sources and safety issues in handling such sources. As documented and referenced by Svec' review [84], porous polymeric stationary phases can be synthesized at any temperature and in any column dimension using high-energy radiation such as γ rays without adding initiators, thus producing monolithic

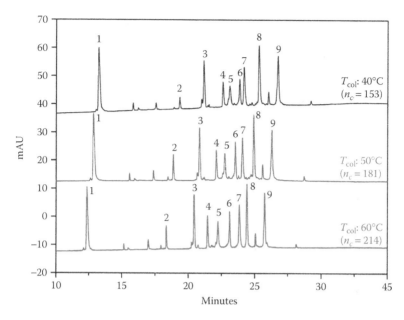

FIGURE 6.27 RP–cLC analysis of nine intact proteins performed at different temperatures. Column: γ-poly-(LMA-*co*-HDDMA) monolith (250 × 0.250 mm id). Conditions: mobile phase A: H_2O/ACN 95:5 v/v + 0.05% TFA, mobile phase B: H_2O/ACN 5:95 v/v + 0.05% TFA; gradient profile: 0–50% B in 30 min; flow rate:15 µL/min, UV at 214 nm, cell: 25 nL; volume injected: 200 nL. Protein mixture: 1, ribonuclease B; 2, lysozyme; 3, α-lactalbumin; 4, myoglobin; 5, creatine phosphokinase type I; 6, β-lactoglobulin B; 7, carbonic anhydrase; 8, enolase; 9, β-amylase. (Reproduced from Simone, P. et al., *J. Sep. Sci.*, 39, 264, 2010. With permission.)

columns with high chemical purity and reproducibility. Taking advantages of this particularity, a polymethacrylate-based poly(lauryl methacrylate-*co*-1,6-hexanediol dimethacrylate) [poly(LMA-*co*-HDDMA)] monolithic column has been prepared by γ-radiation-induced polymerization [85]. The γ-radiated monolithic columns exhibited high permeability and mechanical stability even at elevated temperatures and low pH values, which allowed the use of longer columns to obtain separations with high resolution. This poly(LMA-*co*-HDDMA) monolith was successfully used to analyze intact proteins under RPC conditions (Figure 6.27). As can be seen in this figure, increasing column temperature from 40°C to 60°C increased the *apparent column peak capacity* (as it is measured in gradient elution) from 153 to 214.

Photoinitiated thiol–acrylate polymerization was exploited in the preparation of monolithic columns whereby the acrylate are not only homopolymerized but also coupled with thiol-containing monomers offering an alternative way to fabricate various porous polymer monoliths. For instance, a simple approach has been developed for the rapid preparation of polymeric monolithic columns in UV-transparent fused silica capillaries *via* photoinitiated thiol–acrylate polymerization of PEDAS and trimethylolpropane tris(3-mercaptopropionate) (TPTM) within 10 min, in which the acrylate homopolymerized and copolymerized with thiol simultaneously [86].

Compared to poly(PEDAS) monolith prepared solely from PEDAS monomer, poly(PEDAS-*co*-TPTM) monolith possessed better permeability and better separation efficiencies. The authors believed that the formation of poly(PEDAS-*co*-TPTM) monolith was dominantly controlled by thiol–acrylate addition reaction mechanism, changing the pore structure and morphology of poly(PEDAS)-based monoliths, which favored an enhanced permeability. The chromatographic properties of poly(PEDAS-*co*-TPTM) monolith was evaluated by micro liquid chromatography (μLC) separation of alkylbenzenes, phenols, and basic compounds under RPC conditions. Besides the separation of these small molecules, the peptide mapping of the tryptic digest of four proteins (BSA, myoglobin, ovalbumin, and α-casein) was performed on poly(PEDAS-*co*-TPTM) monolith by μLC-MS/MS.

Step-growth polymerization is another novel polymerization technique, which refers to a type of polymerization process in which difunctional or multifunctional monomers react to form first dimers, then trimers, and so on until a high-molecular weight polymer is obtained. This step-growth polymerization approach led to monoliths with entirely distinct morphologies as compared to the globular structure generated by chain-growth polymerization (i.e., free radical polymerization), opening a great potential for the fabrication of polymer monoliths with high separation efficiencies for small molecules. This was demonstrated by using thiol-yne click polymerization to synthesize monoliths with higher cross-linking density because each alkynyl group could ideally react with two thiol groups *via* a two-step process. Two monolithic polymer columns have been prepared in UV-transparent fused silica capillaries *via* photoinitiated thiol-yne click polymerization of 1,7-octadiyne (ODY) with a dithiol (1,6-hexanedithiol [HDT] or a tetrathiol pentaerythritol tetrakis(3-mercaptopropionate [PTM]) within 15 min [87], yielding two monoliths referred to as O$_2$SH and O$_4$SH, respectively (Figure 6.28). The 2,2-dimethoxy-2-phenylacetophenone was used as the photoinitiator for the polymerization reaction and two porogenic systems of diethylene glycol diethyl ether (DEGDE)/tetrahydrofuran (THF) and DEGDE/PEG 200 were found to effectively control the porous properties of the two kinds of polymeric monoliths, namely O$_2$SH and O$_4$SH, respectively. The authors reported that both monoliths featured relatively homogeneous porous structure and stable structure morphology and chromatographic performance even after 3 months of usage. In addition, the prepared monolithic columns showed good linear relationships for back pressure versus flow rate below the pressure of 40 MPa.

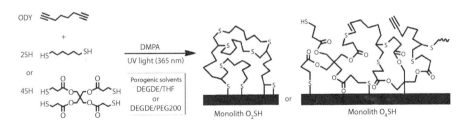

FIGURE 6.28 Preparation of porous polymer monoliths *via* thiol-yne click polymerization. (Reproduced from Liu, Z. et al., *Anal. Chem.*, 86, 12334, 2014. With permission.)

The baseline separation of alkylbenzenes using a mobile phase at 75% ACN revealed the strong hydrophobicity of the monolithic column O_2SH. In addition, O_4SH was also evaluated with the alkylbenzenes test solutes and satisfactory results were achieved. To further apply the monolithic columns in the RPC mode, bovine serum albumin (BSA) tryptic digest was separated on the O_2SH column by cLC–MS/MS. Furthermore, these columns were successfully used to separate mixtures of natural products and standard proteins under RPC conditions.

The same research group reported a novel organic monolith *via* thiol-ene click polymerization reaction of 1,2,4-trivinylcyclohexane (TVCH) and PTM within 10 min [88]. Although other thiol containing monomers, HDT and trimethylolpropane tris (3-mercaptopropionate) were also used for monolithic column preparation, the resulting monoliths were either too dense or too easily detached from the capillary wall. Authors claimed that compared to free-radical polymerization, the approach adopted in this report is more convenient and time-saving, offering a good alternative for organic monolith fabrication using multi-ene and multi-thiol as precursors. In addition, the resulting organic monoliths exhibited RPC mechanism and were reported useful for the separations of small molecules such as alkylbenzenes, pesticides, and basic compounds in cLC.

Low-conversion polymerization (i.e., short-term polymerization) has been applied for the preparation of poly(*N*-vinylcarbazole-*co*-1,4-divinylbenzene) using NVC as the functional monomer and DVB as the cross-linker [89]. Applying this approach in combination with variation in the radical initiator content proved to have a noticeable impact on the porous properties of the newly developed poly(NVC-*co*-DVB) stationary phase. Micro- and mesopores contributed greatly to the overall surface area, but only mesopores are important for the chromatographic separation process of small molecules. An ideal bimodal pore size distribution profile with a good fraction of mesopores was obtained by applying a polymerization time of 90 min, whereas a monomodal distribution of macropores was obtained by polymerization carried out for 270 min. Furthermore, it was demonstrated that the fraction of mesopores is significantly reduced by an increase in polymerization time and hence the chromatographic performance of the support toward the separation of small molecules was reported to be reduced. Moreover, the specific surface area of the monolithic column examined by multipoint BET nitrogen adsorption measurements was reported to be 418 m^2/g when using a polymerization time of 90 min, whereas a specific surface area of 160 m^2/g could be achieved by applying long-term polymerization. In addition, the specific surface area was decreased from 418 to 340 m^2/g on increasing the radical initiator content from 2% to 3%. The repeatability results during 1000 consecutive runs demonstrated the high stability of the chromatographic supports without any noticeable performance reduction (Figure 6.29). Furthermore, chromatographic behavior of the newly developed NVC/DVB monolithic stationary phase was examined by gradient separation of a variety of low-molecular weight compounds, including alkylbenzenes, beta blockers, flavonoids, parabens, and phenols. The highest number of theoretical plates per meter was found to be 116,000, which calculates to a minimum theoretical plate height value of 8.6 µm, showing the importance of understanding and controlling the polymerization conditions that lead to the desired pore size distribution profile.

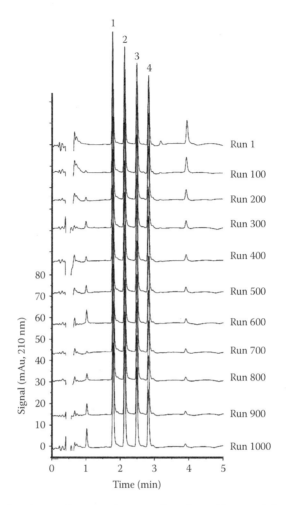

FIGURE 6.29 One thousand consecutive runs of parabens: 80 mm × 0.2 mm i.d. capillary column 1 (Table 6.1, entry 1). Separation conditions: 1%–99% B in 5 min, 15 µL/min, 70°C, UV 210 nm, inj. 1 µL, 1 ng each paraben:1 methylparaben, 2 ethylparaben, 3 propylparaben, 4 butylparaben. (Reproduced from Koeck, R. et al., *Anal. Bioanal. Chem.*, 406, 5897, 2014. With permission.)

6.3.4 Postpolymerization Functionalization

Organic polymer-based monoliths are known for their low specific surface areas, which make them not suitable for the separation of small molecules. One way to remedy this shortcoming has been to subject the monoliths in question to postpolymerization hypercrosslinking functionalization, a process that results in substantially increasing the specific surface areas when compared to that of the precursor or parent monoliths [90,91]. Hypercrosslinked monoliths date back to 1973 to the work of Davankov [92], which used Friedel–Crafts alkylation to further connect the free ends of styrene and chloromethyl styrene monomers.

Two hypercrosslinked monolithic columns, namely poly(styrene-*co*-vinylbenzyl chloride (VBC)-*co*-DVB) and poly(4-methylstyrene(MST)-*co*-VBC-*co*-DVB) were prepared and used in CEC [93]. The hypercrosslinking was achieved by flushing the two monolithic columns with 1,2-dichloroethane in the presence of Fe^{3+} as the catalyst. These monoliths showed enhanced specific surface areas when compared to the parent monolith (i.e., without hypercrosslinking), which resulted in increasing the EOF by 34-fold for hypercrosslinked poly(styrene-*co*-VBC-*co*-DVB) monoliths and by 21-fold for the hypercrosslinked poly(MST-*co*-VBC-*co*-DVB). As reported by the authors, augmenting the specific surface area on hypercrosslinking would increase the adsorption of ions from the mobile phase, and in turn generates increased EOF. This led to speeding up the separation of six alkylbenzenes in less than 8 min and 6 min on poly (styrene-*co*-VBC-*co*-DVB) and poly (MST-*co*-VBC-*co*-DVB), respectively.

In another approach, hypercrosslinked poly(styrene-*co*-DVB) monoliths were formed by using a Fe^{3+} catalyzed Friedel–Crafts alkylation using three external cross-linkers [94] (1) 4,4′-bis(chloromethyl)-1,1′-biphenyl (BCMBP), (2) α,α′-dichloro-*p*-xylene (DCX), and (3) formaldehyde dimethyl acetal. Of the three external cross-linkers, 4,4′-bis(chloromethyl)-1,1′-biphenyl was found to produce monoliths with the best chromatographic performance. Monoliths with extremely large surface areas reaching up to 900 m^2/g were obtained using a precursor monolith polymerized for only 2.5 h, and hypercrosslinked with 4,4′-bis(chloromethyl)-1,10-biphenyl. The hypercrosslinked monoliths were evaluated in capillary RPC using a test mixture composed of acetone and six alkylbenzenes. Under isocratic elution conditions, the authors reported separation efficiencies exceeding 70,000 plates/m were readily achieved for retained analytes.

A reactive polymer monolithic column has been prepared from glycerol carbonate methacrylate (GCMA) functional monomer and EDMA cross-linker followed by further functionalization with different ligands [95]. Cyclic carbonates can undergo ring opening in the presence of nucleophilic compounds such as primary amines for further functionalization of the monoliths with organic compounds of interest. Therefore, the surface modification of the pendant cyclic carbonates with allylamine yielded alkene functionalized monoliths. A successive radical phototriggered thiol-ene *click* addition allowed for the efficient grafting of 1-octane thiol or mercaptobutyric acid onto the alkene-functionalized monolith.

The poly(HPMA-Cl-*co*-EDMA) monolith described earlier in the section on polar monoliths has also been successfully used to prepare RPC monolithic columns through the PPF process. The 3-chloro-2-hydroxypropyl groups on this monolith were conveniently used to covalently attach various ligands *via* one-step reactions. In one case, poly(HPMA-Cl-*co*-EDMA) was functionalized with *N,N*-methyl-*N*-dodecylamine (DMDA) and the resulting monolith was evaluated in anion-exchange/hydrophobic mixed-mode interaction using nano-LC [96]. In a second case, poly(HPMA-Cl-*co*-EDMA) monolith was treated with sodium bisulfite in order to anchor sulfonic acid functionalities as charge-bearing groups to generate EOF during CEC separations [97]. Alkylbenzenes, phenols, and benzoic acids were separated on this monolith *via* typical RPC behavior in which the *k* values for alkylbenzenes decreased with increasing the %ACN content of the mobile phase.

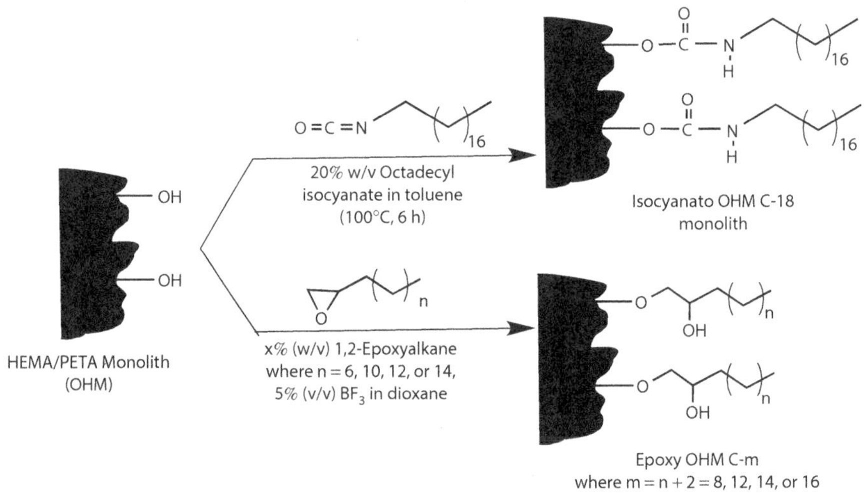

FIGURE 6.30 Schematic of the reaction of the precursor poly(HEMA-*co*-PETA) monolith (i.e., OHM) with 1,2-epoxyalkanes at varying alkyl chain length and with octadecyl isocyanate. (Reproduced from Khadka, S. and El Rassi, Z., *Electrophoresis*, 37, 2016. With permission.)

A novel precursor monolithic capillary column referred to as *hydroxy monolith* or OHM was prepared by the *in situ* copolymerization of HEMA with PETA yielding the neutral poly(HEMA-*co*-PETA) monolith. This neutral precursor OHM capillary was subjected to PPF of the hydroxyl functional groups present on its surface with 1,2-epoxyalkanes catalyzed by boron trifluoride (BF_3) ultimately providing Epoxy OHM C-m capillary column at varying alkyl chain lengths where m = 8, 12, 14, and 16 for RP–CEC (Figure 6.30). In addition, the same precursor OHM was grafted with octadecyl isocyanate yielding isocyanato OHM C18 column (Figure 6.30) to provide an insight into the effect of the nature of the linkage to the surface hydroxyl groups of the OHM precursor. Although the epoxide reaction leaves on the surface of the OHM precursor hydroxy–ether linkages, the isocyanato reaction leaves carbamate linkages on the same surface of the OHM precursor. This study revealed that changing the alkyl chain length resulted in changing the column phase ratio (ϕ) and also the solute distribution constant (K). Although increasing the surface alkyl chain length increased steeply the solute hydrophobic selectivity, that is, methylene group selectivity, the nature of the ligand linkage produced different retention for the same solutes and affected the selectivity of slightly polar solutes. The various monoliths proved very useful for RP–CEC of different small solutes at varying polarity over a wide range of mobile-phase composition.

In a second part of the series of investigations involving the PPF of the OHM capillary column described earlier, the surface hydroxyl groups were reacted with epoxy biphenyl, thus yielding the so-called biphenyl OHM capillary column. The modification involved the epoxy ring opening of the 2-biphenylyl glycidyl ether catalyzed by BF_3 and its subsequent reaction with the hydroxyl groups on the OHM

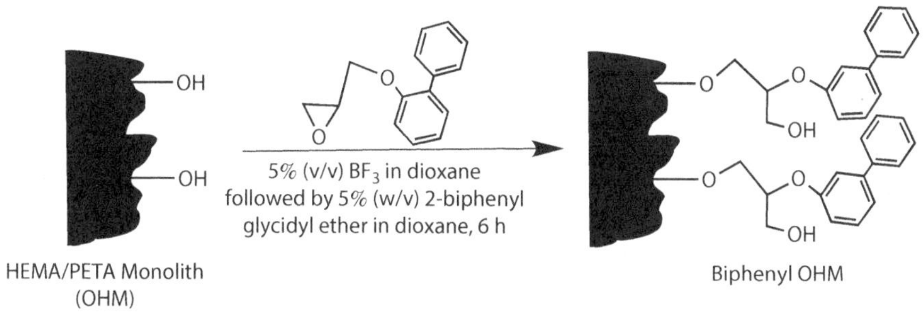

FIGURE 6.31 Postpolymerization modification of the poly(HEMA-*co*-PETA) precursor monolith with 2-biphenylglycidyl ether to yield the Biphenyl OHM HPLC column and the Biphenyl OHM capillary column. (Reproduced from Khadka, S. and El Rassi, Z., *Electrophoresis*, 37, 2016. With permission.)

precursor (Figure 6.31). The biphenyl OHM capillary column thus obtained exhibited the typical reversed-phase behavior by primarily hydrophobic interactions vis-à-vis the homologous series of alkylbenzenes and in addition by π–π interactions toward nitroalkane homologous series *via* their π-electron rich nitro groups. This dual retention mechanism was very distinctly observed with a set of PAH solutes in the sense that the *k* values of the PAH solutes were comparable to those obtained on a more nonpolar stationary phase, namely the epoxy OHM C16 reported in the preceding section. Other aromatic solutes showed the dual retention mechanism on the biphenyl OHM capillary, including phenols, anilines derivatives, and phenoxy acid herbicides.

6.3.5 Monoliths with Incorporated Nanoparticles

As mentioned earlier in Section 6.2 on polar monoliths, because of the nanosize of their particles, nanoentities must be held on a good support with favorable flow characteristics that can be readily prepared and confined in columns. Polymer monolithic columns are the preferred support media for incorporating nanoparticles because they are readily prepared *via in situ* polymerization in columns and channels of all sizes, can be tailor-made from many available monomers, can trap or incorporate covalently nanoparticles, and have very favorable flow characteristics.

Carbon nanotubes were incorporated in poly(GMA-*co*-EDMA) monoliths in order to improve the retention abilities of the monolith toward small molecules [98]. Entrapment of the multiwalled carbon nanotubes (MWCNT) in the methacrylate monoliths significantly improved the column efficiency to around 35,000 plates/m when compared to the parent GMA–EDMA monolith, which was on the order of 1800 plates/m. In a second approach, the authors reported the attachment of the MWCNT onto the surface of the monolith by passing the oxidized carbon nanotubes onto an amino modified GMA–EDMA monolith. Six alkylbenzenes were separated on this new MWCNT modified monolithic column with a separation efficiency of 44,000 plates/m.

In a different investigation, MWCNT was incorporated in a benzyl methacrylate-co-EDMA monolithic column in capillary LC [99]. The amount of nanotubes was varied between 0 and 0.4 mg/mL in the polymerization solution to prepare capillaries of 0.32 mm i.d. filled with the monolith. The test solutes alkylphenones and phenols yielded significant improvement in separation efficiencies between the parent monolith and the monolith incorporated with MWCNT.

The separations of small and large biomolecules by HPLC were performed onto monolithic columns with incorporated MWCNT [100], which were prepared *via* two different approaches pertaining to the investigation of carbon nanotubes either (1) as entities to modulate solute retention on monolithic columns bearing well-defined retentive ligands or (2) as entities that constitute the stationary phase responsible for solute retention and separation. In approach (1), the carbon nanotubes were incorporated into ODM columns while in approach (2), an ideal monolithic support was coated with carbon nanotubes to yield a real *carbon nanotube stationary phase* for the HPLC separation of a wide range of solutes. First, an ODM column based on the *in situ* polymerization of ODA and TRIM was optimized for use in HPLC separations of small and large solutes (e.g., proteins). To further modulate the retention and separation of proteins, small amounts of carbon nanotubes were incorporated into the octadecyl monolith column. In approach (2), an inert, relatively polar monolith based on the *in situ* polymerization of GMM and EDMA proved to be the most suitable support for the preparation of "carbon nanotube stationary phase." This carbon nanotube *coated* monolith proved useful in the HPLC separation of a wide range of small solutes, including enantiomers. In approach (2), a more homogeneous incorporation of carbon nanotubes into the diol monolithic columns (i.e., GMM/EDMA) was achieved when hydroxyl functionalized carbon nanotubes were incorporated into the GMM/EDMA monolithic support. In addition, high-power sonication for a short time enhanced further the homogeneity of the monolith incorporated with nanotubes. In all cases, nonpolar and π interactions were responsible for solute retention on the monolith incorporated carbon nanotubes.

Graphene oxide (GO) nanoparticles, which are characterized by their high surface area and π–π stacking ability were incorporated into a monolithic column for CEC separations [101]. The incorporation of GO into the poly(MAA-*co*-EDMA) monolith introduced oxide functional groups for the efficient EOF generation for CEC separations. This monolith yielded enhanced retention and resolution of some neutral and polar aromatic compounds than its parent monolith (Figure 6.32). Furthermore, the separation of some aniline compounds was also demonstrated on the GO incorporated monolithic column. Along the same lines, another column with GO incorporated into the poly(GMA-*co*-EDMA) monolith was made, which facilitated the separation of small molecules *via* RPC than the parent monolith [102]. The GO was modified with 3-(trimethoxysilyl)propyl methacrylate and polymerized along with GMA and EDMA in a 50.0 mm × 4.6 mm i.d. stainless steel column. Separations of some steroids and aniline compounds were demonstrated on this column.

Recently, Aydoğan and El Rassi have incorporated surface modified fumed silica nanoparticles (FSNPs) into polymethacrylate-based monolithic columns for use in RPC separation of small molecules and proteins [103]. First, FSNPs were modified with 3-(trimethoxysilyl)propylmethacrylate to yield the *hybrid* methacryloyl fumed

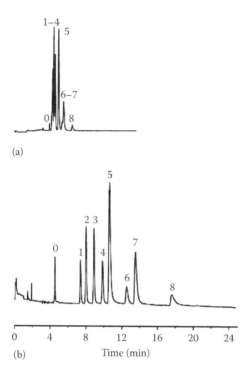

FIGURE 6.32 Electrochromatograms for the separation of test neutral compounds on (a) poly MAA–EDMA monolith; (b) poly GO–MAA–EDMA monolith. Separation conditions: mobile phase, 70% ACN/30% acetate buffer (12.5 mM, pH 5.6); applied voltage, 15 kV; DAD detection at 214 nm; temperature, 25°C. Peak identity: 0, thiourea; 1, benzene; 2, toluene; 3, ethylbenzene; 4, isopropylbenzene; 5, naphthalene; 6, acenaphthalene; 7, fluorine; 8, anthracene. (Reproduced from Wang, M.-M. and Yan, X.-P., *Anal. Chem.*, 84, 39, 2011. With permission.)

silica nanoparticles (MFSNP) monomer. The resulting MFSNP was then mixed with GMM and EDMA in a binary porogenic solvent composed of cyclohexanol and dodecanol. Thereafter, the *in situ* copolymerization of MFSNP, GMM, and EDMA was performed in a stainless steel column of 4.6 mm i.d. The silanol groups of the hybrid monolith thus obtained were grafted with octadecyl ligands by perfusing the hybrid monolithic column with a solution of 4% (w/v) of dimethyloctadecylchlorosilane in toluene; whereas the column was maintained at 110°C for 6 h (Figure. 6.33a). This hybrid poly(GMA-EDMA-MFSNP) having surface-bound octadecyl ligands exhibited hydrophobic interactions under RPC elution conditions and various probe solutes namely, alkylbenzenes, phenol, and aniline derivatives were well separated. Furthermore, this C18–FNSP monolithic column demonstrated high selectivity in the separation of six standard proteins using linear gradient elution at increasing ACN concentration in the mobile phase (Figure 6.33b).

Although the stability of the chemical bond between organosilanes and iron oxide magnetic nanoparticles (MNPs) is not known and assessed yet, vinylized MNPs

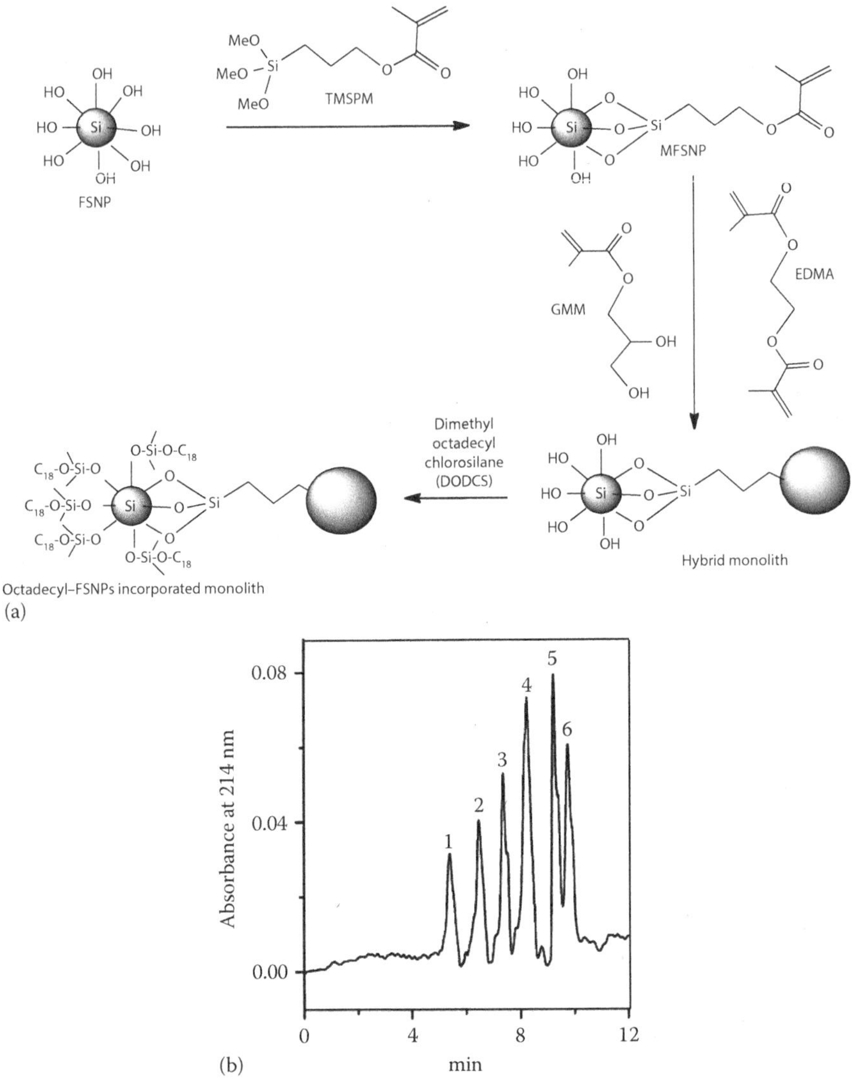

FIGURE 6.33 (a) Schematic diagram of the synthesis of the GMM/EDMA/MFSNP monolith followed by grafting the surface of the hybrid monolith with octadecyl ligands. (b) Chromatogram of six standard proteins obtained on the monolithic column (10 cm × 4.6 mm i.d.) with incorporated FSNP; linear gradient elution at increasing ACN content in the mobile phase from 100% mobile phase A to 100% mobile phase B in 10 min; mobile phase A, 90% water: 10% ACN at 0.1% (v/v) TFA; mobile phase B, 10% water: 90% ACN at 0.1 % (v/v) TFA; flow rate, 1 mL/min; detection wavelength, 214 nm. Peaks order, (1) ribonuclease A, (2) cytochrome C, (3) carbonic anhydrase, (4) lysozyme, (5) myoglobin, and (6) α-chymotrypsinogen A. (Reproduced from Aydoğan, C. and El Rassi, Z., *J. Chromatogr. A*, 1445, 62, 2016. With permission.)

(VMNPs) were prepared *via* reaction of MNPs with 3-(trimethoxysilyl) propylmethacrylate. The VMNPs were then incorporated into polymethacrylate monolithic columns in order to develop novel stationary phases with enhanced performances [104]. In this work, hybrid polymeric monoliths containing different amounts of VMNPs were prepared by copolymerizing GMA, EDMA, META, and VMNPs together in the presence of cyclohexanol and dodecanol as porogens. The synthesized monolith was applied to the CEC separations of alkylbenzenes and organophosphorous pesticides. The results demonstrated increased retention and high column efficiency in comparison to the control monolith (i.e., 0 wt% VMNPs). The higher solute retention seems to be due to the increase of the surface area of the monolith arising from the incorporated VMNPs.

Metal-organic frameworks (MOFs) are viewed as a new class of highly porous separation media made of inorganic ions or clusters coordinated to organic ligands. In fact, the MOF (HKUST-1) nanoparticles were incorporated into poly(GMA-*co*-EDMA) monoliths (Figure 6.34) to afford stationary phases with enhanced performance in the separation of small molecules by RPC using cLC [105]. The MOF (i.e., HKUST-1) is made from dimmer Cu paddle wheels linked by 1,3,5-benzenetricarboxylates with a three-dimensional square-shaped channel system. It was reported that increasing HKUST-1 nanoparticle amount in the polymerization mixture improved the separation and column efficiency. Using a hydrophobic mixture of five PAHs to investigate the retention mechanism of the HKUST-1-poly(GMA-*co*-EDMA) monolith yielded retention and elution order that increased with the hydrophobicity and aromaticities of the analytes, which

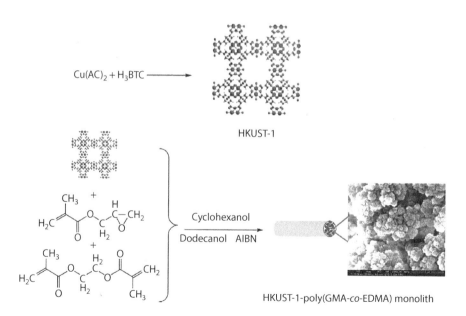

FIGURE 6.34 Synthesis processes of HKUST-1-poly(GMA-*co*-EDMA) monolith. (Reproduced from Yang, S. et al., *J. Chromatogr. A*, 1360, 143, 2014. With permission.)

indicated that the retention mechanism was based on hydrophobic as well as π–π interactions. It was also observed that solute retention decreased with increasing ACN content in the mobile phase, a trend typical to RPC behavior. The HKUST-1-poly(GMA-*co*-EDMA) monolith yielded baseline resolution in the separation of ethylbenzene, styrene, some aromatic acids, and some phenols.

A monolithic column containing a composite of MOFs has been reported [106]. It involved the incorporation of MIL-53 (Al) into poly(HMA-*co*-EDMA), and was demonstrated by the separation of small aromatic compounds using cLC. The presence of 1,4-benzenedicarboxylate moieties within the structure of MIL-53 (Al) as an organic linker was shown to facilitate the separation of aromatic mixtures through π–π interactions. The stability studies that were undertaken by the authors showed excellent mechanical stability of the MIL-53(Al)-(HMA-*co*-EDMA) composite column in cLC separations as well as reproducible results. Furthermore, baseline resolution was obtained for a series of alkylbenzenes in less than 8 min, and a RPC separation mechanism was assessed as the solute retention factor decreased with increasing the ACN content of the mobile phase.

Another MOF has been incorporated into poly(BMA-*co*-EDMA) monolith and evaluated by CEC [107]. It was a lanthanide-based MOF referred to as [Eu$_2$(ABTC)$_{1.5}$ (H$_2$O)$_3$(DMA)] (NKU-1). A binary porogenic system consisting of DMF and 1-butyl-3-methylimidazolium tetrafluoroborate [BMIM]BF$_4$, which is a room temperature IL, was used to provide homogeneous distribution of NKU-1 in the polymerization mixture. The resulting monolithic column exhibited a relatively high separation efficiency (up to 210,000 plates/m) in the analysis of four groups of small molecules, including alkylbenzenes, PAHs, aniline series, and naphthyl substitutes under RPC conditions (Figure 6.35). The results indicated the involvement of π–π interactions in addition to hydrophobic interactions during the separation of these aromatic compounds.

6.3.6 Ionic Liquid Immobilized Monoliths

A few articles have been published regarding the incorporation of IL into porous polymer monoliths in the aim of improving the separation of small molecules with monolithic columns [108]. ILs are nonmolecular ionic solvents known for possessing excellent solvation qualities, a wide viscosity range, excellent chemical and thermal stability, and a negligible vapor pressure. The incorporation of the specific features of IL into the porous structure of polymer monoliths should in principle provide monolithic columns with improved performance. For the structures of the ILs incorporated into monolithic columns, see Table 6.3.

An IL incorporated polymer monolith has been developed by copolymerizing 1-allyl-3-methylimidazolium chloride (AMIM$^+$Cl$^-$) and TMPTA functional monomers with the cross-linker EDMA in the presence of a binary porogenic mixture consisting of 1-propanol and 1,4-butanediol [108]. It was reported that the monoliths without IL exhibited poor resolution and low separation efficiency, whereas the monolith with added IL provided enhanced resolution and separation efficiency for small molecules. Owing to its good mechanical stability and permeability, the IL containing monolith proved useful for the HPLC separation of aromatic hydrocarbons,

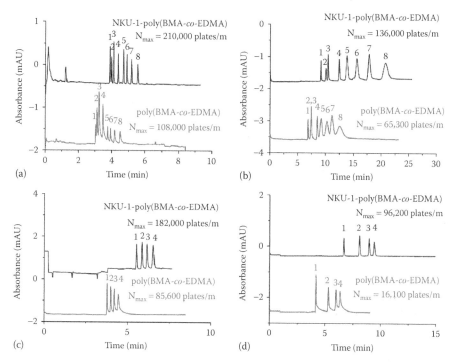

FIGURE 6.35 Electrochromatograms on NKU-1-poly (BMA-*co*-EDMA) monolith 5 and poly (BMA-*co*-EDMA) monolith 5. Conditions: capillary, 100 μm inner diameter, 41.5 cm total length and 32.5 cm effective length; separation voltage, 25 kV; temperature, 25°C; UV-vis detector, 254 nm. Mobile phase: acetonitrile/0.01 mol/L acetate (pH 6.0)(75:25, v/v) (a, c, and d); acetonitrile/0.01 mol/L acetate (pH 6.0) (70:30, v/v) (b). (a) Analytes–alkylbenzenes: (1) acetone, (2) 2,5-dihydroxyacetophenone, (3) acetophenone, (4) butyrophenone, (5) toluene, (6) ethylbenzene, (7) propylbenzene, (8) butylbenzene; (b) Analytes–PAHs: (1) naphthalene, (2) acenaphthylene, (3) fluorene, (4) anthracene, (5) fluoranthene, (6) pyrene, (7) benzo (a) anthracene, (8) benzo (b) fluoranthene; (c) Analytes–aniline series: (1) acetanilide, (2) 4-fluoroaniline, (3) 2-nitroanline, (4) 1-naphthylamine; (d) Analytes–naphthyl substitutes: (1) 1-naphthol, (2) 1-methylnaphthalene, (3) 1-chloronaphthalene, (4) 1-bromonaphthalene. (Reproduced from Zhang, L.S. et al., *J. Chromatogr. A*, 1461, 171, 2016. With permission.)

phenylenediamine isomers, and benzene and its homologues under RPC mobilephase conditions.

The aforementioned research group prepared a similar poly(TMPTA-*co*-EDMA) monolith by including 1-butyl-3-methylimidazolium chloride (BMIM$^+$Cl$^-$) IL in the polymerization solution [109]. The authors claimed in the current work that IL acted as the porogen in the polymerization process. In addition, and as reported earlier the addition of IL to the monolith enhanced the resolution and column efficiency compared to the monolith prepared without IL. The authors reported that the good solvent properties of ILs in the prepolymerization solution are the reason for the high porosity and surface area of the monolith, leading to better separation efficiencies for small molecules. The back pressure of the column was linearly increased with increasing the mobile-phase flow rate, indicating a good mechanical stability of the monolith. The poly(TMPTA-*co*-EDMA) with incorporated BMIM⁺Cl⁻ was demonstrated in the separation of some acidic and basic compounds, some homologues of benzene and toluene derivatives under RPC separation conditions.

A monolithic column has been prepared in 4.6 mm i.d. stainless steel column for HPLC separation using the IL 1-vinyl-3-butylimidazolium chloride (VBC-ILs), 1-dodecene (C12), and TMPTA as the comonomer with EDMA as the cross-linker [110]. Aniline, p-xylene, naphthalene, diphenylamine, and triphenylamine were eluted in the order of decreasing polarity and the retention of these compounds increased with decreasing methanol concentration in the mobile phase showing a typical RPC mechanism. Similarly, benzene, naphthalene, biphenyl, and anthracene eluted in accordance to their decreasing polarities.

6.3.7 NONPOLAR HYBRID ORGANIC-SILICA MONOLITHS

As mentioned above in the section on polar monoliths, organic-silica hybrid monolithic columns offer some advantages over silica-based monoliths and polymer-based monoliths because they combine the best of both types of monoliths, including mechanical stability, simple preparation procedures, and better pH stability. Considering these advantages, a hybrid phenyl-modified silica monolith was prepared *via* the polycondensation of phenyltrimethoxysilane (PTMS) and TMOS precursors in the presence of PEG as the porogen [111]. By treatment with an alkaline solution, mesopores were then introduced into the silica skeleton to yield the PTMS–TMOS-based macroporous–mesoporous hybrid silica monolith. The phenyl groups on the monolithic surface provided π–π interactions in addition to hydrophobic interactions for the separation of a wide range of analytes under RPC conditions. A set of model analytes, including propylbenzene, butylbenzene, and anthracene were baseline separated in a shorter time on the PTMS–TMOS hybrid monolith with higher separation efficiencies when compared to phenyl-modified silica particles packed in a column, under otherwise identical conditions.

Due to the lack of commercially available trialkoxysilanes with the desired functional groups, the PPF of organic-silica precursor monoliths containing vinyl, amino, chloropropyl, and thiol groups has been practiced very often to introduce various functional groups into the monolithic surface. The *one-pot* approach is one of the most convenient approaches for this purpose, which involves two thermal treatments, including the sol-gel process that is carried out at lower temperature to achieve uniform porous monolithic matrix and the subsequent copolymerization at a relatively high temperature to ensure the incorporation of a given monomer into the monolithic matrix. Along this strategy, a novel (3-sulfopropyl methacrylate potassium) silica hybrid monolithic column has been prepared by a simple one-pot approach [112] based on the thiol-ene chemistry. In this process, a reactive precursor monolith was synthesized by polycondensation of TMOS and 3-mercaptopropyltrimethoxysilane (MPTMS) and at the same time *in situ* thiol-ene click reaction between the precondensed siloxane and the organic monomer 3-sulfopropyl methacrylate potassium (SPMA) occurred in a one-pot manner. The SPMA-silica hybrid monolithic capillary column with its negatively charged sulfonate groups on its surface generated a cathodal EOF in CEC runs. This column was demonstrated for the CEC separation of different kinds of compounds, including anilines, alkylbenzenes, and phenols under RPC conditions. Using thiourea (EOF tracer) as the test solute the SPMA–silica hybrid monolith exhibited a plate height of about 3.9 μm corresponding to 250 000 plates/m.

Another hybrid organic-silica monolith based on a one-pot approach has been a C18 reversed-phase organic-silica cationic monolithic column with an IL as the organic monomer [113]. In the synthetic process of the precursor monolithic matrix *via* the polycondensation of tetraethyl orthosilicate and triethoxyvinylsilane, the IL 1-vinyl-3-octadecylimidazolium bromide ($VC_{18}HIm^+Br^-$) was simultaneously anchored into the monolithic matrix. This hybrid monolith showed strong anodal EOF in a wide range of pH (3.0–11.0), which is due to the existence of a strong cationic imidazole group. It also showed good separation repeatability and satisfactory reproducibility. A mixture of alkylbenzenes was well separated on the $VC18HIm^+Br^-$ hybrid monolithic capillary column and the analytes were eluted according to their decreased polarity indicating a typical RPC retention mechanism. A mixture of standard amino acids was also baseline separated and the retention pattern of these charged compounds seemed to be the resultant of the combination of electrophoretic mobility, ionic exchange, and hydrophobic interactions.

Organic-silica hybrid monoliths containing polyhedral oligomeric silsesquioxanes with specific cage structures referred to as POSS have received some interest. As documented in an early review article on this topic [114], the POSS oligomeric monomers, which have been known for many years, are inorganic–organic hybrid architecture made up of an inner inorganic framework of silicone and oxygen $(SiO_{1.5})_x$ that is externally covered by organic substituents. These external substituents can be entirely nonpolar hydrocarbonaceous in nature or they can incorporate a range of polar structures and various functional groups making POSS ideal alternatives to alkoxysilanes for synthesizing a broad range of organic-silica hybrid copolymers [114]. By considering the attractive properties of POSS, various hybrid monolithic columns have been prepared in UV-transparent fused silica capillaries *via* photoinitiated thiol–acrylate polymerization of an acrylopropyl polyhedral oligomeric silsesquioxane (acryl-POSS) and monothiol monomer (1-octadecanethiol [ODT] or sodium 3-mercapto-1-propanesulfonate) within 5 min [115]. In this work, a *one-pot* or *two-step* approaches were used for the preparation of C18-functionalized hybrid monolithic columns *via* photoinitiated thiol–acrylate polymerization as shown in Figure 6.36. The acryl–POSS containing several acrylate groups was either (1) homopolymerized and simultaneously functionalized with ODT or sodium 3-mercapto-1-propanesulfonate in a *one-pot* manner *via* thiol-ene addition reaction, or (2) the resulting monolithic column from the homopolymerization acryl–POSS was thereafter grafted with ODT or sodium 3-mercapto-1-propanesulfonate *via* also thiol-ene addition reaction in a *two-step* approach. Five alkylbenzenes were baseline separated on the C18-functionalized hybrid monolithic column with a plate count of 60,000–73,000 plates/m. Due to the better permeability and stronger hydrophobicity of the C18-functionalized hybrid monolith prepared, mixtures of basic and phenolic compounds were readily separated on the acryl–POSS column. In addition, mixtures of four protein digests were separated in cLC using gradient elution.

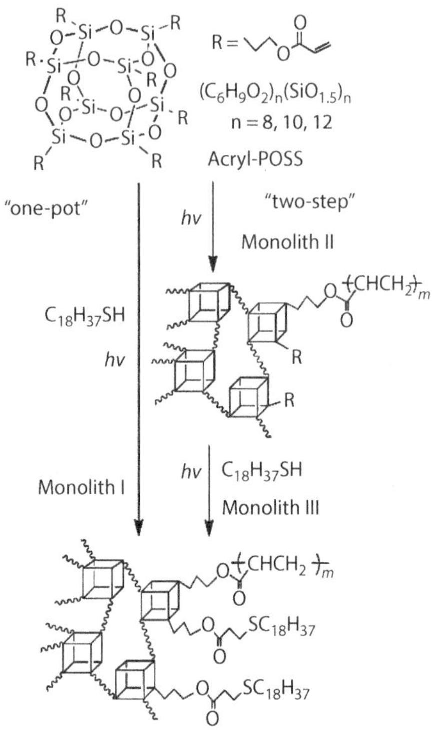

FIGURE 6.36 *One-pot* and *two-step* approaches for the preparation of C18-functionalized hybrid monolithic columns via photoinitiated thiol acrylate polymerization. (Reproduced from Zhang, H. et al., *Anal. Chem.*, 87, 8789, 2015. With permission.)

The sulfonate containing hybrid monolith, which was prepared using sodium 3-mercapto-1-propanesulfonate instead of ODT, exhibited reversed-phase and ion-exchange mixed-mode retention mechanism.

Two additional contributions from the same earlier research group have appeared. One report described the preparation of three hybrid monoliths *via* thiol–methacrylate click polymerization using methacrylate–polyhedral oligomeric silsesquioxane and three multithiol monomers, including 1,6-hexanedithiol (HDT), trimethylolpropane tris(3-mercaptopropionate) (TPTM), and pentaerythritol tetrakis(3-mercaptopropionate) (PTM) in the presence of n-propanol and PEG 200 as the porogenic solvents and dimethylphenylphosphine as the catalyst [116]. These copolymerization yielded hybrid monoliths with hydrophobic surfaces that exhibited RPC behaviors in the separations by cLC of mixtures of small and large molecules such as alkylbenzenes, PAHs, phenols, anilines, and BSA tryptic digest (Figure 6.37).

In a second report, the same research group described the preparation of two kinds of hybrid monoliths *via* thiol–epoxy click polymerization using a multiepoxy monomer, octaglycidyldimethylsilyl POSS, and two multithiols, including TPTM and PTM [117]. Ethanol and PEG 10,000 were chosen as the porogens, whereas KOH

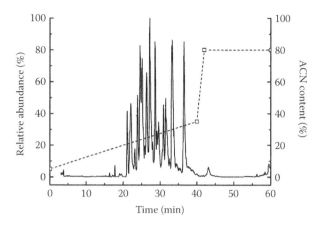

FIGURE 6.37 Separations of BSA digest on the hybrid monolithic column POSS–PTM in cLC–MS/MS. Experimental conditions: column dimensions 30.0 cm × 75 μm i.d.; gradient, 95% water (containing 0.1% formic acid) to 35% ACN (containing 0.1% formic acid) in 40 min, then to 80% ACN (containing 0.1% formic acid) in 2 min; flow rate, 200 nL/min (after split); injection volumes, 0.5 μg. (Reproduced from Lin, H. et al., *J. Chromatogr. A*, 1379, 34, 2015. With permission.)

served as the catalyst in the preparation of these hybrid monoliths, which exhibited RPC behaviors as manifested by the decrease in the retention of alkylbenzenes on increasing the ACN content of the mobile phase. These monoliths performed well in the analysis of alkylbenzenes, PAHs, phenols, dipeptides, intact proteins, and BSA tryptic digest. The mechanical stabilities of the two monoliths were confirmed by the pressure resistance test, as the measured back pressure linearly increased (R > 0.998) to even more than 41.4 MPa on increasing the mobile-phase flow rate.

6.4 CONCLUSION

This chapter provided a review of the progress made in the past 5–6 years in the area of organic polymer-based monolithic columns for their use in HPLC and CEC. It summarized the evolvement of both polar and nonpolar organic monolithic columns and their use in HILIC/HICEC and RPC/RP–CEC, respectively. By evaluating the results reported in a total of 117 references, one can say that some major efforts have been invested in optimizing monolithic columns for the separation of small molecules by augmenting the plate number *via* increasing the specific surface areas of monoliths by novel approaches, including hypercrosslinking, low-conversion polymerization (i.e., short term polymerization), and relatively speaking by single cross-linkers–based monoliths. Despite these attempts, further progress is still badly needed to produce monolithic columns that are more suitable for the separation of small molecules. On another front, monoliths with incorporated nanoparticles and nanostructured materials are believed to enhance the separation of small molecules by providing different and unique selectivity. Finally, despite the suitability of monolithic columns for the separation of large macromolecules, the monolithic media are far from being fully exploited in this important area of the life sciences.

REFERENCES

1. F. Svec, *J. Sep. Sci.* 32 (2009) 3.
2. F. Svec, Y. Lv, *Anal. Chem.* 87 (2014) 250.
3. J. Urban, *J. Sep. Sci.* 39 (2016) 51.
4. M. Jonnada, R. Rathnasekara, Z. El Rassi, *Electrophoresis* 36 (2015) 76.
5. S. Karenga, Z. El Rassi, *Electrophoresis* 32 (2011) 90.
6. A.J. Alpert, *J. Chromatogr. A* 499 (1990) 177.
7. X. Chen, H.D. Tolley, M.L. Lee, *J. Sep. Sci.* 34 (2011) 2088.
8. M.-L. Chen, L.-M. Li, B.-F. Yuan, Q. Ma, Y.-Q. Feng, *J. Chromatogr. A* 1230 (2012) 54.
9. J. Lin, S. Liu, J. Lin, X. Lin, Z. Xie, *J. Chromatogr. A* 1218 (2011) 4671.
10. Z. Lin, H. Huang, X. Sun, Y. Lin, L. Zhang, G. Chen, *J. Chromatogr. A* 1246 (2012) 90.
11. J. Cheng, X. Chen, Y. Cai, Y. He, Z. Chen, Z. Lin, L. Zhang, *Electrophoresis* 34 (2013) 1189.
12. R. Yu, W. Hu, G. Lin, Q. Xiao, J. Zheng, Z. Lin, *RSC Advances* 5 (2015) 9828.
13. J. Guo, Q. Zhang, Z. Yao, X. Zhao, D. Ran, J. Crommen, Z. Jiang, *J. Sep. Sci.* 37 (2014) 1720.
14. D.N. Gunasena, Z. El Rassi, *J. Chromatogr. A* 1317 (2013) 77.
15. X. Lin, S. Feng, W. Jia, K. Ding, Z. Xie, *J. Chromatogr. A* 1316 (2013) 104.
16. X. Lin, Y. Li, D. Xu, C. Yang, Z. Xie, *Analyst* 138 (2013) 635.
17. N. Li, Y. Shen, L. Qi, Z. Li, J. Qiao, Y. Chen, *RSC Advances* 5 (2015) 61436.
18. Z. Lin, R. Yu, W. Hu, J. Zheng, P. Tong, H. Zhao, Z. Cai, *Analyst* 140 (2015) 4626.
19. M.-L. Chen, S.-S. Wei, B.-F. Yuan, Y.-Q. Feng, *J. Chromatogr. A* 1228 (2012) 183.
20. H. Jiang, H. Yuan, Y. Qu, Y. Liang, B. Jiang, Q. Wu, N. Deng, Z. Liang, L. Zhang, Y. Zhang, *Talanta* 146 (2016) 225.
21. Z. Liu, Y. Peng, T. Wang, G. Yuan, Q. Zhang, J. Guo, Z. Jiang, *J. Sep. Sci.* 36 (2013) 262.
22. H.C. Foo, J. Heaton, N.W. Smith, S. Stanley, *Talanta* 100 (2012) 344.
23. G. Yuan, Y. Peng, Z. Liu, J. Hong, Y. Xiao, J. Guo, N.W. Smith, J. Crommen, Z. Jiang, *J. Chromatogr. A* 1301 (2013) 88.
24. M. Staňková, P. Jandera, V. Škeříková, J. Urban, *J. Chromatogr. A* 1289 (2013) 47.
25. P. Jandera, M. Staňková, T. Hájek, *J. Sep. Sci.* 36 (2013) 2430.
26. D. Qiu, F. Li, M. Zhang, J. Kang, *Electrophoresis* 37 (2016) 1725.
27. C. Liu, W. Chen, G. Yuan, Y. Xiao, J. Crommen, S. Xu, Z. Jiang, *J. Chromatogr. A* 1373 (2014) 73.
28. V. Škeříková, J. Urban, *J. Sep. Sci.* 36 (2013) 2806.
29. Y. Lv, Z. Lin, F. Svec, *Anal. Chem.* 84 (2012) 8457.
30. I. Tijunelyte, J. Babinot, M. Guerrouache, G. Valincius, B. Carbonnier, *Polymer* 53 (2012) 29.
31. M. Guerrouache, S. Mahouche-Chergui, M.M. Chehimi, B. Carbonnier, *Chem. Comm.* 48 (2012) 7486.
32. Y. Lv, Z. Lin, F. Svec, *Analyst* 137 (2012) 4114.
33. H.C. Kolb, M.G. Finn, K.B. Sharpless, *Angew. Chem. Int. Ed.* 40 (2001) 2004.
34. C.E. Hoyle, C.N. Bowman, *Angew. Chem. Int. Ed.* 49 (2010) 1540.
35. P.N. Nair, M. Podgórski, S. Chatani, T. Gong, W. Xi, C.R. Fenoli, C.N. Bowman, *Chem. Mater.* 26 (2014) 724.
36. S. Belbekhouche, M. Guerrouache, B. Carbonnier, *Macromol. Chem. Phys.* 217 (2016) 997.
37. Ç. Kip, D. Erkakan, A. Gökaltun, B. Çelebi, A. Tuncel, *J. Chromatogr. A* 1396 (2015) 86.

38. S. Khadka, Z. El Rassi, *Electrophoresis* 37 (2016) 3178.
39. Z. Chen, Q. Ye, L. Liu, H. Dong, *J. Chromatogr. Sci.* (2015) 1.
40. C. Aydoğan, B. Çelebi, A. Bayraktar, *Polym. Adv. Technol.* 25 (2014) 777.
41. S. Janků, V. Škeříková, J. Urban, *J. Chromatogr. A* 1388 (2015) 151.
42. S. Currivan, J.M. Macak, P. Jandera, *J. Chromatogr. A* 1402 (2015) 82.
43. C. Aydoğan, Z. El Rassi, *J. Chromatogr. A* 1445 (2016) 55.
44. Y. Liang, C. Wu, Q. Zhao, Q. Wu, B. Jiang, Y. Weng, Z. Liang, L. Zhang, Y. Zhang, *Anal. Chim. Acta* 900 (2015) 83.
45. S. Tang, S. Liu, Y. Guo, X. Liu, S. Jiang, *J. Chromatogr. A* 1357 (2014) 147.
46. X. Shi, L. Qiao, G. Xu, *J. Chromatogr. A* 1420 (2015) 1.
47. T. Wang, Y. Chen, J. Ma, X. Zhang, L. Zhang, Y. Zhang, *Analyst* 140 (2015) 5585.
48. J. Wang, F. Wu, R. Xia, Q. Zhao, X. Lin, Z. Xie, *J. Chromatogr. A* 1449 (2016) 100.
49. N. Tanaka, H. Kobayashi, N. Ishizuka, H. Minakuchi, K. Hosoya, T. Ikegami, *J. Chromatogr. A* 965 (2002) 35.
50. Z. Lin, X. Tan, R. Yu, J. Lin, X. Yin, L. Zhang, H. Yang, *J. Chromatogr. A* 1355 (2014) 228.
51. M.-L. Chen, J. Zhang, Z. Zhang, B.-F. Yuan, Q.-W. Yu, Y.-Q. Feng, *J. Chromatogr. A* 1284 (2013) 118.
52. L. Zhao, L. Yang, Q. Wang, *J. Chromatogr. A* 1446 (2016) 125.
53. J. Urban, P. Jandera, P. Langmaier, *J. Sep. Sci.* 34 (2011) 2054.
54. C. Puangpila, T. Nhujak, Z.E. Rassi, *Electrophoresis* 33 (2012) 1431.
55. F.M. Okanda, Z. El Rassi, *Electrophoresis* 26 (2005) 1988.
56. S. Karenga, Z. El Rassi, *J. Sep. Sci.* 31 (2008) 2677.
57. S.-L. Lin, Y.-R. Wu, T.-Y. Lin, M.-R. Fuh, *J. Chromatogr. A* 1298 (2013) 35.
58. Ç. Kip, A. Tuncel, *Electrophoresis* 36 (2015) 945.
59. P. Jandera, M. Staňková, V. Škeříková, J. Urban, *J. Chromatogr. A* 1274 (2013) 97.
60. W. Alshitari, C.L. Quigley, N. Smith, *Talanta* 141 (2015) 103.
61. H. Zhang, J. Ou, Y. Wei, H. Wang, Z. Liu, L. Chen, H. Zou, *Anal. Chim. Acta* 883 (2015) 90.
62. S.D. Chambers, T.W. Holcombe, F. Svec, J.M.J. Frechet, *Anal. Chem.* 83 (2011) 9478.
63. M. Jonnada, Z. El Rassi, *J. Chromatogr. A* 1409 (2015) 166.
64. Q. Duan, C. Liu, Z. Liu, Z. Zhou, W. Chen, Q. Wang, J. Crommen, Z. Jiang, *J. Chromatogr. A* 1345 (2014) 174.
65. Z. Jiang, N.W. Smith, P. David Ferguson, M. Robert Taylor, *J. Biochem. Biophys. Meth.* 70 (2007) 39.
66. S. Karenga, Z. El Rassi, *Electrophoresis* 32 (2011) 1044.
67. S. Karenga, Z. El Rassi, *Electrophoresis* 32 (2011) 1033.
68. H. Liu, X. Bai, D. Wei, G. Yang, *J. Chromatogr. A* 1324 (2014) 128.
69. X. Bai, H. Liu, D. Wei, G. Yang, *Talanta* 119 (2014) 479.
70. W. Niu, L. Wang, L. Bai, G. Yang, *J. Chromatogr. A* 1297 (2013) 131.
71. S. Kositarat, N.W. Smith, D. Nacapricha, P. Wilairat, P. Chaisuwan, *Talanta* 84 (2011) 1374.
72. E.J. Carrasco-Correa, F. Vela-Soria, O. Ballesteros, G. Ramis-Ramos, J.M. Herrero-Martínez, *J. Chromatogr. A* 1379 (2015) 65.
73. A. Svobodova, T. Krizek, J. Sirc, P. Salek, E. Tesarova, P. Coufal, K. Stulik, *J. Chromatogr. A* 1218 (2011) 1544.
74. G. Hasegawa, K. Kanamori, N. Ishizuka, K. Nakanishi, *ACS Appl. Mater. Interfaces* 4 (2012) 2343.
75. R. Koeck, R. Bakry, R. Tessadri, G.K. Bonn, *Analyst* 138 (2013) 5089.
76. M. Bedair, Z. El Rassi, *Electrophopresis* 23 (2002) 2938.
77. M. Bedair, Z. El Rassi, *J. Chromatogr. A* 1013 (2003) 35.
78. Y. Li, H.D. Tolley, M.L. Lee, *J. Chromatogr. A* 1218 (2011) 1399.
79. K. Liu, H.D. Tolley, M.L. Lee, *J. Chromatogr. A* 1227 (2012) 96.

80. K. Liu, H.D. Tolley, J.S. Lawson, M.L. Lee, *J. Chromatogr. A* 1321 (2013) 80.
81. K. Liu, P. Aggarwal, H.D. Tolley, J.S. Lawson, M.L. Lee, *J. Chromatogr. A* 1367 (2014) 90.
82. P. Aggarwal, J.S. Lawson, H.D. Tolley, M.L. Lee, *J. Chromatogr. A* 1364 (2014) 96.
83. J. Zhong, M. Hao, R. Li, L. Bai, G. Yang, *J. Chromatogr. A* 1333 (2014) 79.
84. F. Svec, *J. Chromatogr. A* 1217 (2010) 902.
85. P. Simone, G. Pierri, P. Foglia, F. Gasparrini, G. Mazzoccanti, A.L. Capriotti, O. Ursini, A. Ciogli, A. Laganà, *J. Sep. Sci.* 39 (2016) 264.
86. H. Wang, J. Ou, J. Bai, Z. Liu, Y. Yao, L. Chen, X. Peng, H. Zou, *J. Chromatogr. A* 1436 (2016) 100.
87. Z. Liu, J. Ou, H. Lin, H. Wang, Z. Liu, J. Dong, H. Zou, *Anal. Chem.* 86 (2014) 12334.
88. L. Chen, J. Ou, Z. Liu, H. Lin, H. Wang, J. Dong, H. Zou, *J. Chromatogr. A* 1394 (2015) 103.
89. R. Koeck, M. Fischnaller, R. Bakry, R. Tessadri, G.K. Bonn, *Anal. Bioanal. Chem.* 406 (2014) 5897.
90. J. Urban, F. Svec, J.M. Fréchet, *J. Chromatogr. A* 1217 (2010) 8212.
91. J. Urban, F. Svec, J.M. Fréchet, *Anal. Chem.* 82 (2010) 1621.
92. V. Davankov, S. Rogozhin, M. Tsyurupa, US Patent Appl 3 (1971) 729.
93. X.-J. Chen, N.P. Dinh, J. Zhao, Y.-T. Wang, S.-P. Li, F. Svec, *J. Sep. Sci.* 35 (2012) 1502.
94. F. Maya, F. Svec, Polymer 55 (2014) 340.
95. R. Poupart, D.N. El Houda, D. Chellapermal, M. Guerrouache, B. Carbonnier, B. Le Droumaguet, *RSC Advances* 6 (2016) 13614.
96. C. Aydoğan, *J. Chromatogr. A* 1392 (2015) 63.
97. A. Gökaltun, C. Aydoğan, B. Çelebi, A. Denizli, A. Tuncel, *Chromatographia* 77 (2014) 459.
98. S.D. Chambers, F. Svec, J.M.J. Frechet, *J. Chromatogr. A* 1218 (2011) 2546.
99. A. Aqel, K. Yusuf, Z.A. Al-Othman, A.Y. Badjah-Hadj-Ahmed, A.A. Alwarthan, *Analyst* 137 (2012) 4309.
100. E. Mayadunne, Z. El Rassi, *Talanta* 129 (2014) 565.
101. M.-M. Wang, X.-P. Yan, *Anal. Chem.* 84 (2011) 39.
102. Y. Li, L. Qi, H. Ma, *Analyst* 138 (2013) 5470.
103. C. Aydoğan, Z. El Rassi, *J. Chromatogr. A* 1445 (2016) 62.
104. E.J. Carrasco-Correa, G. Ramis-Ramos, J.M. Herrero-Martínez, *J. Chromatogr. A* 1385 (2015) 77.
105. S. Yang, F. Ye, Q. Lv, C. Zhang, S. Shen, S. Zhao, *J. Chromatogr. A* 1360 (2014) 143.
106. K. Yusuf, A.Y. Badjah-Hadj-Ahmed, A. Aqel, Z.A. ALOthman, *J. Sep. Sci.* 39 (2016) 880.
107. L.S. Zhang, P.-Y. Du, W. Gu, Q.-L. Zhao, Y.-P. Huang, *J. Chromatogr. A* 1461 (2016) 171.
108. J. Wang, L. Bai, Z. Wei, J. Qin, Y. Ma, H. Liu, *J. Sep. Sci.* 38 (2015) 2101.
109. J. Wang, X. Jiang, H. Zhang, S. Liu, L. Bai, H. Liu, *Anal. Meth.* 7 (2015) 7879.
110. J. Qin, L. Bai, J. Wang, Y. Ma, H. Liu, S. He, T. Li, Y. An, *Anal. Meth.* 7 (2015) 218.
111. R. Meinusch, K. Hormann, R. Hakim, U. Tallarek, B.M. Smarsly, *RSC Advances* 5 (2015) 20283.
112. H. Yang, Y. Chen, Y. Liu, L. Nie, S. Yao, *Electrophoresis* 34 (2013) 510.
113. G. Fang, H. Qian, Q. Deng, X. Ran, Y. Yang, C. Liu, S. Wang, *RSC Advances* 4 (2014) 15518.
114. G. Li, L. Wang, H. Ni, C.U. Pittman Jr, *J. Inorg. Organomet. P.* 11 (2001) 123.
115. H. Zhang, J. Ou, Z. Liu, H. Wang, Y. Wei, H. Zou, *Anal. Chem.* 87 (2015) 8789.
116. H. Lin, J. Ou, Z. Liu, H. Wang, J. Dong, H. Zou, *J. Chromatogr. A* 1379 (2015) 34.
117. H. Lin, L. Chen, J. Ou, Z. Liu, H. Wang, J. Dong, H. Zou, *J. Chromatogr. A* 1416 (2015) 74.

7 Solid-Core or Fully Porous Columns in Ultra High-Performance Liquid Chromatography — Which Way to Go for Better Efficiency of the Separation?

Shulamit Levin

CONTENTS

7.1 INTRODUCTION

High-performance liquid chromatography (HPLC) has been evolving and maturing during the past five decades into a major analytical technique for discovery, qualitative, and quantitative analysis in all areas of life. This is in large due to the many features offered by state-of-the-art HPLC instrumentation in terms of operation, understanding of separation principles, high sensitivity, and selectivity. The users of HPLC have been constantly striving for higher efficiency of the separation to get higher sensitivity and resolution. Getting higher efficiency by HPLC can be achieved by decreasing diffusion caused by the broadening effects of the analytes while migrating through the column and the system.

7.2 ULTRA HIGH-PERFORMANCE LIQUID CHROMATOGRAPHY EMERGENCE

On 2004, a groundbreaking instrumentation was introduced to the world of liquid chromatography (LC), Ultra-Performance Liquid Chromatography (UPLC™),[1,2] which became a trademark by Waters Corporation. The term UHPLC is the more generic one, which refers to separations performed at pressures above 400 bar with lower dispersion. These systems were aimed to increase efficiency of the separations, thereby increasing peak capacity,[3–5] reducing analysis time,[6] improving sensitivity,[7] resolution,[8] and throughput.[9] The driving force for this evolution were the columns that were packed with sub-2 μm particles.[10,11]

While maintaining retention and capacity similar to HPLC, the UPLC™ system was created to accommodate newly designed sub-2 μm fully porous particles that can withstand high pressures, around 15,000 psi, which traditional silica-based particles could not endure. The groundbreaking performance of UPLC™ was reached thanks to these columns which have enhanced mechanical stability, and being the second generation of hybrid silica particles, based on bridged ethylene hybrid (BEH) technology.[2] The development of such sub-2 μm particles was a significant challenge.[2] Moreover, packing these 1.7 μm particles into reproducible and rugged columns were also a challenge that needed to be overcome. Requirements included a smoother interior surface of the column hardware, and redesigning the end frits to retain the small particles and resist clogging. Packed bed uniformity was also critical, especially if shorter columns were to maintain the same resolution as HPLC, while accomplishing the goal of faster separations. Moreover, at high pressures, frictional heating of the mobile phase could be quite significant and had to be considered.[2] With column diameters typically used in HPLC (3.0 to 4.6 mm), the consequence of frictional heating would be a loss of performance due to heat-induced nonuniform flow. To minimize the effects of frictional heating, smaller diameter columns (1–2.1 mm) were typically used for UPLC.[2]

Figure 7.1 illustrates schematically a typical chromatogram in traditional HPLC (e.g., 4.6 × 150 mm 5 μm particles column) in comparison to a typical UPLC

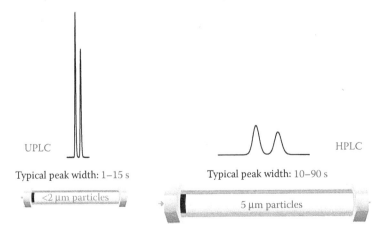

FIGURE 7.1 A scheme showing typical values obtained in HPLC and in UPLC.

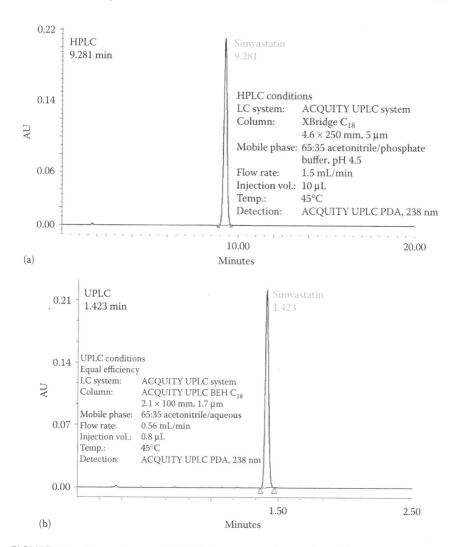

FIGURE 7.2 Chromatograms obtained with a conventional column (a) a sub-2 μm column (b). (Reproduced with the permission of Waters Corporation.)

(e.g., 2.1 × 75 mm 1.7 μm particles column), whereas Figure 7.2 brings an actual assay of Simvastatin, in which a one peak chromatogram obtained in a run by HPLC (A) is compared to the same sample of one peak, running in UPLC™ system. The ability to obtain higher efficiency is taken advantage of here to shorten run time for better productivity.

The estimated pressure drop across a 15 cm long column packed with 1.7 μm particles is about 15,000 psi when operated at a typical flow rate. Therefore, a pump capable of delivering the solvent smoothly and reproducibly at this pressure range, that can compensate for the solvent's rapidly varying compressibility, and can operate precisely in both the gradient and isocratic separation modes, was needed.[1]

Sample introduction was also critical. Conventional HPLC injection valves, either automated or manual, were not designed and hardened to work at extreme pressures. To protect the column from experiencing extreme pressure fluctuations, the injection process had to be relatively pulse-free. Injection volumes were required to be reduced below 5 µL, as the methods were scaled down to the 2 mm ID columns, and not to create volume overload. A fast injection cycle time was needed to fully benefit from the speed attained by UPLC™, which, in turn, allowed a high sample throughput. Better injector washing was needed for minimal carry over due to the newly gained sensitivity.

With 1.7 µm particles, half-height peak widths of less than 1 s could be obtained in LC for the first time, posing significant challenges for the detectors. In order to accurately and reproducibly integrate a peak, the detector sampling rate had to be high to collect enough data points across the peak. Moreover, the detector cell had to have minimal contribution to band broadening, so that the newly gained column's intrinsic separation efficiency will be preserved outside it as much as possible. In general, signal-to-noise values for UPLC detection could be at least 2–3 times higher than with HPLC separations, depending on the detection technique that was used. Conventional absorbance-based optical detectors are concentration-sensitive detectors and, for UPLC use, the flow cell volume would have had to be reduced in standard UV/visible detectors to maintain concentration and signal within Beers' Law limitations.

Figure 7.3 brings experimental comparison between HPLC and UPLC™ runs of a standard peptide mixture, in which the benefits of the enhanced performance of the instrumentation allowed the increase in resolution and sensitivity simultaneously, keeping the run times the same in both LC systems, to gain information that was not

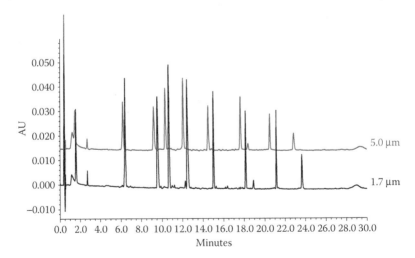

FIGURE 7.3   Column: 2.1 × 50 mm 1.7 µm Acquity C18 and XBridge C18 2.1 × 50 mm 5.0 µm; Flow: 0.3 mL/min; Temperature: 38°C; Gradient (30 min): 5%–50% ACN with 0.018% TFA; Detector: UV at 214 nm, 20 points/sec, $Tc = 0.1$; Injection: 10 µL; Backpressure: 4500 PSI (1.7 µm), 1400 PSI (5.0 µm); Sample: MassPREP™ Peptide Mixture: Allantoin, RASG-1, Angiotensin frag. 1-7, Bradykinin, Angiotensin II, Angiotensin I, Renin Substrate, Enolase T35, Enolase T37, Melittin.

obtained before, thanks to the enhanced performance of the UHPLC system. This is a different approach to scaling down to UHPLC, not looking for saving of time, unlike the work shown in Figure 7.2.

Early adoption of UHPLC technology was demonstrated by the pharmaceutical sciences.[5,6,11] The drive for higher efficiency or smaller height equivalent to theoretical plate (HETP), originated from an ever increasing need for better sensitivity, expressed as signal-to-noise, or as higher peak capacity for complex chromatograms in discovery by metabolomics/metabonomics and other *omics* fields.[5] Figure 7.4 brings an example for such a high peak capacity mass spectrometry total ion chromatogram (TIC) trace, produced from a rat urine, on a UPLC–QTof™ system. It can be understood from the strikingly higher peak capacity obtained here how UHPLC, which has shown a significant improvement in the separation capabilities of these types of samples, became the inlet of choice for LC-MS/MS in such studies. Note the high temperature used here, which will be discussed later on.

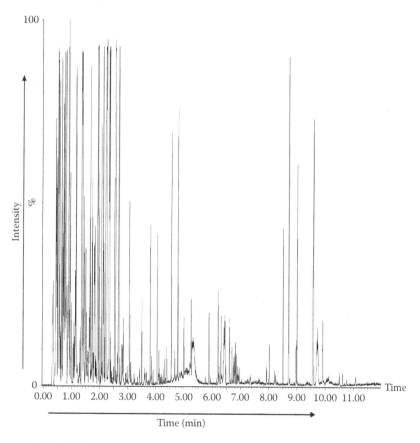

FIGURE 7.4   TIC of a reversed-phase gradient UPLC separation of rat urine on a 2.1 × 150 mm, 1.7 µm *Acquity BEH C18 column over 10 min* at 90°C. (Reprinted with permission from Plumb, R.S. et al., *Anal. Chem.*, 78, 20, 7278–7283, 2006. Copyright 2006 American Chemical Society.)

7.3   EFFICIENCY OF THE SEPARATION

The studies of increased efficiency of the modern columns with smaller particles brought back the extensive use of the van Deemter equation, which describes the relation between efficiency, expressed as the height equivalent to a theoretical plate H, to linear velocity (u).[12] In general, the broadening of a chromatographic peak is expressed as a variance, σ_c^2 and is the sum of the broadening effects that occur within the column during migration of the sample zone through the column.

$$H = \frac{\sigma_c^2}{L} = A + \frac{B}{u} + C_s \cdot u + C_m \cdot u \qquad (7.1)$$

where:

H is the HETP

σ_c^2 is the length variance of the peak in the column in the absence of extra-column broadening

L is the column length

A represents the dispersive contribution from the flow profile or Eddy diffusion, which has been originally considered to be independent of the linear velocity u in the practical range of LC operation

B arose from the axial molecular diffusion and is proportional to the inverse of the linear velocity, and the C terms or mass transfer terms were proportional to velocity

C_m arose from slowness of equilibration or mass transfer in the mobile phase

C_s arose from slowness of equilibration or mass transfer in the stationary phase

Giddings[13] and Knox[14] both have extended the model, and they used a reduced plate height $h = H/d_p$ and reduced linear velocity $v = ud_p/D_m$ (where D_m is the diffusion and d_p is particle diameter):

$$h = \frac{B}{v} + A^{0.33} + C \cdot v \qquad (7.2)$$

Equations 7.1 and 7.2 are empirical formulae that describe the relationship between linear velocity (related to flow rate) and column's efficiency of the separation, as measured by the height equivalent to theoretical plate, HETP or its reduced term h.

The particle size is one of the significantly affecting parameters on HETP, so the van Deemter curve has been used to investigate and compare columns' chromatographic performance.[15,16] Using these equations the study of column's efficiency showed that it depends on particle size and analyte's diffusion coefficient:

$$\frac{\sigma_c^2}{L} = H = A \cdot d_p + \frac{B \cdot D_m}{u_0} + C \cdot \frac{d_p^2}{D_m} \cdot u_0 \qquad (7.3)$$

where:

H is the theoretical plate height

u_0 is the linear velocity

D_m is the diffusion coefficient of the analyte
d_p is the particle size
A, B, and C are the coefficients of the reduced van Deemter equation

It should be noted here that the linear velocity evaluation is based on the migration velocity of the unretained peak.

Figure 7.5 shows the effect of particle diameter on the van Deemter equation,[2] where the optimal HETP (minimal) point in each curve is lower, and shifts to higher velocities as particles get smaller. Moreover, the curve levels off at relatively high linear velocities for the sub-2 μm particles. This leveling off indicates that the new technology of sub-2 μm particles results in a lower contribution of the term C in the van Deemter equation. This is the basis for the newly gained chromatographic performance: The capability of reducing analysis times by increasing flow rates with minimal sacrifice of the efficiency of the separation at higher velocities.

The increase in efficiency of the separations came at a cost. The narrow-bore sub-2 μm particles columns created significant back pressure according to this equation[12]:

$$\Delta P = \Phi \frac{\eta \cdot L \cdot u}{d_p^2} \qquad (7.4)$$

where:
L is the column length
u is the average mobile-phase velocity

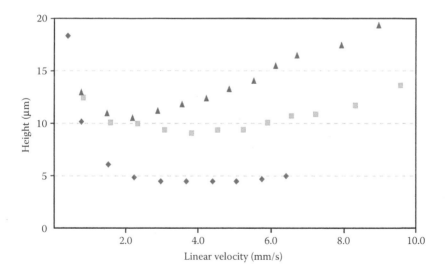

FIGURE 7.5 Van Deemter curves for hexylbenzene on BEH C18 columns packed with particles that are 1.7 μm (red diamonds), 3.5 μm (green squares), and 5.0 μm (blue triangles) in diameter. Column dimensions: 2.1 × 50 mm; mobile phase, acetonitrile:water 7:3 (v/v); temperature, 25.0°C. (Reprinted with permission from Jeffrey, R.M. et al., *Anal. Chem.*, 77, 23, 460A–467A, 2005. Copyright 2005 American Chemical Society.)

η is the mobile-phase viscosity
Φ is the flow resistance
d_p is the particle size

As d_p was decreased, when increasing u, a back pressure higher than 400 bar could be generated with columns longer than 3 cm, when packed with sub-2 μm particles.

A possible direction to overcome this obstacle was the suggestion to use high temperatures, as discussed by Nguyen et al.[9] They showed experimentally that the decrease in the eluent viscosity (η) at high temperatures led to a decrease in back pressures and therefore allowed the use of higher flow rates. They investigated the efficiency measured by the ratio of plate count N to time unit N/t_0, to describe the compromise between high efficiency and fast LC analysis. They could show experimentally that according to the following Equation 7.5, the simultaneous use of high temperatures and sub-2 μm particles could significantly increase the plate number per time unit, N/t_0, due to the concomitant increase of diffusion coefficient (D_m) and reduction of particle size diameter (d_p).

$$\frac{N}{t_0} = \frac{\nu \cdot D_m}{h \cdot d_p^2} \tag{7.5}$$

The study discussed before, shown in Figure 7.4, also describes a work done at elevated temperatures, using relatively high linear velocity for shorter run times, getting ultrahigh peak capacity. The columns used in this study were 2.1 × 150 mm in length, which can produce high back pressure even for UHPLC systems and required elevation of the temperature to be able to operate the column at flow rates appropriate for enhanced efficiency. Connection of two such columns in the series was also possible to increase peak capacity (not shown here).

As the technology of columns' particles and their packing has improved, it became more and more difficult experimentally to get close to their theoretical plate count, due to extra-column band broadening by the internal voids of the chromatographic systems, which was a significant limitation and hindrance to the evolution of ultra high-performance liquid chromatographic applications.[8]

Adding various volumetric and time-related contributions to extra-column band broadening from the system, HETP becomes the following equation[8]:

$$H = A \cdot d_p + \frac{B \cdot D_m}{u_0} + C \cdot \frac{d_p^2}{D_m} \cdot u_0 + \frac{1}{L} \cdot \frac{\sigma_v^2}{\pi^2 \cdot r^4} \cdot \frac{1}{\varepsilon_t^2 \cdot (k'+1)^2} + \frac{1}{L} \cdot \frac{\tau^2 \cdot u_0^2}{(k'+1)^2} \tag{7.6}$$

where:
  σ_v^2 is volumetric in nature and derives from the injection volume, the detector volume, and the volume of the connection tubing between the injector and detector
  r is the column radius
  ε_t is the total porosity, that is, the mobile-phase volume divided by the empty column volume
  k' is the retention factor

τ^2 is the time-related events contributing to the dispersion such as sampling rate and response time

L is the column length

Newly designed LC instrumentation for narrow-bore columns packed with sub-2–3 µm particles was proven to be essential to get values as close as possible to the theoretical plate count, while enduring their increasingly higher back pressures.[8,17,18] First and foremost, dispersion of the eluting narrower peaks had to be minimized by redesigning and polishing the internal surfaces of the tubing and connections. Then, pumps had to be able to withstand higher back pressure range and more precise short blast gradient times. Sample injection ports had to be able to deliver precisely lower injection volumes, typically 1–5 µL injections, with extensive washing to reduce the more easily detected carry over. Detector tubing and flow cell had to be redesigned for minimum dispersion, while keeping the same optical path, as well as having faster response times and higher sampling rates.[8] Extensive work was made to evaluate the effect of an extra-column band broadening on the separation efficiency of sub-2 µm particle columns in UHPLC systems by various researchers,[18–26] most of whom used highly optimized UHPLC systems for these studies.

Detailed discussion on the instrumental requirements in order to get the utmost efficiency of the separation is described by Fountain et al.[18] The variance of the system can be measured as described therein. Once it is known, one can evaluate the performance expected from certain geometry of the columns, diameter, and particles size. Figure 7.6, in which the behavior of plate count versus retention factor is measured for columns having different diameters but the same length and particle size both on a well-performing HPLC system with a variance of $\sigma_v^2 = 52$ µL^2 and on a UPLC with a variance of $\sigma_v^2 = 1$ µL^2. It can be noted in this figure that only columns of ID $\geq$ 3 mm can be operated with such an HPLC system with minimal loss of efficiency as long as $k' > 2.5$. On the other hand, columns below 3 mm ID tend to lose efficiency significantly in the lower k' values, in accordance with Equation 7.4, which shows how the efficiency rapidly decreases with increasing column radius to the fourth power. It is clear from Figure 7.6 that using narrow-bore columns of 2.1 mm in diameter and below suffers a significant efficiency loss even at high k' values, due to the large extra-column band broadening of a typical HPLC system.

Figure 7.7 shows the van Deemter plots for acenaphthene using columns with different particle sizes on two different LC systems. It is notable that when not considering system contribution to dispersion, any comparison between various columns packed with either sub-2 µm or above is meaningless.[18–22] When a 1.7 µm particle column was compared to a 2.5 µm particle column on a conventional HPLC system (Figure 7.7a), the difference in column performance was not apparent. The maximum plate counts achieved on this system were 3860 for the 1.7 µm particle column (optimum flow rate = 0.2 mL/min) and 3430 for the 2.5 µm particle column (optimum flow rate = 0.2 mL/min). When these same columns were evaluated on a UPLC™ system, the expected difference in performance between both particle sizes was indeed observed (Figure 7.7b). For the 2.5 µm particle column, the maximum plate count was 5880. The maximum plate count achieved on the 1.7 µm particle column increased by more than 130% to 8990 plates (optimum flow rate = 0.6 mL/min).

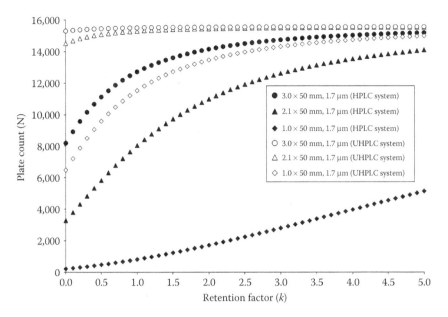

FIGURE 7.6 Calculated plate count (N) as a function of column diameter and retention factor. The extra-column band spreading for HPLC was measured on an Alliance 2695 system (Waters); the variance is $\sigma_v^2 = 52\ \mu L^2$. The extra-column band spreading for UPLC was measured on an Acquity UPLC I-Class system (Waters) with a fixed-loop injector and 5 μL loop; the variance is $\sigma_v^2 = 1\ \mu L^2$. Sample was a 0.5 μL injection of 160 mg/mL caffeine in 50:50 ratio of ACN/water. The mobile phase was 50:50 ACN/water pumped at 0.5 mL/min. UV light was at 273 nm at the highest possible data rate (80 Hz) and lowest filter time constant setting (zero if available). Band spreading was measured at 5σ. A 10 cm length piece of 50 μm PEEKSil tubing was used in place of the analytical column. (Fountain, K.J. and Iraneta, P.C., Instrumentation and columns for UHPLC separations, In Guillarme, D. et al. (Eds.), *UHPLC in Life Sciences*, Royal Society of Chemistry, London, UK, pp. 1–16, 2012. Reproduced by permission of The Royal Society of Chemistry.)

There were two reasons for this improvement in column efficiency on the described UPLC™ system. First, the system contribution to band spreading is significantly larger on the conventional HPLC system, which had more than 2.5 times the band spreading (in μL) of the UPLC system. Due to the smaller peak volumes associated with the higher-performing column, columns packed with 1.7 μm particles are proportionally more affected by extra-column effects than columns packed with larger particles ($\geq$2.5 μm), especially for analytes with small k' values (Equation 7.6). When calculating column efficiency, it is important to remember that the extra-column variance (square of the measured band spreading value) is incorporated in the calculated value and reduces the measured plate count.

Second, conventional HPLC systems are limited to operating pressures of approximately 345–415 bar (5000–6000 psi). For small molecular weight analytes, the optimum linear velocity for sub-2 μm particle columns typically generates pressures that are above this limit. Therefore, one must operate these columns at suboptimal flow rates in order to remain within the pressure constraints of traditional

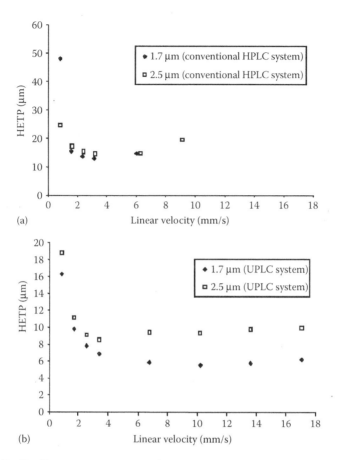

FIGURE 7.7 Van Deemter measurements for 1.7 and 2.5-μm particle columns on (a) a conventional HPLC system, and (b) a UPLC system. Acenaphthene was used as the test probe (*k* = 3.6). Plate counts were not corrected for extra-column effects in order to show system differences. Peak width was measured from 4σ. (Reproduced with the permission of Waters Corporation.)

HPLC systems, resulting in measured column efficiencies that are much lower than expected. This can be seen in Figure 7.7, where the optimum linear velocity for the 1.7 μm particle column is ~3 mm/s on the conventional HPLC system. However, this same column run on the UHPLC system showed maximum efficiency at ~10 mm/s, which is more than three times the linear velocity that can be used with traditional HPLC. The pressure required to achieve this flow rate on a 50 mm length column was ~375 bar (5600 psi), either beyond the limit or close to the limit of classical HPLC systems. It is clear that sub-2 μm particle columns must be used with low dispersion, high-pressure instrumentation in order to achieve the full performance potential of these particles.

The pharmaceutical industry was very interested in the new technology[6,11] thanks to high throughput of runs in the isocratic mode, as the use of the narrow bore sub-2 μm particle columns resulted in a 5- to 10-fold reduction in analysis time with limited influence on efficiency and resolution. The run time could be further reduced by up to 30-fold with the shorter available columns (i.e., 30 mm length). Moreover, in the gradient mode, the separation of complex mixtures, containing an active pharmaceutical compound and related impurities, has been significantly improved with column length equal to 100 mm, to increase peak capacity and resolution relative to their existing HPLC methods as shown in Figure 7.5.

7.4   REEMERGENCE OF THE SOLID-CORE COLUMNS

The need for special instrumentation as well as method validation of every transferred method from HPLC to UHPLC held back the pharmaceutical users from adopting methods of UHPLC up until August 2014, when the United State Pharmacopoeia allowed the scaling down of isocratic methods to UHPLC without full revalidation (USP37-NF32 S1, August 2014). In the meantime an alternative was suggested; one which did not necessarily require revalidation of each and every scaled-down method, the solid-core columns.[24] On 2006 the format of solid-core or superficially porous particles-based stationary phases was revived, a concept which was presented in the 1970s as pellicular particles: Corasil (Waters Inc) and Zipax (Du-Pont Inc).[26,27] The format of solid-core particles was resurrected with a significantly improved sub-3 μm pure silica particles with solid/fused core and porous shell around them, as can be seen in the schematic Figure 7.8. These solid-core particles have been developed to be highly efficient, enable fast separations with modest operating pressures.[25,28–41]

Figure 7.8 shows a scheme of a solid-core type of particle. The ratio between the thickness of the porous shell and the solid core and its relationship with chromatographic behavior was studied by DeStefano et al.[28] This study showed that a reduction in shell thickness for superficially porous particles results in an efficiency better than

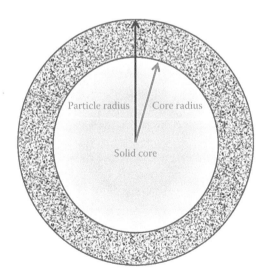

FIGURE 7.8   A scheme showing a solid-core particle with porous shell around it.

for totally porous particles, as might also be expected from theory, showing lower loss in efficiency as the mobile-phase velocity increased. However, reducing the shell thickness on a solid-core particle initiated a compromise, as thinner porous shells resulted in reduced sample loading capability (due to lower accessible surface area).

The use of these columns on traditional HPLC instrumentation was considered at first as an effective answer to the compromise between the need for high efficiency, analysis speed, and operating pressure, as they showed efficiencies not previously reported for totally porous particles of equivalent size in traditional HPLC systems.[25] As a result, high mobile-phase velocities could be used for very fast separations with only modest loss in column efficiency. The lower back pressure even allowed the connection of columns in series to enhance plate count per total length, relative to the sub-2 μm columns, which omitted the need for specially designed lower dispersion LC systems according to the authors.[30]

However, soon enough it turned out that the increase in efficiency was even more dramatic when the solid-core columns were operated on the more optimized UHPLC systems, bringing them even closer to their theoretical higher plate count.[19] At this stage, it became a challenge for the theoreticians to explain the reason behind the dramatic improvement in efficiency, in terms of HETP.[32]

7.5 THEORETICAL STUDIES OF THE EXTENDED EFFICIENCY OF SOLID-CORE COLUMNS

The initial rationale behind using solid-core stationary phases was that a lower diffusion path of the analyte inside the particles of the stationary phase, will create a lower mass transfer resistance in van Deemter plot, but this was proven not to be a dominant parameter by the comprehensive theoretical analysis critical overview of the solid-core columns by Guiochon and Gritti.[32] The overview began with a detailed history, then presented theoretical analysis covering the following parameters: Influence of the shell thickness on retention times and the saturation capacity of the column; Influence of particle size on the column permeability; Band broadening in chromatographic columns and the plate height equation; Column efficiency and the mass transfer kinetics; Influence of the shell thickness on the mass transfer resistance across the particles, and so on. An important discussion in this chapter covered the instrument's contributions to band broadening for these columns, especially the narrow-bore ones. The conclusion was that the modern low-dispersion UHPLC platform is required to operate the solid-core columns as well as the fully porous ones to get as close as possible to their extraordinary potential efficiency.[19]

Strikingly, this work showed that the exceptional performance of 4.6 mm I.D. columns, packed with the recent generation of sub-3 μm solid-core particles, could not be mainly attributed to the reduction of the C term in van Deemter equation as a result of reduced diffusion within the stationary phase of the particles, as was initially suggested by its early manufacturers. This work proved that this specific parameter had a marginal role in the gain of column efficiency,[32,33] and that the actual advantages of columns packed with the recent core–shell particles originate mainly from the reduction of both the longitudinal diffusion B coefficient (-20% to -30%) and the eddy diffusion A term (-40%). The decrease of the B coefficient was expected, because a significant part of the column volume is now occupied by the solid material, through which analytes cannot axially diffuse. However, the significant reduction in the eddy diffusion term was unexpected. It remained uncertain whether the significant decrease of the *A* term was caused by the tighter size distribution of solid-core (5%–7%) versus fully porous (15%–20%) particles (decrease of the short-range interparticle velocity deviations) or by the decrease of the cross-column velocity deviations caused by the smoothness of the external surface area of the solid-core particles.

Experimental results presented in this overview, showed some of the chromatographic properties and performance of the recent solid-core sub-3 μm particles columns.[32] Plate heights remarkably as low as 3 μm have been observed. However, such achievement could not have been obtained using a standard HPLC systems (<400 bar), due to overall band broadening, which is seriously deteriorating the quality of the separations.[18-24] It was clearly demonstrated all over again that LC instruments must show new highly precise injection modules, narrower connecting internal tubing, and smaller detection cell volumes with the standard optical path length, to get the utmost performance from these sub-3 μm particles solid-core columns, as well as was needed for the fully porous columns packed with sub-2 μm particles columns.

An interesting work[19] described how the measurement of intrinsic HETP of very efficient, short, and narrow-bore columns packed with sub-2 μm solid-core particles could be accurately measured, even in highly dispersive UHPLC, without removing the columns. Figure 7.9 shows results from this study.[19] The proposed protocol was based on using plots of the apparent HETP (which include the total contributions of the column and the instrument) of a few homologous compounds as a function of the reciprocal of $(1 + k')$.[2] These analytes have very similar structures and differed only by the presence of a few methylene groups. The graphs were truly linear and their intercept provided the value of intrinsic column-only HETP, so that the instrument variance could be deduced from them without removing the column. This way they could learn more about the contribution of the instruments to the overall efficiency of the separation.

In a critical standpoint paper,[33] showing theoretical as well as experimental results, Gritti and Guiochon explained facts and currently propagated four *legends* in HPLC, which were harnessed in widespread presentations, to explain the dramatic increase of efficiency of the separation by the columns packed with solid-core particles.[33] To better understand these claims it is recommended to read through Grushka's classic explanation from the 1970s[42] of the sources of band broadening inside the chromatographic column.

In this chapter,[33] the relative contributions of the longitudinal diffusion (B/u), the eddy dispersion (A), and C_u terms of the van Deemter Equation 7.1 were reevaluated and compared to those of columns packed with standard fully porous particles. Chromatographic myths or legends such as the relative impact of the B/u and C_u HETP terms on the overall column performance or the influence of the particle size distribution (PSD) on column performance are discussed and revisited, describing four common *legends* in the literature of chromatography.

Legend 1 discussed the contribution of axial diffusion to *B* term in the van Deemter equation. The claim was that the contribution of the axial diffusion term to the optimum column HETP is systematically neglected in the scientific literature. Although this fact is rarely explicitly mentioned, it accounts for nearly 25% of the total band

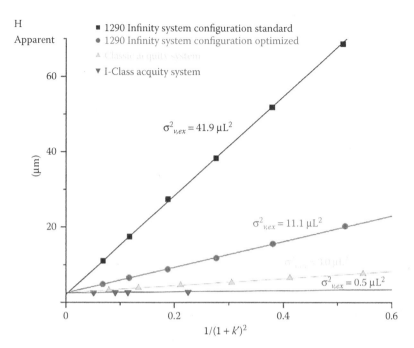

FIGURE 7.9 Plots of the apparent HETP's of homologous compounds versus the reciprocal of $(1 + k')^2$ for four different UHPLC instruments. Column: 2.1 mm × 100 mm packed with 1.6 μm prototype RPLC-C18 core–shell particles. Eluent: acetonitrile/water, 75/25 (v/v). $T = 24°C$. Despite the large differences in instrument performance and in extra-column volume variances (see values of $\sigma^2_{v,ex}$ in the graph for each instrument), the extrapolated values of H Apparent (2.55 ± 0.05 μm) when $k' \to 0$ is the same for all UHPLC systems. (Adapted from Gritti, F. and Guiochon, G., *J. Chromatogr. A*, 1327, 49–56, 2014.)

broadening of highly retained compounds on reversed-phase liquid chromatography (RPLC) columns, used at optimum velocity and packed with fully porous particles. On account of the volume occupied by the solid cores of shell particles represents typically ~25% of the column volume, one can anticipate even a smaller *B* term and greater column efficiency for columns packed with solid-core particles than for those packed with conventional particles. The author suggested that the measurement of the *B* term and its quantitative contribution to the overall plate height (at a constant ratio of shell diffusivity to bulk sample diffusion) need to be revisited.

Legend 2 concerns the common assumption that because solid-core particles have diffusion path length effectively shorter than through fully porous particles, this parameter is significant in improving the efficiency. They criticized strongly the statements such as "This shorter diffusion path allows for faster mass transfer. The result is less band broadening for higher peak efficiency comparable to or better than sub-2 μm porous particles." The authors claimed that although the diffusion path length is indeed shorter, its effect on reduction of the *C* term of the van Deemter equation is not as significant as originally thought, and needs still to be proven theoretically and experimentally.

Legend 3 discussed the eddy diffusion term (*A*) for small molecules. In many textbooks, scientific papers, and company reports, this term is assumed to be independent of the eluent's velocity, which is in profound contradiction with the general theory of eddy diffusion of Giddings.[13] The authors showed that the eddy diffusion was not constant in very small flow rates, and as the velocity, corrected for particle diameter and diffusion coefficient, gets as high as possible technically, the eddy diffusion term is significantly higher in the fully porous particles compared to the solid-core ones, contributing significantly to the overall broadening. Legend 4 also relates to the term A, is widespread in the chromatographic literature, stating that narrow particle size distribution of packing materials provides less sample dispersion in the interstitial eluent. Although it is true that particle size distribution in solid-core particles columns are narrower (RSD ~5%) than that of conventional fully porous particles (~20%), there is no experimental proof and no theoretical consideration to support this claim that tight particle size distribution is indeed a major reason behind the significant increased chromatographic efficiency of the solid-core columns.

LC users adopted the solid-core particle columns with enthusiasm, and a recent publication[41] even suggested using a ballistic gradient in one generic method, using narrow bore short solid-core column when there is a need to quickly develop LC method for a new chemical entity (NCE) to be considered as active pharmaceutical ingredient (API) in the pharmaceutical industry. Figure 7.10 shows chromatograms of a mixture

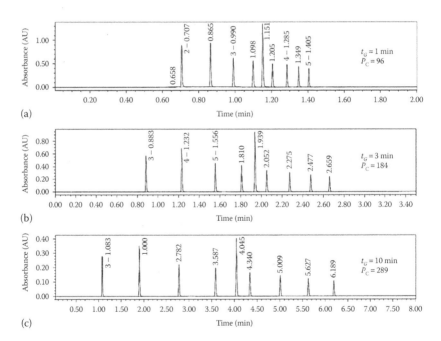

FIGURE 7.10 Comparative gradient UHPLC chromatograms obtained with gradient times of (a) 1 min, (b) 3 min, and (c) 10 min showing the effect on peak capacity. Column: 50 mm × 2.1 mm, 1.6 μm Cortecs C18+; mobile-phase A: 0.05% formic acid; mobile-phase B: acetonitrile; gradient: 30%–100% B; flow rate: 1 mL/min; temperature: 40°C; pressure: 12,000 psi; detection: UV at 254 nm; sample: a test mixture of nine alkylphenones. Peak capacities were measured by dividing the gradient time by the average measured base peak widths. (Adapted from Dong, M.W., *LCGC N. Am.*, 34, 6, 408, 2016.)

at three gradient times, displaying the remarkable peak capacity and speed of the separation that could be achieved with solid core columns in an optimized UHPLC system. The method was also compatible with traditional HPLC system, although much better performance and shorter run times could be achieved by using UHPLC.

7.6 CONCLUSION

Solid-core stationary phases were developed in the 1970s to provide enhanced efficiency of the separation in LC and were abandoned until the emergence of UHPLC UPLC™ system, which operated narrow-bore columns with sub-2 μm particles. These columns produced strikingly efficient LC, but required entirely new design of the internal parts of the LC technology. This was a drive for column manufacturers to reintroduce solid-core type of particles in the format of sub-3 μm, based on ultrapure silica, to enhance efficiency of the separation in an alternative way, without the need for newly designed LC systems. The reasoning was based on the theoretical predictions of reduced band broadening due to the reduced path of the molecules inside the stationary phase as a result of the solid core. Nevertheless, soon enough it was shown in many publications that running these columns in ultra high-performance liquid chromatographic systems resulted in efficiency closer to their theoretically calculated values, fulfilling their great potential.

Thanks to the success of UHPLC instrumentation, manufacturers of such solid-core particles proceeded toward the next challenges: creating sub-2 μm solid-core particles, then packing them into narrow-bore columns. These enhanced formats of solid-core columns truly require fully optimized LC systems for the dramatically enhanced efficiency of the separations. Theoretical works described here attempted to explain this extra efficiency and revisit some wrong widespread notions of possible reasons for it.

REFERENCES

1. Swartz, M.E., 2005. UPLC™: An introduction and review. *Journal of Liquid Chromatography & Related Technologies*, 28(7–8), 1253–1263.
2. Jeffrey, R.M., Uwe, D.N., Marianna, K. and Robert, S.P. 2005. A new separation technique takes advantage of sub-2-μm porous particles, *Analytical Chemistry*, 77(23), 460A–467A.
3. Wren, S.A., 2005. Peak capacity in gradient ultra performance liquid chromatography (UPLC). *Journal of Pharmaceutical and Biomedical Analysis*, 38(2), 337–343.
4. Neue, U.D., 2008. Peak capacity in unidimensional chromatography. *Journal of Chromatography A*, 1184(1), 107–130.
5. Plumb, R.S., Rainville, P., Smith, B.W., Johnson, K.A., Castro-Perez, J., Wilson, I.D. and Nicholson, J.K., 2006. Generation of ultrahigh peak capacity LC separations via elevated temperatures and high linear mobile-phase velocities. *Analytical Chemistry*, 78(20), 7278–7283.
6. Novakova, L., Matysova, L. and Solich, P., 2006. Advantages of application of UPLC in pharmaceutical analysis. *Talanta*, 68(3), 908–918.
7. Churchwell, M.I., Twaddle, N.C., Meeker, L.R. and Doerge, D.R., 2005. Improving LC–MS sensitivity through increases in chromatographic performance: Comparisons of UPLC–ES/MS/MS to HPLC–ES/MS/MS. *Journal of Chromatography B*, 825(2), 134–143.

8. Fountain, K.J., Neue, U.D., Grumbach, E.S. and Diehl, D.M., 2009. Effects of extra-column band spreading, liquid chromatography system operating pressure, and column temperature on the performance of sub-2-μm porous particles. *Journal of Chromatography A*, 1216(32), 5979–5988.

9. Nguyen, D.T.T., Guillarme, D., Heinisch, S., Barrioulet, M.P., Rocca, J.L., Rudaz, S. and Veuthey, J.L., 2007. High throughput liquid chromatography with sub-2μm particles at high pressure and high temperature. *Journal of Chromatography A*, 1167(1), 76–84.

10. Nguyen, D.T.T., Guillarme, D., Rudaz, S. and Veuthey, J.L., 2006. Chromatographic behaviour and comparison of column packed with sub-2μm stationary phases in liquid chromatography. *Journal of Chromatography A*, 1128(1), 105–113.

11. Russo, R., Guillarme, D., Nguyen, D.T.T., Bicchi, C., Rudaz, S. and Veuthey, J.L., 2008. Pharmaceutical applications on columns packed with sub-2 μm particles. *Journal of Chromatographic Science*, 46(3), 199–208.

12. Van Deemter, J.J., Zuiderweg, F.J. and Klinkenberg, A.V., 1956. Longitudinal diffusion and resistance to mass transfer as causes of nonideality in chromatography. *Chemical Engineering Science*, 5(6), 271–289.

13. Giddings, J.C., 2002. *Dynamics of Chromatography: Principles and Theory*. CRC Press: Boca Raton, FL.

14. Bristow, P.A. and Knox, J.H., 1977. Standardization of test conditions for high performance liquid chromatography columns. *Chromatographia*, 10(6), 279–289.

15. De Villiers, A., Lestremau, F., Szucs, R., Gélébart, S., David, F. and Sandra, P., 2006. Evaluation of ultra performance liquid chromatography: Part I. Possibilities and limitations. *Journal of Chromatography A*, 1127(1), 60–69.

16. Nguyen, D.T.T., Guillarme, D., Rudaz, S. and Veuthey, J.L., 2006. Chromatographic behaviour and comparison of column packed with sub-2μm stationary phases in liquid chromatography. *Journal of Chromatography A*, 1128(1), 105–113.

17. Guillarme, D., Veuthey, J.L. and Smith, R.M. (Eds.) 2012. *UHPLC in Life Sciences* (No. 16). Royal Society of Chemistry: London, UK.

18. Fountain, K.J. and Iraneta, P.C. Instrumentation and columns for UHPLC separations. In *UHPLC in Life Sciences*, Guillarme, D. et al. (Eds.), Royal Society of Chemistry, London, UK, pp. 1–16, 2012.

19. Gritti, F. and Guiochon, G., 2014. Accurate measurements of the true column efficiency and of the instrument band broadening contributions in the presence of a chromatographic column. *Journal of Chromatography A*, 1327, 49–56.

20. Fekete, S. and Fekete, J., 2011. The impact of extra-column band broadening on the chromatographic efficiency of 5 cm long narrow-bore very efficient columns. *Journal of Chromatography A*, 1218(31), 5286–5291.

21. Gritti, F. and Guiochon, G., 2010. On the extra-column band-broadening contributions of modern, very high pressure liquid chromatographs using 2.1 mm ID columns packed with sub-2μm particles. *Journal of Chromatography A*, 1217(49), 7677–7689.

22. Gritti, F. and Guiochon, G., 2011. On the minimization of the band-broadening contributions of a modern, very high pressure liquid chromatograph. *Journal of Chromatography A*, 1218(29), 4632–4648.

23. Stankovich, J.J., Gritti, F., Stevenson, P.G. and Guiochon, G., 2013. The impact of column connection on band broadening in very high pressure liquid chromatography. *Journal of Separation Science*, 36(17), 2709–2717.

24. Wu, N., Bradley, A.C., Welch, C.J. and Zhang, L., 2012. Effect of extra-column volume on practical chromatographic parameters of sub-2-μm particle-packed columns in ultra-high pressure liquid chromatography. *Journal of Separation Science*, 35(16), 2018–2025.

25. Fekete, S., Oláh, E. and Fekete, J., 2012. Fast liquid chromatography: The domination of core–shell and very fine particles. *Journal of Chromatography A*, 1228, 57–71.
26. Kirkland, J.J., 1971. Columns for modern analytical liquid chromatography. *Analytical Chemistry*, 43(12), 36A–48A.
27. Kirkland, J.J., 1992. Superficially porous silica microspheres for the fast high-performance liquid chromatography of macromolecules. *Analytical Chemistry*, 64(11), 1239–1245.
28. DeStefano, J.J., Langlois, T.J. and Kirkland, J.J., 2008. Characteristics of superficially-porous silica particles for fast HPLC: Some performance comparisons with sub-2-μm particles. *Journal of Chromatographic Science*, 46(3), 254–260.
29. Gritti, F. and Guiochon, G., 2012. Mass transfer kinetics, band broadening and column efficiency. *Journal of Chromatography A*, 1221, 2–40.
30. Cunliffe, J.M. and Maloney, T.D., 2007. Fused-core particle technology as an alternative to sub-2-μm particles to achieve high separation efficiency with low backpressure. *Journal of Separation Science*, 30(18), 3104–3109.
31. Felinger, A., 2011. Diffusion time in core–shell packing materials. *Journal of Chromatography A*, 1218(15), 1939–1941.
32. Guiochon, G. and Gritti, F., 2011. Shell particles, trials, tribulations and triumphs. *Journal of Chromatography A*, 1218(15), 1915–1938.
33. Gritti, F. and Guiochon, G., 2012. Facts and legends on columns packed with sub-3-μm core-shell particles. *LC GC North America*, 30, 586–595.
34. Gritti, F., Shiner, S., Fairchild, J.N. and Guiochon, G., 2014. Characterization and kinetic performance of 2.1 × 100 mm production columns packed with new 1.6 μm superficially porous particles. *Journal of Separation Science*, 37(23), 3418–3425.
35. Fekete, S., Kohler, I., Rudaz, S. and Guillarme, D., 2014. Importance of instrumentation for fast liquid chromatography in pharmaceutical analysis. *Journal of Pharmaceutical and Biomedical Analysis*, 87, 105–119.
36. Walter, T.H. and Andrews, R.W., 2014. Recent innovations in UHPLC columns and instrumentation. *TrAC Trends in Analytical Chemistry*, 63, 14–20.
37. Omamogho, J.O., Hanrahan, J.P., Tobin, J. and Glennon, J.D., 2011. Structural variation of solid core and thickness of porous shell of 1.7 μm core–shell silica particles on chromatographic performance: Narrow bore columns. *Journal of Chromatography A*, 1218(15), 1942–1953.
38. Gritti, F., Leonardis, I., Abia, J. and Guiochon, G., 2010. Physical properties and structure of fine core–shell particles used as packing materials for chromatography: Relationships between particle characteristics and column performance. *Journal of Chromatography A*, 1217(24), 3819–3843.
39. Liekens, A., Denayer, J. and Desmet, G., 2011. Experimental investigation of the difference in B-term dominated band broadening between fully porous and porous-shell particles for liquid chromatography using the effective medium theory. *Journal of Chromatography A*, 1218(28), 4406–4416.
40. DeStefano, J.J., Schuster, S.A., Lawhorn, J.M. and Kirkland, J.J., 2012. Performance characteristics of new superficially porous particles. *Journal of Chromatography A*, 1258, 76–83.
41. Dong, M.W., 2016. A universal reversed-phase HPLC method for pharmaceutical analysis. *LCGC North America*, 34(6), 408.
42. Grushka, E., 1974. Band spreading in liquid chromatography. *Analytical Chemistry*, 46(6), 510A.

8 Inverse Size-Exclusion Chromatography

Annamária Sepsey, Ivett Bacskay,
and Attila Felinger

CONTENTS

8.1 INTRODUCTION

The most preferred method to separate sample molecules based on their size relative to the pore size is size exclusion chromatography (SEC) (also referred to as gel-permeation, gel-filtration, molecular sieve, or simply gel chromatography) because using a strong solvent, there will be no interaction between the solute molecules and the stationary phase. The inverse version of SEC, inverse size-exclusion chromatography (ISEC) was also described in the middle of the 1970s [1,2], where the pore sizes were determined in the knowledge of the molecule size. Some sources attributed the first description of the ISEC technology incorrectly to Ogston [3] or to Aggerbrandt [4].

There is a plethora of information on the theoretical and experimental aspects of SEC and ISEC and also a number of great reviews have been published. The porous structure of the chromatographic particles is of great complexity and that has a number of consequences during the separation process. The physicochemical properties of high-performance liquid chromatography (HPLC) stationary phases play an important role on column performance and efficiency. The proper characterization of the pore structure and the pore size distribution (PSD) is relevant, because the mass transfer across the particles is greatly affected by the nature of the pores.

SEC allows getting a more accurate picture of the impact of the distributions (pore size distribution and polydispersity) on the separation efficiency from ISEC measurements using a proper model. In this chapter, we summarize the most important theories and the newest applications regarding ISEC.

8.2 BAND BROADENING PROCESSES IN LIQUID CHROMATOGRAPHY

In chromatography, the sample components become distributed over the separation path as discrete zones and these zones expand continuously as the separation process advances. The result of the separation depends on whether the zones are narrow enough, so that we can keep avoiding overlapping and cross-contamination with adjacent zones [5]. Therefore, the major aim of chromatography is the limitation of zone spreading. To fulfill this, one needs to understand the processes underlying the zone formation and spreading.

The events contributing to zone spreading in separation systems are of a random nature. The central limit theorem of probability theory guarantees that a sequence of random events of the most general kind will lead to a Gaussian distribution function if the single random displacements are independent and small compared to the final mean displacement [6]. If this is fulfilled, the variances of the independent processes (σ_i^2) contributing to the zone spreading are additive. Thus the total variance can be written as

$$\sigma_{\text{total}}^2 = \sigma_1^2 + \sigma_2^2 + \sigma_3^2 + \cdots = \sum_{i=1}^{n} \sigma_i^2 \tag{8.1}$$

As the apparent diffusion coefficients (D_i) are directly proportional to the respective variances and inversely proportional to the time (t) as

$$D_i = \frac{\sigma_i^2}{2t}, \tag{8.2}$$

they are additive in the same manner ($D_{\text{total}} = \sum D_i$). With each increment in zone spreading, there is obviously a corresponding loss of separation efficiency. Figure 8.1 illustrates the contributions to overall band spreading obtainable in a chromatographic system.

The spreading caused by the HPLC system (injection needle, needle seat capillary, injection valve, connecting tubes, detector cell [7]), and column hardware can be measured and optimized easily. The dead volume contains only these effects and we can remove them the easiest way by deconvolution of an inert, unretained component's peak from the analyte's peak we are interested in. By this operation—which is also called as normalization to the system volume—we achieve that the initial zone of the analyte can be approximated at relatively low concentrations as a Dirac delta function, which is an infinitely narrow zone of unit area. The diffusion and diffusion-like processes in the column convert even such a delta function to a Gaussian one with the passage of time (Figure 8.2a). However, if the initial zone was already

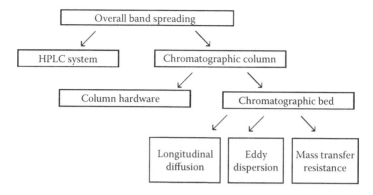

FIGURE 8.1   Contributions to overall band spreading in a chromatographic system.

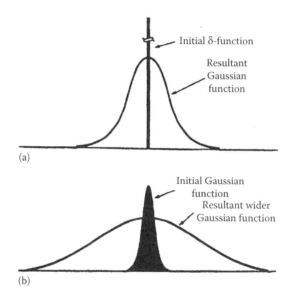

FIGURE 8.2   Formation of concentration profiles (a) from initial Dirac δ function and (b) from initial Gaussian function. (From Giddings, J.C.: *Unified Separation Science.* 1991. Copyright Wiley-VCH Verlag GmbH & Co. KGaA. Reproduced with permission.)

Gaussian, it becomes broader and lower as time passes (Figure 8.2b). If the initial zone cannot be approximated by a delta function, the final zone can be obtained as a sum of Gaussian functions [5].

In the bed, while separating the components from each other, several processes occur: the analyte molecules travel with the mobile phase along the column at high pressure, whereas several adsorption/desorption steps occur in adsorption chromatography and pore entrance/release steps in SEC. We can distinguish three major sources of zone dispersion inside the chromatographic bed [8,9]:

- Longitudinal diffusion along the mobile phase percolating through the bed, which is inversely proportional to the mobile phase velocity because the lesser time the band spends in the column, the lesser time it has to diffuse.
- Multipath (eddy) dispersion, due to the anastomosis of the channels conveying the sample band across the bed of particles.
- Mass transfer resistance, due to the time molecules need to diffuse in and out of the particles and through their pores and due to slow adsorption–desorption kinetics.

Those three contributions account for the band broadening due to the mass transfer processes encountered in any type of chromatographic column (packed or monolithic columns), independent of the physical state of the mobile phase (gas, liquid, or supercritical fluid), the flow regime (laminar to turbulent), or the nature of the stationary phase (liquid, solid, or narrow pores in which molecular movements are hindered) [10].

8.3   DISTRIBUTIONS CONTRIBUTING TO BAND BROADENING

In the previous section, the classical, long known peak broadening factors were discussed. However, other contributions can also affect the peak shape and thus separation efficiency. In this section, the pore size distribution of the stationary phase particles and the distribution of the molecule size affecting the separation efficiency will be discussed. Other parameters such as temperature-induced effects will not be addressed.

8.3.1   PARTICLE SIZE DISTRIBUTION

The physical characteristics of particles are important and manufacturers monitor the particle properties in various ways. The size of the particles is especially important, as this parameter largely determines the size of the flow channels between particles and thus the efficiency of the packed columns. The particle size, however, cannot be defined by a single value; it is governed by a distribution that may cover a wide range [11]. The particle size and its distribution can be obtained either with electron microscopy, or with special instruments such as Coulter counter which provides a more convenient and quantitative information. The typical relative standard deviation of the particle size is 5% for superficially porous particles and 10%–20% for totally porous particles but it strongly depends on the method used to characterize it [12].

The importance of particle diameter distribution in influencing separation has been discussed several times in practice and by chromatographic models as well [11,13–20]. There are several contradictory consequences that were drawn on the efficiency of the separations. According to one of the most recent papers, there is no evident correlation between the particle size distribution and the column efficiency [19] and this observation is in good agreement with the measured data. Besides, a narrow particle size distribution provides a lower pressure drop for comparable efficiency.

8.3.2 Pore Size Distribution

Porous particles can be characterized by their specific surface area, specific pore volume, and pore diameter, whose properties depend strongly on each other. The pores provide the surface area necessary for separations in adsorption or partition chromatography and the molecules are separated according to their size relative to that of the pore diameter in SEC. The pore diameter, however—similarly to the particle size—is not a constant, well-definable value; the pores of modern porous stationary phases may exhibit a momentous PSD, which is an important feature of the adsorbent [21]. Experimental data confirm that the pore sizes can cover a rather wide range, including pores in the molecular size range as well as macroscopic fissures and cracks, thus the pore size should preferably be presented on a logarithmic scale [11,22]. This is shown in Figure 8.3 where the PSD of three commercially available HPLC stationary phases are presented when measured with low temperature nitrogen adsorption (LTNA).

The nature and the breadth of the PSD have significant impact on the mass transfer properties of the stationary phase thus on the efficiency of the separation, because the hindered diffusion of the molecules in the pore network gives a critical contribution to band broadening. There could be large molecules that enter the pores only a few times during the separation process and their individual residence time in the pore is huge compared to the small molecules that can enter the pores without hindrance [23].

There are a number of methods for determining relevant information about the porous media. Particle surface area and PSD typically are measured by LTNA at 77 K using the Brunauer–Emmett–Teller procedure (BET) or by mercury intrusion [24–26].

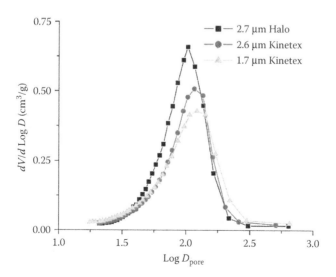

FIGURE 8.3 Pore size distribution of three superficially porous particles obtained from low temperature nitrogen adsorption data.

The former is based on the physisorption of gases, whereas the latter is based on the controlled penetration of fluids. Both methods are based on certain assumptions, for example, on the pore shape. LTNA measurements are used for microporous and mesoporous silica but mercury intrusion is only reliable for mesoporous and macroporous silica. For mesoporous silica, nearly the same pore volume distributions were found by means of porosimetry and nitrogen sorption for mean pore diameters between 4 and 40 nm [24]; however, in other cases porosimetry gives slightly smaller values for the pore diameter [27]. Mercury intrusion does not work well for fragile particles (with a large pore volume and also for monolithic columns) or for soft polymer particles.

These techniques are either too expensive or they destroy the chromatographic column so that it cannot be used for any further analysis. It is also possible that they give irrelevant information because fake assumptions are made while deriving values of parameters. For example, it was long neglected that the size of the nitrogen molecule is much smaller than the size of the analyte molecules used in chromatographic practice, so that the pore volume derived from such measurements and the conclusions that were drawn could have been quite misleading [11].

The surface modification should also be taken into account. It was usually assumed that the molecular area of nitrogen is 1.62 nm^2 [28] but it was shown that nitrogen adsorption is less localized on hydrophobic surfaces due to weaker adsorbate–surface interactions and has more freedom for lateral movement, thus effectively occupying a larger area of the surface. Therefore the use of the 1.62 nm^2 value would lead to an underestimation of the actual surface area of hydrophobic surfaces and we should rather use 2.05 nm^2 [29].

8.3.3 POLYDISPERSITY

Size-exclusion chromatography is more often used to obtain molecular size or molecular weight than for separating polymers from each other [30]. This determination relies on a calibration process in which a number of special standard polymers with well-known molecular weight are eluted and a calibration curve is drawn by their partition coefficient (ln M_W vs. K). The retention properties of the unknown sample are then compared with the calibration data to determine the weight or size of the investigated molecule. Even though we say that the standards used for the calibration process have well-defined molecular weights, they do have a distribution. Manufacturers use a quantity to characterize the fitness-to-purpose of these polymers, the polydispersity (P).

The polydispersity of a polymer sample is defined as the ratio of the weight- and number-averaged relative molecular weights of the polymer sample [31]:

$$P = \frac{M_W}{M_n} \tag{8.3}$$

Although there is a new recommended International Union of Pure and Applied Chemistry (IUPAC) terminology for polydispersity [32] where *degree-of-polymerization dispersity* and shortly the word *dispersity* is recommended for use, we rather use the term polydispersity here because this is still the common way how polymer scientists refer to the distribution of polymer molecules.

The knowledge and the effect of polydispersity—that it increases the peak width and leads to a decrease in separation efficiency—is almost of the same age as SEC [33–36]. It is very difficult, if feasible at all, to obtain experimentally the real polydispersity by SEC. The first equations that described the contribution of polydispersity to the height equivalent to a theoretical plate (HETP) value, however, greatly underestimated the effect of polydispersity [33,34], it was proven that similarly to a new equation given by Knox [31], the data given by the manufacturers overestimate it as well [37,38].

As indicated in Equation 8.1, the variances of the independent events can be divided into contributors and we can distinguish the contribution of the kinetic events (σ^2_{kin}) and that of the polydispersity (σ^2_P) to the total band broadening as

$$\sigma^2 = \sigma^2_{kin} + \sigma^2_P \tag{8.4}$$

The elimination of the first term of Equation 8.4, so the measurement of σ^2_P is experimentally impossible. Although recent improvements in experimental detection techniques in SEC made precise determination of molecular-weight distribution possible, Netopilík showed that for samples with low polydispersity, the errors due to the band broadening and due to the interdetector volume increases as polydispersity decreases [39]. The effect of polydispersity is in fact so large that even with $P = 1.01$, it is difficult to make accurate determinations of σ^2_{kin} because polydispersity of even a relatively narrow polymer fraction will contribute substantially to the total width of the eluted peak [31].

8.4   SIZE-EXCLUSION CHROMATOGRAPHY

SEC is defined as the differential elution of solutes from the porous stationary phase caused by different degrees of steric exclusion [40]. The stationary phase used in SEC is always a mechanically stable porous media. The molecules traveling along the column in the mobile phase can enter the pores if the size of the pore is larger than that of the molecule. However, it is still uncertain which exact size parameter determines the separation, in general it is accepted that the gyration radius or diameter of the molecule is used to determine whether the molecule can enter the pore [41].

In an ideal SEC, there is no interaction between the molecules and neither the stationary nor the mobile phase particles, so that the separation is entropy driven. There are however other mechanisms that are responsible for the migration of the molecules along the column (hydrodynamic and stress-induced diffusion, the polarization effect, multipath, enthalpic and soft-body interactions, and the so-called wall effect), they can be ignored in almost every case. The main mechanism responsible for retention of micromolecules in SEC is the size-exclusion effect [42] and for macromolecules it is the combination of the size-exclusion and the hydrodynamic effects [43].

There are two types of molecules that are of major importance in SEC based on their size: the completely permeable particle (indicated by a subscript *perm* in the equations) that is small enough to visit all the pores, and the barely excluded particle

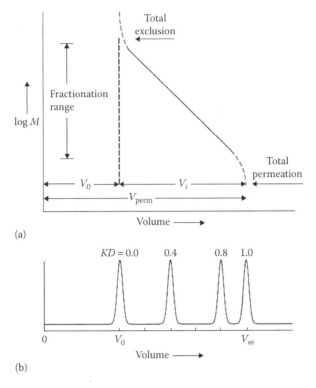

FIGURE 8.4 Hypothetical SEC calibration curve (a) and chromatogram (b). (Snyder, L.R. et al.: Introduction to Modern Liquid Chromatography. 2010. Copyright Wiley-VCH Verlag GmbH & Co. KGaA. Reproduced with permission.)

that is too large to enter the pores and will be excluded (indicated by a subscript *excl* in the equations). The former molecules have access to the stagnant and moving zone of the mobile phase as well, whereas for the latter molecules only the interstitial volume between the stationary phase particles is available.

The operational aspects of SEC are presented in Figure 8.4, where V_0 indicates the column void volume obtained by the retention volume of an excluded molecule, and V_{perm} is the volume available for the totally permeable molecules. Solutes of intermediate range have hydrodynamic radii between that of excluded and totally permeable molecules.

The partition coefficient (K_{SEC}) can easily be calculated by means of the retention volumes of the abovementioned and the unknown molecules by

$$K_{SEC} = \frac{V_i - V_{excl}}{V_{perm} - V_{excl}} = \frac{V_p}{V_{p,perm}} = \frac{t_p}{t_{p,perm}} \tag{8.5}$$

where after multiplication by the mobile phase velocity, the numerator stands for time spent by the investigated molecule in the pores of the stationary phase particles and the denominator indicates the residence time spent by the completely permeable particle in the pores (subscript p refers to pore).

As mentioned earlier, the retention of macromolecules arises from either size-exclusion effect, hydrodynamic effect, or a combination of both according to their size relative to the pore size of the stationary phase [43]. Therefore, the retention volume (V_R) is written as

$$V_R = V_p K_{SEC} + V_0 K_{HDC} + V_{system} \qquad (8.6)$$

where:

K_{HDC} accounts for the hydrodynamic chromatography contribution
V_{system} is the volumetric contribution of the system [43,45]

K_{HDC} is defined as

$$K_{HDC} = \frac{1}{1 + 2(r_{eff}/r_0) - C(r_{eff}/r_0)^2} \qquad (8.7)$$

where:

C is a constant that depends on the packing–carrier–polymer system (here $C = 2.698$ [46])
r_0 is the hydraulic radius of the packing material
r_{eff} is the effective size of macromolecule

$$r_{eff} = \frac{\sqrt{\pi}}{2} r_G \qquad (8.8)$$

where r_G is the mean gyration radius of the molecule.

It was shown [45] that by combining the Kirkwood–Riseman relationship with the Stokes–Einstein equation, one gets a direct relation of the gyration radius to the molecular weight (M_W) as

$$r_G = \frac{k_B T M_W^B}{4 \pi \eta K} \qquad (8.9)$$

where:

k_B is the Boltzmann constant
T is the temperature
η is the viscosity of the mobile phase
B is the exponent of the power law relating the diffusion coefficient of the polymer to the reciprocal of its molecular weight
K is a constant

As K and B are tabulated for various polymer–solvent systems, r_G can safely be used. The typical value for B is 0.549 and for K is 3×10^{-8} m^2/s [45].

In a widely used retention model [30,45,47] the ratio of the gyration radius of the molecule (r_G) to the radius of the pore opening (r_p) is a unique size parameter:

$$\rho = \frac{r_G}{r_p} \qquad (8.10)$$

which is used to define the partition coefficient as

$$K_{SEC} = \begin{cases} (1-\rho)^m & \text{if } 0 \le \rho \le 1 \\ 0 & \text{if } \rho > 1 \end{cases} \qquad (8.11)$$

where m is a constant whose value depends only on the pore geometry.

8.5 INVERSE SIZE-EXCLUSION CHROMATOGRAPHY

Size-exclusion chromatography can be used to determine the molecular weight or size of a molecule by a calibration process, and by using the inverse way, ISEC is an appropriate tool to investigate the properties of the stationary phase (pore geometry, porosity, surface area, and PSD): by measuring molecules of known size, the size of the pores can be determined. For the effective use of ISEC, however, a proper theory is needed that relates PSD to the retention time of polymers.

ISEC is a very comfortable and preferred method to derive information about the structure of the porous packing material, such as the interstitial porosity and the surface area of the packed bed or PSD of various porous particles and monoliths [1,2,27,43,45–95]. ISEC has several advantages in comparison to other classical or novel methods for determining the porous structure. Although in mercury intrusion and nitrogen absorption, dry experimental environment with high pressure and low temperature is necessary that causes the structural damage of the porous material, in ISEC intact structural information of materials can be achieved without morphological changes in a nondestructive way. In addition, ISEC is operated under typical chromatographic conditions, so it is a relatively convenient approach [81]. It only needs an HPLC instrument, a good solvent, and a series of standard molecules with defined size; it is very gentle with the stationary phases and enables the determination of the PSD within the entire range of 1–400 nm of various types of stationary phases [2].

8.5.1 HISTORICAL PERSPECTIVES

During the past four decades several assumptions were made to obtain the structural information from SEC data. Halász et al. plotted the sum of residues (i.e., the pore volume formed by pores with diameter greater than a specific value) against the pore diameter to get a so-called integral distribution curve and took the value at 50% accessibility of the pore volume as the mean pore diameter [2,49,50,52,55]. They got smaller mean pore diameters compared to the independent mercury porosimetry results and made up a relationship between the PSD and the plate number [49,53]. Shortly after Halász et al. introduced their idea, Freeman et al. realized that a theory needs to be related to the measured chromatographic data [48,54] to account for the size effects observed in SEC and used a relationship between the partition coefficient and pore size parameters. Chiantore commented that for flexible chain molecules the radius of the equivalent hard sphere should be used in their relation instead of the mean external length of the molecule [96]. They used two existing models to derive the molecule size and plotted the partition coefficient versus the solute size to represent the PSD of the stationary phase.

Knox and Scott have developed a rather sophisticated model to determine the PSD from SEC data [58,62]. Their model has two assumptions: the pores are cylindrical and the sample molecules are spherical. They realized that the partition coefficient depends only on the size and shape of the solute molecules and that of the pores of the bed and defined it as

$$
K_{SEC} = \begin{cases} \left(\dfrac{R-r}{R}\right)^2 = \left(1-\dfrac{r}{R}\right)^2 & \text{if } \dfrac{r}{R} < 1 \\ 0 & \text{if } \dfrac{r}{R} > 1 \end{cases}
\tag{8.12}
$$

where:
r is the radius of the hard sphere molecule
R corresponds to the radius of the pore assumed as a cylinder

For a continuous distribution of pore diameters, they obtained the cumulative pore size distribution function [$G(R)$ where R stands for the pore cylinder's radius] by a differential procedure as

$$
G(R) = K_{SEC} - \frac{3}{2}\left[\frac{dK_{SEC}}{d\ln r_G}\right] + \frac{1}{2}\left[\frac{d^2 K_{SEC}}{d(\ln r_G)^2}\right]
\tag{8.13}
$$

where:
K_{SEC} is the partition coefficient in SEC
r_G is the radius of the sample molecule

Later, Hagel et al. pointed out that in the work of Knox et al. they made "an erroneous assumption that the selectivity curve is equal to the pore size distribution curve" [73]. Many other theories and applications were made in that time for ISEC [27,45,56,57,59,60,63,65–72,74–79,82,83,85,86,88,89,91,92]. It was also demonstrated that in some cases ISEC was used incorrectly, which should be avoided [61].

There is no consensus regarding the calculation of the molecule size, which is illustrated in Figure 8.5 where the molecular radius is shown as calculated by different approximations [1,2,45,91,107]. In general, the gyration radius of the molecule determines whether or not the molecule can enter the pore; thus in this work we use Equation 8.9 to account for the size of the molecule. Though, it is still uncertain which exact size parameter determines the separation [41].

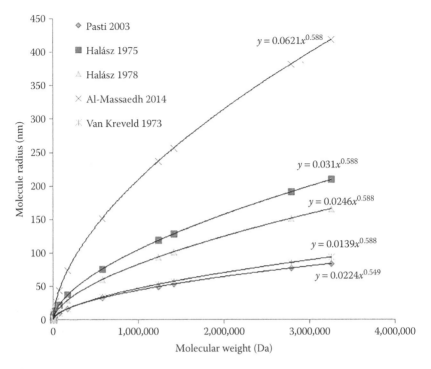

FIGURE 8.5 Molecular radius calculations by different approximations used in [1,2,45,91,107] for polystyrene standards.

8.5.2 Novel Model Based on the Stochastic Theory of Size-Exclusion Chromatography

The stochastic theory seems rather suitable for describing SEC among the basic theories of chromatography because it describes the chromatographic process at the molecular level. It assumes that the number of pore entrapment and release steps are determined by a Poisson process and the time that a molecule spends inside the pore of the stationary phase (residence or sojourn time) is determined by an exponential distribution [97–99]. This approach may become very complex to work with for situations except the simplest cases, so it was not conveniently used until the characteristic function (CF) approach was introduced to this field [47,100].

The stochastic theory of SEC assumes that a molecule of a certain size enters and leaves the pores n_p times on average during the migration along the column and spends τ_p time on average in a single pore. After leaving a pore, the molecule spends τ_m time on average in the mobile phase before entering another pore. All these variables are random quantities. The moments of the chromatographic peak (the first moment is the mean residence time, the second central moment is the variance of the peak, and the third central moment gives information about the peak symmetry) can be determined by

$$\mu_1 = n_p \tau_p \tag{8.14}$$

$$\mu_2' = 2n_p \tau_p^2 \tag{8.15}$$

$$\mu_3' = 6n_p \tau_p^3 \tag{8.16}$$

The aforementioned equations are only valid if we assume that the pore size in the stationary phase particles is uniform.

Based on the work of Cavazzini et al. [101], Sepsey et al. improved the stochastic theory of SEC to account for the pore size distribution [93] where they derived the partition coefficient as

$$K_{SEC} = \frac{1}{2} \sum_{k=0}^{\varphi} (-\rho)^k e^{\frac{k^2 \sigma^2}{2}} \binom{\varphi}{k} \text{erfc}\left(\frac{k\sigma^2 + \ln\rho}{\sqrt{2}\sigma} \right) \tag{8.17}$$

where:
  ρ is the size ratio defined in Equation 8.10
  σ refers to the standard deviation of the log-normal PSD and which strongly depends on the pore shape (for the calculation of the first absolute moment $\varphi = 1, 2$, or even 3 means that the pore is slit shaped, cylindrical, or conical/ spherical, respectively)

The PSD has important consequences on the retention time and peak shape. We obtain the relative pore accessibility by plotting the partition coefficient against the size ratio, ρ, when no pore geometry is assumed, that is, $\varphi = 0$ in Equation 8.17 and a certain value of the standard deviation of the PSD is substituted. This is illustrated in Figure 8.6a. In the case of uniform pores ($\sigma = 0$), there is a sharp distinction between the molecules that can visit the pores and the ones that are excluded. All the molecules that are small enough to fulfill $\rho < 1$, that is, molecules for which $r_G < r_p$ can visit all the pores. On the other hand, every molecule for which $\rho > 1$ is excluded from all the pores. When a range of pore sizes are present in the stationary phase, so that $\sigma > 0$, there are pores that are accessible for the molecules larger than $r_{p,0}$ too. The broader the PSD, the smoother is the transition between inclusion and exclusion. The rest of Figure 8.6 shows the effect of ρ and σ on the partition coefficient in case of slit-shaped pore geometry, cylindrical pore geometry, and conical or spherical pore geometry, respectively. The correctness of the model becomes clear when Figure 8.6 is compared against Figure 2 in Reference 47 or against experimental results such as Figure 4 in Reference 64.

The parameters affecting the retention properties were changed individually to investigate their influence. It was shown that the effect of the model parameters studied and that of the quantities characterizing the separation (number of theoretical plates, relative resolution) are the most intensive in the case of conical pore geometry. The pore size distribution has minor influence on the retention properties of small molecules compared to the larger molecules.

Based on the stochastic model for pore size distribution, Bacskay et al. determined the pore structure (pore volume, PSD, and pore geometry) of nine commercially

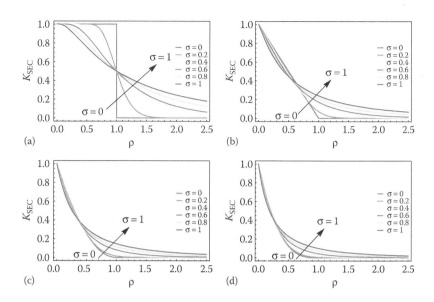

FIGURE 8.6 The influence of the pore shape and the size of the molecule relative to the pore size (ρ), on the partition coefficient for (a) relative pore accessibility, (b) slit-shaped pores, (c) cylindrical pores, and (d) conical or spherical pores. (Reprinted from *J. Chromatogr. A*, 1331, Sepsey, A. et al., Molecular theory of size exclusion chromatography for widepore size distributions, 52–60, Copyright 2014, with permission from Elsevier.)

available C_{18} columns out which three were packed with totally porous particles (TPP), four were packed with superficially porous particles (SPP), one of them was first generation monoliths (1G), and one was second generation monoliths (2G), respectively [94,95]. All measurements were executed at the same conditions, that is, on the same instrument, with neat tetrahydrofuran (THF) eluent, at the same temperature and flow rate, using polystyrene standards of low polydispersity. It was demonstrated that the pore geometry can be described as a transition between cylindrical and conical geometry.

By plotting V_R against r_G, the void volume (V_0), the pore volume (V_p), and the pore size (r_p) can be obtained by nonlinear fitting of Equation 8.6 using Equation 8.11 to the retention data of the polystyrene standards. This is shown in Figure 8.7 with solid orange line for data obtained on a Phenomenex Aeris Widepore column. To obtain the void volume, V_0, the pore volume, V_p, the mean pore size, $r_{p,0}$, and the breadth of the PSD, σ from experimental data, we should fit Equation 8.6 combining with the K_{SEC} derived in the novel model (Equation 8.17). The fitting is shown for the same experimental data points in Figure 8.7 with a blue dashed line. The goodness of fit is significantly better at 5% significance level in the latter case ($F = 3.40$ to 21.61, $F_{critical} = 2.85$).

The difference of the pore size distribution of totally porous and superficially porous stationary phases is obvious in Figure 8.8 where the probability density functions are shown for the investigated stationary phases. The technologies to produce TPPs and SPPs are completely different. The wide PSD of superficially porous particles arise from the method of the shell synthesis [102].

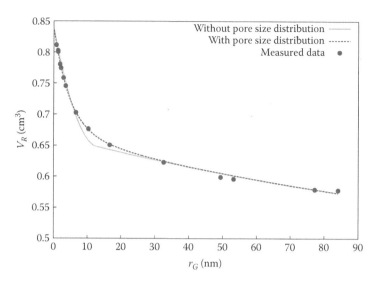

FIGURE 8.7   SEC calibration curve obtained on a Phenomenex Aeris Widepore column. V_R is corrected for the extracolumn volume; r_G is calculated using Equation 8.9. (Reprinted from *J. Chromatogr. A*, 1339, Bacskay, I. et al., Determination of the pore size distribution of high performance liquid chromatography stationary phases via inverse size exclusion chromatography, 110–117, Copyright 2014, with permission from Elsevier.)

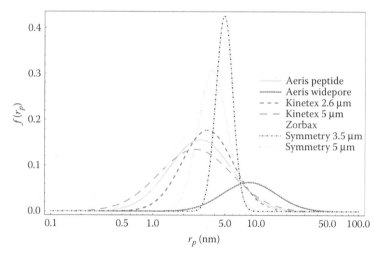

FIGURE 8.8   Comparison of the pore size distribution of the various columns; pore sizes and PSDs are estimated by the stochastic theory assuming cylindrical pore shapes. $r_{p,0}$ and σ obtained from Reference 94, substituted to the log-normal distribution.

The results derived by the stochastic model were compared to the Knox model [58,62] by applying it to the experimental data and reproducing the log-normal distribution from that. The results are presented in Table 8.1.

The stochastic model allows characterizing the porous structure of the monolithic columns as well [95]. Figure 8.9 represents the calibration curves obtained with the

TABLE 8.1
Numerical Comparison of the Pore Size Distributions Calculated with the Stochastic and with the Knox Model with the Pore Size Given by Manufacturers; $r_{p,0}$ Is the Mean and σ Is the Standard Deviation of the Distribution

Column	Manufacturer Data	Stochastic Model		Knox Model	
	r_p	$r_{p,0}$	σ	$r_{p,0}$	σ
Aeris peptide	5	4.671	0.696	6.282	0.748
Aeris widepore	10	12.434	0.603	14.977	0.653
Kinetex 2.6 μm	5	4.657	0.567	5.427	0.601
Kinetex 5 μm	5	5.042	0.817	8.045	0.870
Zorbax	4.75	4.158	0.635	5.228	0.668
Symmetry 3.5 μm	5	5.089	0.187	4.622	0.230
Symmetry 5 μm	5	4.203	0.325	4.052	0.351

Source: Bacskay, I. et al., *J. Chromatogr. A*, 1339, 110–117, 2014.
Note: Data are presented in nanometers.

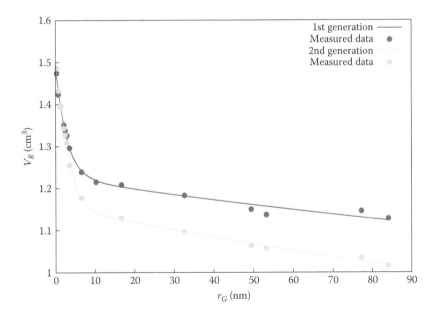

FIGURE 8.9 SEC calibration curves obtained for the 1G and 2G monolithic columns. Equation 8.6 with K_{SEC} expressed by Equation 8.17 was fitted to the measured retention volumes through the parameters V_0, r_0, V_p, $r_{p,0}$, and σ. V_R is corrected for the extracolumn volume and r_G was calculated using Equation 8.9. (Reprinted from *J. Chromatogr. A*, 1359, Bacskay, I. et al., The pore size distribution of the first and the second generation of silica monolithic stationary phases, 112–116, Copyright 2014, with permission from Elsevier.)

1G and 2G monolithic columns. One can see that the standards that can enter the mesopores have rather similar retention volumes on both columns and the difference between the columns is more pronounced for excluded polystyrenes, that is, in the range of hydrodynamic chromatography. The lines in Figure 8.9 show the results of nonlinear fitting of Equation 8.6 to the V_R versus r_G data with K_{SEC} expressed by Equation 8.17. Parameters V_0, r_0, V_p, $r_{p,0}$, and σ were also fitted. The numerical results are listed in Table 8.2. In case of monolithic stationary phases, V_0 and r_0 refer to the volume of macropore space and the radius of macropore, respectively.

The 22.7% increased mesopore volume, V_p, of the 2G chromolith can be explained with the well-described highly structured shape of the monolith [103–105].

However, the mean pore size of the macropores decreased significantly, the mean mesopore size is increased, but its standard deviation decreased considerably. Due to the asymmetrical log-normal distribution of mesopore sizes, $r_{p,0}$ is not the real average value of the pore sizes but that can be obtained with calculation of the first moment of log-normal distribution. Practically, there is no difference between the two generations of monolithic when the real average mesopore sizes are compared, the μ_{1,r_p} values coincide (Table 8.2).

It must be noted that the effect of the polydispersity was also considered using the stochastic theory of SEC [106] and it was concluded that experimentally it is not possible to distinguish the distribution of pore size from that of molecule size, and one is

TABLE 8.2

Parameters Obtained by the Stochastic Model of SEC with Wide Pore Size Distribution. The Macropore Volume (V_0), the Macropore Radius (r_0), the Mesopore Volume (V_p), the Mesopore Radius ($r_{p,0}$), and the Standard Deviation of the PSD (σ), Were Obtained by Nonlinear Fitting of Equation 8.6 with K_{SEC} Expressed by Equation 8.17. The Average Value of the Pore Radius (μ_{1,r_p}) Was Calculated Using the First Moment of the Log-Normal Distribution

	Monolithic Column	
	1G	2G
V_0 (cm³)	1.225 ± 0.011	1.160 ± 0.007
r_0 (µm)	1.545 ± 0.280	0.949 ± 0.075
V_p (cm³)	0.265 ± 0.007	0.343 ± 0.008
$r_{p,0}$ (nm)	7.446 ± 0.326	8.297 ± 0.318
σ	0.580 ± 0.093	0.369 ± 0.120
χ^2	3.465×10^{-5}	6.659×10^{-5}
μ_{1,r_p} (nm)	8.810	8.882

Source: Bacskay, I. et al., *J. Chromatogr. A*, 1359, 112–116, 2014.

not able to tell whether a molecule of an exact size visits a number of pores of different sizes or a bunch of molecules of different size enter in a pore of well-defined size.

8.6 CONCLUSION

ISEC is a powerful tool to determine the size and the distribution of the stationary phase pores in liquid chromatography.

Using the stochastic theory of SEC, not only the PSD of the stationary phase, but also the polydispersity of the sample can be determined [93,106]. When assuming a pore geometry (slit shaped, cylindrical, and conical or spherical), the statistical moments of the peak profiles can easily be calculated and the experimentally recorded chromatograms can therefore be used for the calculations.

Both the PSD and the polydispersity have strong influence on the retention properties (retention time, peak width, and peak shape) of macromolecules. For the separation of macromolecules, the wide PSD will increase the retention and efficiency. Therefore, in all modes of liquid chromatography, the efficient separation of macromolecules calls for a broad PSD.

The model accounting for PSD can be used to develop SEC measurements and so to obtain relevant information from the pore structure by a nondestructive way on the basis of containing information about both the pore geometry and the distribution of the pore sizes [94,95].

Using the stochastic theory, ISEC has become a more accurate method to investigate the structure of a porous HPLC packing material (pore size and its distribution) without destroying the column.

The pore size distribution of superficially porous particles is always broader as compared to the totally porous particles. That difference arises from the synthesis method of the particles.

Furthermore, one can characterize the interstitial zone of the columns with ISEC. Thus a significant difference between the 1st and 2nd generation silica monolithic columns was identified in the hydrodynamic range, that is, for the molecules excluded from the mesopores. From the hydrodynamic effect, the sizes of the macropores of the monoliths can be determined.

REFERENCES

1. I. Halász, K. Martin, Bestimmung der Porenverteilung (10–4000 Å) von Festkörpern mit der Methode der Ausschluß-Chromatographie, *Ber. Bunsenges Phys. Chem.*, 79 (1975), 731–732.
2. I. Halász, K. Martin, Pore sizes of solids, *Angew. Chem., Int. Ed. Engl.*, 17 (1978), 901–908.
3. A. G. Ogston, The spaces in a uniform random suspension of fibres, *Trans. Far. Soc.*, 54 (1958), 1754–1757.
4. L. G. Aggerbrandt, O. Samuelson, Penetration of water-soluble polymers into cellulose fibers, *J. Appl. Plym. Sci.*, 8 (1964), 2801–2812.
5. J. C. Giddings, *Unified Separation Science*, John Wiley & Sons, New York, 1991.
6. W. Feller, *An Introduction to Probability Theory and its Application*, 3rd ed, Vol. I, John Wiley & Sons, New York, 1968.

7. F. Gritti, C. A. Sanchez, T. Farkas, G. Guiochon, Achieving the full performance of highly efficient columns by optimizing conventional benchmark high-performance liquid chromatography instruments, *J. Chromatogr. A*, 1217 (2010), 3000–3012.

8. L. Lapidus, N. R. Amundson, Mathematics of adsorption in beds. VI. The effect of longitudinal diffusion in ion exchange and chromatographic columns, *J. Phys. Chem.*, 56 (1952), 984–988.

9. J. J. van Deemter, F. J. Zuiderweg, A. Klinkenberg, Longitudinal diffusion and resistance to mass transfer as causes of non ideality in chromatography, *Chem. Eng. Sci.*, 5 (1956), 271–289.

10. F. Gritti, G. Guiochon, Perspectives on the evolution of the column efficiency in liquid chromatography, *Anal. Chem.*, 85 (2013), 3017–3035.

11. F. Gritti, I. Leonardis, J. Abia, G. Guiochon, Physical properties and structure of fine core–shell particles used as packing materials for chromatography: Relationships between particle characteristics and column performance, *J. Chromatogr. A*, 1217 (2010), 3819–3843.

12. F. Gritti, G. Guiochon, Mass transfer kinetics, band broadening and column efficiency, *J. Chromatogr. A*, 1221 (2012), 2–40.

13. I. Halász, M. Naefe, Influence of column parameters on peak broadening in high-pressure liquid chromatography, *Anal. Chem.*, 44 (1972), 76–84.

14. R. Endele, I. Halász, K. Unger, Influence of the particle size (5–35 um) of spherical silica on column efficiency in hplc, *J. Chromatogr.*, 99 (1974), 377.

15. C. Dewaele, M. Verzele, Influence of the particle size distribution of the packing material in reversed-phase high-performance liquid chromatography, *J. Chromatogr.*, 260 (1983), 13–21.

16. J. Billen, D. Guillarme, S. Rudaz, J.-L. Veuthey, H. Ritchie, B. Grady, G. Desmet, Relation between the particle size distribution and the kinetic performance of packed columns. Application to a commercial sub-2-μ-particle material, *J. Chromatogr. A*, 1161 (2007), 224–233.

17. D. Cabooter, J. Billen, H. Terryn, F. Lynen, P. Sandra, G. Desmet, Kinetic plot and particle size distribution analysis to discuss the performance limits of sub-2 μm and supra-2 μm particle columns, *J. Chromatogr. A*, 1204 (2008), 1–10.

18. A. Daneyko, A. Höltzel, S. Khirevich, U. Tallarek, Influence of the particle size distribution on hydraulic permeability and eddy dispersion in bulk packings, *Anal. Chem.*, 83 (2011), 3903–3910.

19. F. Gritti, T. Farkas, J. Heng, G. Guiochon, On the relationship between band broadening and the particle-size distribution of the packing material in liquid chromatography: Theory and practice, *J. Chromatogr. A*, 1218 (2011), 8209–8221.

20. K. Horváth, D. Lukács, A. Sepsey, A. Felinger, Effect of particle size distribution on the separation efficiency in liquid chromatography, *J. Chromatogr. A*, 1361 (2014), 203–208.

21. F. Rouquerol, J. Rouquerol, K. Sing, *Adsorption by Powders and Porous Solids*, Academic Press, New York, 1999.

22. B. M. Wagner, S. A. Schuster, B. E. Boyes, J. J. Kirkland, Superficially porous silica particles with wide pores for biomacromolecular separations, *J. Chromatogr. A*, 1264 (2012), 22–30.

23. A. Felinger, L. Pasti, F. Dondi, M. van Hulst, P. J. Schoenmakers, M. Martin, Stochastic theory of size exclusion chromatography: Peak shape analysis on single columns, *Anal. Chem.*, 77 (2005), 3138–3148.

24. K. K. Unger, *Porous Silica*, Elsevier, Amsterdam, the Netherlands, 1979.

25. S. Brunauer, P. H. Emmett, E. Teller, Adsorption of gases in multimolecular layers, *J. Amer. Chem. Soc.*, 60 (1938), 309–319.

26. H. L. Ritter, L. C. Drake. Pressure porosimeter and determination of complete macropore-size distributions, *Ind. Eng. Chem. Anal. Ed.*, 17 (1945), 782–786.

27. H. Guan, G. Guiochon, Study of physico-chemical properties of some packing materials: I. measurements of the external porosity of packed columns by inverse size-exclusion chromatography, *J. Chromatogr. A*, 731 (1996), 27–40.

28. S. J. Gregg, K. S. W. Sing, *Adsorption, Surface Area and Porosity*, Academic Press, London, UK, 1982.

29. A. Giaquinto, Z. Liu, A. Bach, Y. Kazakevich, Surface area of reversed-phase HPLC columns, *Anal. Chem.*, 80 (2008), 6358–6364.

30. W. W. Yau, J. J. Kirkland, D. D. Bly, *Modern Size-Exclusion Liquid Chromatography*, Wiley, New York, 1979.

31. J. H. Knox, F. McLennan, Allowance for polydispersity in the determination of the true plate height in GPC, *Chromatographia*, 10 (1977), 75–78.

32. R. F. T. Stepto, Dispersity in polymer science, *Pure Appl. Chem.*, 8 (2009), 351–353.

33. D. D. Bly, Determination of theoretical plates in gel permeation chromatography by using polydisperse materials (polymers), *J. Polym. Sci. Part A-1*, 6 (1968), 2085–2089.

34. D. D. Bly, Properties and uses of a linear, semilog calibration in gel permeation chromatography, *Anal. Chem.*, 41 (1969), 477–480.

35. K. K. Unger, R. Kern, M. C. Ninou, K.-F. Krebs, High-performance gel permeation chromatography with a new type of silica packing material, *J. Chromatogr. A*, 99 (1974), 435–443.

36. K. K. Unger, R. Kern, Column parameters controlling resolution in high performance gel permeation chromatography, *J. Chromatogr. A*, 122 (1976), 345–354.

37. Y. V. Heyden, S.-T. Popovici, B. Staal, P. J. Schoenmakers, Contribution of the polymer standards' polydispersity to the observed band broadening in size-exclusion chromatography, *J. Chromatogr. A*, 986 (2003), 1–15.

38. M. Netopilík, S. Podzimek, P. Kratochvíl, Estimation of width of narrow molecular-weight distributions by size-exclusion chromatography with concentration and light scattering detection, *J. Chromatogr. A*, 922 (2001), 25–36.

39. M. Netopilík, Effect of local polydispersity in size exclusion chromatography with dual detection, *J. Chromatogr. A*, 793 (1998), 21–30.

40. J. Silberring, M. Kowalczuk, J. Bergquist, A. Kraj, P. Suder, T. Dylag, M. Smoluch, J.-P. Chervet, J. Ekman, Size-exclusion chromatography, In Heftmann, E. (Ed.), *Fundamentals and Applications of Chromatography and Related Differential Migration Methods Fundamentals and Techniques Part A*, Elsevier, Amsterdam, the Netherlands, 2004, pp. 213–252.

41. I. Teraoka, Calibration of retention volume in size exclusion chromatograhy by hydrodynamic radius, *Macromolecules*, 37 (2004), 6632–6639.

42. M. Potschka, A soft-body theory of size exclusion chromatography, In Potschka, M., Dubin, P.L. (Eds.), *Strategies in Size Exclusion Chromatography*, ACS, Washington, DC, 1996, pp. 67–87.

43. G. Stegeman, J. C. Kraak, H. Poppe, Hydrodinamic and size-exclusion chromatography of polymers on porous particles, *J. Chromatogr.*, 550 (1991), 721–739.

44. L. R. Snyder, J. J. Kirkland, J. W. Dolan, *Introduction to Modern Liquid Chromatography*, Wiley, Hoboken, NJ, 2010.

45. L. Pasti, F. Dondi, M. van Hulst, P. J. Schoenmakers, M. Martin, A. Felinger, Experimental validation of the stochastic theory of size exclusion chromatography: 1. Retention on single and coupled columns, *Chromatographia*, 57 (2003), S171–S186.

46. G. Stegeman, J. C. Kraak, H. Poppe, Dispersion in packed-column hydrodynamic chromatography, *J. Chromatogr.*, 634 (1993), 149–159.

47. F. Dondi, A. Cavazzini, M. Remelli, A. Felinger, M. Martin, Stochastic theory of size exclusion chromatography by the characteristic function approach, *J. Chromatogr. A*, 943 (2002), 185–207.

48. D. H. Freeman, I. C. Poinescu, Particle porosimetry by inverse gel permeation chromatography, *Anal. Chem.*, 49 (1977), 1183–1188.

49. I. Halász, P. Vogtel, Determination of morphological properties of swellable solids by size exclusion chromatography, *Angew. Chem., Int. Ed. Engl.*, 19 (1980), 24–28.

50. R. Nikolov, W. Werner, I. Halász, Pore size distribution of 'in situ' coated silica gels determined by exclusion chromatography, *J. Chromatogr. Sci.*, 18 (1980), 207–216.

51. S. B. Schram, D. H. Freeman, Characterization of LC stationary phases by inverse size exclusion chromatography, *J. Liq. Chromatogr.*, 3 (1980), 403–417.

52. W. Werner, I. Halász, Pore structure of chemically modified silica gels determined by exclusion chromatography, *J. Chromatogr. Sci.*, 18 (1980), 277–283.

53. W. Werner, I. Halász, Relative peak broadening and pore size distribution of the stationary phase in exclusion chromatography, *Chromatographia*, 13 (1980), 271–272.

54. D. H. Freeman, S. B. Schram, Characterization of microporous polystyrene-divinylbenzene copolymer gels by inverse gel permeation chromatography, *Anal. Chem.*, 53 (1981), 1235–1238.

55. T. Crispin, I. Halász, Determination of the pore size distribution, by exclusion chromatography, of ion-exchange polymers which swell in water, *J. Chromatogr.*, 229 (1982), 351–362.

56. B. Haidar, A. Vidal, H. Balard, J. B. Donnet, Structural differences exhibited by networks prepared by chemical and photochemical reactions. III. characterization by inverse GPC, *J. Appl. Polym. Sci.*, 29 (1984), 4309–4320.

57. J. Capillon, R. Audebert, C. Quivoron, Can porosity properties of particles really be determined by inverse gel permeation chromatography?, *Polymer*, 26 (1984), 575–580.

58. J. H. Knox, H. P. Scott, Theoretical models for size-exclusion chromatography and calculation of pore size distribution from size-exclusion chromatography data, *J. Chromatogr. A*, 316 (1984), 311–332.

59. K. Jerabek, Determination of pore volume distribution from size exclusion chromatography data, *Anal. Chem.*, 57 (1985), 1595–1597.

60. K. Jerabek, Characterization of swollen polymer gels using size exclusion chromatography, *Anal. Chem.*, 57 (1985), 1598–1602.

61. K. Jerabek, On the feasibility of determining porosity properties of particles by inverse gel permeation chromatography, *Polymer*, 27 (1986), 971.

62. J. H. Knox, H. J. Ritchie, Determination of pore size distribution curves by size-exclusion chromatography, *J. Chromatogr. A*, 387 (1987), 65–84.

63. K. Jerabek, K. Setinek, An investigation of the morphology of the glycidyl methacrylate copolymers using inverse size-exclusion chromatography, *React. Polym.*, 5 (1987), 151–156.

64. A. A. Gorbunov, L. Y. Solovyova, V. A. Pasechnik, Fundamentals of the theory and practice of polymer gel-permeation chromatography as a method of chromatographic porosimetry, *J. Chromatogr.*, 448 (1988), 307–332.

65. K. Jerabek, K. Setinek, Structure of macronet styrene polymer as studied by inverse steric exclusion chromatography and by selective sulfonation, *J. Polym. Sci.*, 27 (1989), 1619–1623.

66. K. Jerabek, K. Setinek, Strong acidic ion exchangers structure by inverse steric exclusion chromatography, *J. Polym. Sci.*, 28 (1990), 1387–1395.

67. K. Jerabek, K. J. Shea, D. Y. Sasaki, G. J. Stoddard, Accessibility of the gel phase in macroporous network polymers: A comparison of the fluorescence probe and inverse steric exclusion chromatography techniques, *J. Polym. Sci.*, 30 (1992), 605–611.

68. A. Wang, J. Yan, R. Xu, The determination of the pore structural parameters of isoporous resins by inverse GPC, *J. Polym. Sci.*, 44 (1992), 959–964.

69. K. Jerabek, A. Revillon, E. Puccilli, Pore structure characterization of organic-inorganic materials by inverse size exclusion chromatography, *Chromatographia*, 36 (1993), 259–262.

70. A. Revillon, Possibility of chromatographic characterization of porous materials, specially by inverse size exclusion chromatography, In Rouquerol, J., Rodriguez-Reinoso, F., Sing, K.S.W., Unger, K.K. (Eds.), *Characterization of Porous Solids III*, Elsevier, Amsterdam, the Netherlands, 1994, pp. 363–372.

71. A. Kurganov, K. Unger, T. Issaeva, Packings of unidimensional regular pore structure as model packings in size-exclusion and inverse size-exclusion chromatography, *J. Chromatogr. A*, 753 (1996), 177–190.

72. M. Brissova, M. Petro, I. Lacik, A. C. Powers, T. Wang, Evaluation of microcapsule permeability via inverse size exclusion chromatography, *Anal. Biochem.*, 242 (1996), 104–111.

73. L. Hagel, M. Östberg, T. Andersson, Apparent pore size distributions of chromatography media, *J. Chromatogr. A*, 743 (1996), 33–42.

74. M. Goto, B. J. McCoy, Inverse size-exclusion chromatography for distributed pore and solute sizes, *Chem. Eng. Sci.*, 55 (2000), 723–732.

75. M. Ousalem, X. X. Zhu, J. Hradil, Evaluation of the porous structures of new polymer packing materials by inverse size-exclusion chromatography, *J. Chromatogr. A*, 903 (2000), 13–19.

76. P. DePhillips, A. M. Lenhoff, Pore size distributions of cation-exchange adsorbents determined by inverse size-exclusion chromatography, *J. Chromatogr. A*, 883 (2000), 39–54.

77. N. V. Saritha, G. Madras, Modeling the chromatographic response of inverse size-exclusion chromatography, *Chem. Eng. Sci.*, 56 (2001), 6511–6524.

78. M. Al-Bokari, D. Cherrak, G. Guiochon, Determination of the poreosities of monolithic columns by inverse size-exclusion chromatography, *J. Chromatogr. A*, 975 (2002), 275–284.

79. R. Füssler, H. Schaefer, A. Seubert, Effect of the porosity of PS-DVB-copolymers on ion chromatographic behavior in inverse size-exclusion and ion chromatography, *Anal. Bioanal. Chem.*, 372 (2002), 705–711.

80. M. Kele, G. Guiochon, Repeatability and reproducibility of retention data and band profiles on six batches of monolithic columns, *J. Chromatogr. A*, 960 (2002), 19–49.

81. Y. Yao, A. M. Lenhoff, Determination of pore size distributions of porous chromatographic adsorbents by inverse size-exclusion chromatography, *J. Chromatogr. A*, 1037 (2004), 273–282.

82. D. Lubda, W. Lindner, M. Quaglia, C. du Fresne von Hohenesche, K. K. Unger, Comprehensive pore structure characterization of silica monoliths with controlled mesopore size and macropore size by nitrogen sorption, mercury porosimetry, transmission electron microscopy and inverse size exclusion chromatography, *J. Chromatogr. A*, 1083 (2005), 14–22.

83. B. A. Grimes, R. Skudas, K. K. Unger, D. Lubda, Pore structural characterization of monolithic silica columns by inverse size-exclusion chromatography, *J. Chromatogr. A*, 1144 (2007), 14–29.

84. J. Urban, P. Jandera, P. Schoenmakers, Preparation of monolithic columns with target mesopore-size distribution for potential use in size-exclusion chromatography, *J. Chromatogr. A*, 1150 (2007), 279–289.

85. J. Urban, S. Eeltink, P. Jandera, P. J. Schoenmakers, Characterization of polymer-based monolithic capillary columns by inverse size-exclusion chromatography and mercury-intrusion porosimetry, *J. Chromatogr. A*, 1182 (2008), 161–168.

86. M. Thommes, R. Skudas, K. Unger, D. Lubda, Textural characterization of native and n-alky-bonded silica monoliths by mercury intrusion/extrusion, inverse size exclusion chromatography and nitrogen adsorption, *J. Chromatogr. A*, 1191 (2008), 57–66.

87. F. Gritti, I. Leonardis, J. Abia, G. Guiochon, Physical properties and structure of fine core–shell particles used as packing materials for chromatography: Relationships between particle characteristics and column performance, *J. Chromatogr. A*, 1217 (2010), 3819–3843.

88. M. Jackowska, S. Bocian, B. Gawdzik, M. Grochowicz, B. Buszewski, Influence of chemical modification on the porous structure of polymeric adsorbents, *Mater. Chem. Phys.*, 130 (2011), 644–650.

89. E. N. Viktorova, A. A. Korolev, T. R. Ibragimov, A. A. Kurganov, Porosity of monolithic macroporous sorbents: Inverse hydrodynamic and size-exclusion chromatography study, *Polym. Sci.*, 53 (2011), 899–905.

90. D. Gétaz, N. Dogan, N. Forrer, M. Morbidelli, Influence of the pore size of reversed phase materials on peptide purification processes, *J. Chromatogr. A*, 1218 (2011), 2912–2922.

91. A. A. Al-Massaedh, U. Pyell, Adamantyl-group containing mixed-mode acrylamide-based continuous beds for capillary electrochromatography. Part II. characterization of the synthesized monoliths by inverse size exclusion chromatography and scanning electron microscopy, *J. Chromatogr. A*, 1325 (2014), 247–255.

92. Z. Wang, R. K. Marcus, Determination of pore size distributions in capillary-channeled polymer fiber stationary phases by inverse size-exclusion chromatography and implications for fast protein separations, *J. Chromatogr. A*, 1351 (2014), 82–89.

93. A. Sepsey, I. Bacskay, A. Felinger, Molecular theory of size exclusion chromatography for widepore size distributions, *J. Chromatogr. A*, 1331 (2014), 52–60.

94. I. Bacskay, A. Sepsey, A. Felinger, Determination of the pore size distribution of high-performance liquid chromatography stationary phases via inverse size exclusion chromatography, *J. Chromatogr. A*, 1339 (2014), 110–117.

95. I. Bacskay, A. Sepsey, A. Felinger, The pore size distribution of the first and the second generation of silica monolithic stationary phases, *J. Chromatogr. A*, 1359 (2014), 112–116.

96. O. Chiantore, A comment on the solute dimensions to be applied in inverse size exclusion chromatography, *J. Polymer Sci. C: Polymer Lett.*, 21 (1983), 429–431.

97. J. B. Carmichael, Stochastic model for gel permeation separation of polymers, *J. Polym. Sci. Part A-2*, 6 (1968), 517–527.

98. J. B. Carmichael, The relation between pore size distribution and polymer separation in gel permeation chromatography, *Macromolecules*, 1 (1968), 526–529.

99. J. B. Carmichael, Theory of gel filtration separation of biopolymers assuming a Gaussian distribution of pore sizes, *Biopolymers*, 6 (1968), 1497–1499.

100. F. Dondi, M. Remelli, The characteristic function method in the stochastic theory of chromatography, *J. Phys. Chem.*, 90 (1986), 1885–1891.

101. A. Cavazzini, M. Remelli, F. Dondi, A. Felinger, Stochastic theory of multiple-site linear adsorption chromatography, *Anal. Chem.*, 71 (1999), 3453–3462.

102. G. Guiochon, F. Gritti, Shell particles, trials, tribulations and triumphs, *J. Chromatogr. A*, 1218 (2011), 1915–1938.

103. K. Hormann, T. Müllner, S. Bruns, A. Höltzel, U. Tallarek, Morphology and separation efficiency of a new generation of analytical silica monoliths, *J. Chromatogr. A*, 1222 (2012), 46–58.

104. D. Hlushkou, K. Hormann, A. Höltzel, S. Khirevich, A. Seidel-Morgenstern, U. Tallarek, Comparison of first and second generation analytical silica monoliths by pore-scale simulations of eddy dispersion in the bulk region, *J. Chromatogr. A*, 1303 (2013), 28–38.

105. D. Cabooter, K. Broeckhoven, R. Sterken, A. Vanmessen, I. Vandendael, K. Nakanishi, S. Deridder, G. Desmet, Detailed characterization of the kinetic performance of first and second generation silica monolithic columns for reversed-phase chromatography separations, *J. Chromatogr. A*, 1325 (2014), 72–82.

106. A. Sepsey, I. Bacskay, A. Felinger, Polydispersity in size-exclusion chromatography: A stochastic approach, *J. Chromatogr. A*, 1365 (2014), 156–163.

107. M. E. Van Kreveld, N. Van Den Hoed, Mechanism of gel permeation chromatography: Distribution coefficient, *J. Chromatogr. A*, 83 (1973), 111–124.

9 Studies on the Antioxidant Activity of Foods and Food Ingredients by Thin-Layer Chromatography–Direct Bioautography with 2,2′-Diphenyl-1-Picrylhydrazyl Radical (DPPH*)

Joseph Sherma

CONTENTS

9.1 INTRODUCTION

The determination of antioxidant activity of foods is currently a highly researched area because of the benefit of antioxidants for preservation or protection of the foods against oxidative damage, to avoid deleterious changes and loss of commercial and nutritional value [1] and their value in protecting human health when the foods are consumed. For example, AOAC International approved First Action status for a visible range spectrometric (colorimetric) method for analysis of three fruit juices, carrot juice, green tea, wine, rosemary spice, ready to eat cereal, and yogurt based on reaction of methanol–water extracts with the stable 2,2′diphenyl-1-picrylhydrazyl radical (DPPH*; $C_{18}H_{12}N_5O_6$; MW = 394.32: CAS No. 1898-66-4) to measure the total antioxidant activity with DPPH* reported as Trolox (vitamin E analog standard) equivalents [2]. Antioxidant compounds, including simple phenols, phenolic acids, polyphenols, polyenes, tocopherols, indoles, stilbenes, tannins, anthocyanins, proanthrocyanidins, hydroquinones, and flavonoids, scavenge free radicals, thus inhibiting the oxidative mechanism that can lead to neurodegenerative diseases [3]. During reaction with the antioxidant, the purple color of DPPH* (maximum absorption at 517 nm) changes to yellow through the transfer of a proton plus an electron (hydride equivalent) [2] to form the corresponding reduced DPPH-H [3]. Various advances in the DPPH* spectrometric method [1] and its application to estimate antioxidant capacity of food samples [4] were reviewed.

Online high-performance column liquid chromatography bioassay of antioxidant activity has been reported for the analysis of food extracts based on the decrease in absorbance (negative peaks) at 515 nm after postcolumn reaction of separated compounds with DPPH* solution [5]. Thin-layer chromatography (TLC) was combined with direct bioautographic detection using DPPH* starting in the 1960s to take advantage of the simultaneous separation and activity measurement on the same plate. The advantages of the TLC–DPPH* bioassay over HPLC include direct access to the entire sample chromatogram immobilized on the layer prior to detection, flexibility, and high sample throughput because multiple samples can be examined simultaneously, use of a small amount of mobile phase per sample leading to lower cost for purchase and disposal of solvents, speed of method development, ability to identify the most active antiradical component, different mobile-phase development techniques, and the absence of the mobile phase during biodetection; the speed and simplicity of analysis make it suitable for automated systems in screening applications. Another advantage is the need for little or no sample purification because each plate is used only once compared to an HPLC column, which is used for many samples over a period of time and can become fouled by strong sorption of sample constituents that can change column performance and elute as interferences in later samples. Although reviews of TLC–DPPH* in the search for new drugs has been published [3], this is the first review on its use in the analysis of foods and food ingredients such as plant essential oils, which are widely used as food flavors and preservatives and act as natural antioxidants to extend the shelf life of dishes and processed food products.

9.2   TECHNIQUES OF THIN-LAYER CHROMATOGRAPHY–2,2′DIPHENYL-1-PICRYLHYDRAZYL RADICAL*

This section will describe the general techniques that have been used for TLC–DPPH*, and the next section will present examples of selected applications to food analysis with specific experimental details provided in each chapter.

Samples are usually extracted with traditional techniques such as shaking, sonication, Soxhlet, or refluxing prior to TLC. Stepwise extraction with different solvents is sometimes carried out to find active fractions by TLC–DPPH*, followed by isolation and identification of active compounds in the fractions by additional techniques such as preparative layer chromatography (PLC) and spectrometry. However, the hyphenation of the DPPH* test with TLC often allows use of a single extraction because fractionation of the extract occurs on the plate, which saves much time, effort, and expense. Occasionally, a more advanced extraction technique has been applied, such as accelerated solvent extraction (ASE) and supercritical fluid extraction (SFE).

Sample and standard solutions are applied to the plate manually with a micropipet or automatically with a commercial instrument. The stationary phase is usually 20 × 20 or 10 × 20 cm aluminum or glass backed TLC or high-performance TLC (HPTLC) silica gel plates from Merck with or without a fluorescent phosphor (plates with the designation F contain a phosphor that emits green light when irradiated with 254 nm UV radiation in order to facilitate fluorescence quenching detection of compounds that absorb this radiation as dark zones). Other plates such as polyamide and octadecylsilyl (C18 or RP18), cyano (CN), or diol chemically bonded silica gel have been used when the necessary selectivity was not provided by silica gel.

Plate development with the mobile phase has usually been done at laboratory temperature in a large volume N-chamber such as the CAMAG twin trough chamber (TTC) in the one-dimensional ascending mode. The method used for choosing the mobile phase is almost never given in research papers, but it is usually trial and error based on prior experience and literature searching; however, application of the Prisma optimization method based on the Snyder solvent classification was cited in one case for TLC–DPPH*. More than one mobile phase may be needed for development of multiple silica gel plates if the sample contains a wide polarity range of analytes. Horizontal development, two-dimensional (2D) development, and micro-TLC on smaller plates have also been reported. Antioxidant detection has been carried out by spray or dip application of different concentrations of a methanol, chloroform, or hexane solution of DPPH* usually followed by a waiting period; the concentration of DPPH* used is not always given in research publications. The radical scavenging zones appear yellow on a purple layer background and are evaluated at specified times after reagent application. Other more or less specific reagents may be applied to separate plates for chemical detection of compounds in chromatograms. Figure 9.1 shows examples of biological and chemical detection of chromatograms.

Chromatograms are documented by photography and are quantified using image analysis software; only seldom are antioxidant compounds detected with DPPH* and then quantified with a slit-scanning densitometer. Zone identification has been confirmed by hyphenation with mass spectrometry (MS) either off-line or through an online interface such as the CAMAG TLC–MS interface. Identification and quantification have been facilitated by TLC–EDA *in situ* hyphenation with densitometry using an ultraviolet/visible (UV/Vis) detector or fluorescence detector (FLD).

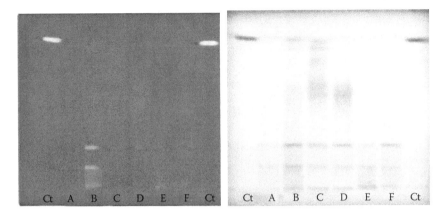

FIGURE 9.1 Flatbed scanner images of chromatograms on silica gel 60 F plates with aceto-nitrile–water–chloroform–formic acid (60:15:10:5) mobile phase. The left plate shows scans of yellow zones on a purple background obtained after dipping in 0.2% DPPH* in methanol or hexane for 5 s and being left in the dark for 30 min, and the right plate shows scans of red–purple zones on a white background after dipping in vanillin–20% methanolic sulfuric acid reagent (1 g:100 mL) and heating at 105°C for 5 min to detect phenols among other compounds. Ct stands for catechin and tracks A–F are *Medemia argun* nut 80% aqueous methanol extracts fractionated on a C18 column before application to the plate. (Courtesy of Dr. Lukasz Ciesla, National Institute on Aging, Baltimore, MD.)

9.3 APPLICATIONS OF THIN-LAYER CHROMATOGRAPHY– 2,2′DIPHENYL-1-PICRYLHYDRAZYL RADICAL*

This section contains selected applications of TLC–DPPH* in the study of antioxidant activity of foods and food ingredients arranged chronologically. In each, the techniques, materials, and instruments that were reported for each step of the TLC process are described.

Yrjonen et al. [6] evaluated the free radical scavenging activity of rapeseed meal, a feed for livestock and poultry and potential new source of natural food antioxidants, using phenolic acids and flavonoids as model compounds. Extraction of the meal was with methanol–water (80:20) using an UltraTurrax T25, samples were applied with a CAMAG Linomat onto RP18 TLC plates, development was in an unsaturated TTC with the mobile-phase methanol–water–*o*-phosphoric acid (55:45:1.1) that was optimized by the PRISMA model, the dried plates were dipped into 0.04% DDPH* in methanol for 5 s using an automated CAMAG Immersion Device 3, and quantification of chromatogram images was made under visible light by measuring the areas of the yellow bands using a Hitachi HV-C20 charge coupled device (CCD) camera and CAMAG Video Scan version 1.01 software 2 min after dipping. A total of 10 separated zones with free radical scavenging activity were detected with R_f values from 0.04 to 0.85, and results based on alpha-tocopherol index showed good correlation between activities measured by TLC–DPPH* and a conventional DPPH* spectrometric method.

Lapornik et al. [7] measured the variation of antioxidant activity for differently prepared extracts of pressed residues (mare) from red wine and black and red currant juice production. Ethanol–water (70:30) was chosen for extracting phenols from samples, and activity was also measured for standards of the anthocyanins present in these beverages. Both spectrometric and TLC–DPPH* methods were used, and the results correlated well for standards (r = 0.979) and slightly less but acceptably for crude extracts (r = 0.857). Antioxidant activity for the extracts was in the order black currant juice > wine > red currant juice, and the strongest anthocyaninin activity was found for cyanidin 3-glucoside. The layer used was 0.25 mm silica gel 60, ethyl acetate–formic acid–water (85:10;15) was the mobile phase for development in an unsaturated TTC, and image analysis with an HP Officejet Pro 1175c flatbed scanner and Videodensitometer Sorbfil 1.1 software were used for zone quantification after plates were treated with methanolic DPPH* reagent.

Antioxidants from *Spirulina platensis* microalga, which is used as a food supplement, food ingredient, or complete food and is often termed a *superfood*, were separated and characterized by Jaime et al. [8] combining ASE with ethanol at 115°C for 15 min in a Dionex ASE 200 instrument; TLC–DPPH* with silica gel 60 F aluminum plates, petroleum ether–acetone (75:25) mobile phase in a closed chamber, and dip application of 0.5 mM methanolic DPPH* reagent; and C18 column HPLC with a diode array detector (DAD). In each experiment two plates were used in parallel, and active zones on the second unsprayed plate were scraped off and eluted corresponding to those detected on the first sprayed plate and then analyzed by HPLC–DAD. It was found that the compounds with the greatest antioxidant activity were carotenoids, phenolics, and degradation products of chlorophyll.

Emami et al. [9] demonstrated the antioxidant activity of the essential oils of different parts of *Juniperus communis* subsp. *Hemisphaerica* and *Juniperus oblonnga*. Fresh leaves of male and female plants as well as fruits were ground in a blender, and the volatile oils were isolated by steam distillation. Spotted samples were developed on silica gel 60 F aluminum plates with toluene–ethyl acetate (97:3), and detection was by spraying with 0.2% methanolic DPPH*. Zones that became yellow within 30 min were considered active. Some potential radical scavenging compounds, for example, vitamin C and quercetin, were used as positive controls. Essential oils from the male leaves of *Juniperus communis* and fruit of *Juniperus oblonga* were most active, and their use in very low concentrations for preserving foods was suggested.

Maksimovic et al. [10] studied the radical scavenging activity of *Thymus glabrescens* Willd. (*Lamiaceae*) essential oil and concluded that it could be used in the food industry as a seasoning agent or for preserving processed foods from oxidative degradation. Essential oil was isolated by hydrodistillation in a clevenger-type apparatus, and activity was screened by TLC–DPPH* with thymol as a standard. Samples and standards were applied manually as spots on two silica gel 60 F aluminum plates, which were developed in toluene–ethyl acetate (93:7) mobile phase. One plate was sprayed with anisaldehyde–sulfuric acid reagent to detect terpene ester, terpene alcohol, and thymol zones, and the other with 0.5 mM methanolic DPPH* and observed immediately. Only the thymol zone reacted with DPPH*. Antioxidant activity was quantified by the spectrometric DPPH* method against butylated hydroxytoluene and thymol as positive controls.

Radical scavenging activity and total phenolic content of extracts of the root bark of *Osyris lanceolata*, which are consumed as a tea and a tonic in soup in some countries, were reported by Yeboah et al. [11]. Exhaustive extractions were performed with methanol, methanol–water (9:1), hexane, and chloroform as well as SFE with CO_2 in a Speed SFE-4.1 instrument and extracts were chromatographed on silica gel 60 F aluminum plates with a different mobile phase having a polarity that matched the polarity of the compounds most likely present in each. Plates were inspected for yellow zones after 5 min, 30 min, and 8 h to take into account the possible variation in the kinetics of the reaction between different compounds and the DPPH* spray reagent. In addition to this visual identification of fractions with antioxidant potential that served as the basis for activity-guided isolation of active compounds, semiquantitative TLC–DPPH* was carried out by visual comparison of zone color changes on a grid of about 1.0 cm line spacing drawn on the plate, and quantification by spectrometric DPPH* assay against ascorbic acid and gallic acid standards. The methanol and methanol–water (9:1) extracts showed several components with high antioxidant activity displaying fast kinetics.

Gu et al. [12] extracted the edible fruit of *Perilla frutescens* var. *acuta* by refluxing with 80% ethanol and subjected this extract to gradient elution C18 HPLC–DAD with collection of fractions. The fractions were analyzed by TLC on silica gel 60 F plates developed in a saturated chamber with hexane–toluene–ethyl acetate–formic acid (2:5:2.5:0.5), the detection reagent was 2.54 mM methanolic DPPH* applied by spraying, and four compounds were subsequently isolated from the active fractions. These compounds were identified as rosmarinic acid, luteolin, apigenin, and chrysoeriol by means of UV spectrometry, nuclear magnetic resonance (NMR) spectrometry, and MS and quantified by C18 HPLC. This study is a prime example of the isolation of antioxidants guided by TLC with DPPH* bioautography.

Hosu et al. [13] determined the antioxidant activity of natural and commercial juices by mixing different concentrations of vitamin C and juice samples with methanolic DPPH* and after 30 min spotting them on a silica gel 60 TLC plate with a microsyringe, photographing the plate using a CAMAG Digistore 2 system, and processing the images using ImageJ computer software freely available online http://www.rsbweb.nih.gov/ij/. A calibration curve was made based on the spot areas and amounts of the vitamin C standard, and the amounts of the antioxidant in the juice spots were interpolated from the curve based on their areas. The authors showed that the results were equivalent to those obtained from parallel analysis with the spectrometric method and claimed that the *in situ* method was more simple, rapid, and inexpensive. They called this a TLC determination, but it is not because there was no layer development with a mobile phase after spotting of samples and standards. This method is included as an example of a so called *dot-blot* test that is often incorrectly called a TLC method.

Rossi et al. [14] used the HPTLC–DPPH* assay to guide the isolation of antioxidant fractions from *Croton lechleri* essential oil. Essential oil was obtained by steam distillation, and HPTLC–DDPH* was carried out by application of methanolic solution in 10 mm bands with a Linomat 4 onto silica gel 60 F plates that were subsequently developed in a chamber with toluene–ethyl acetate–petroleum ether (93:7:20). Three samples were applied to one plate, and after development one chromatogram was sprayed with methanolic DPPH* (20 mg/100 mL) to detect the antioxidant fractions and the other two were useful to obtain corresponding separated active zones to scratch out and extract from the layer with ethyl acetate. The spot extracts were analyzed by gas chromatography/MS. The results suggested the possibility to employ *Croton lechleri* essential oil as a new flavoring protective ingredient for foods or dietary supplements against potential mutagens formed during cooking and/or processing, in general.

Olech et al. [15] investigated the antiradical activity of *Rosa rugosa* Thunb. (Japanese rose), which is cultivated as a food in Japan and elsewhere. Dried and pulverized petals were extracted exhaustively with 80% aqueous methanol, extracts plus four standards (gallic acid, protocatechuic acid, quercetin, and Trolox) were chromatographed on C18 W (water wettable) plates with methanol–water–*o*-phosphoric acid (45:54:1), and zones were detected by dipping into 0.1% methanolic DPPH*. Samples were applied bandwise with a Desaga AS-30 automatic sampler, plates were developed in a horizontal Chromdes DS chamber, chromatograms were photographed with a Canon EOS 350D digital camera with standard 18–35 lens, and the pictures were processed to give areas of negative peaks using the ImageJ 1.43u image processing program. Calibration curves were prepared as plots of peak area response against the applied amount of standard antioxidant, and sample amounts were interpolated from the curves; results were reported as standard activity coefficient relating standard and sample amounts giving the same antiradical activity (SAC). This is a very important chapter because it details the image processing methodology for ImageJ software better than any other source in the literature so far.

Profiles of antioxidants from the *superfruit* blueberries were obtained by Kusznierewicz et al. [16] by applying methanol–formic acid (99:1) sonication extracts of berry fruits on silica gel 60 F plates with a glass capillary, developing with ethyl acetate–formic acid–water (6:1:1), and spraying with 1 mg/mL methanolic DPPH*. The profiles indicated that anthocyanins were the major antioxidants in all the berries that were studied.

Micro-2D TLC was used by Hawryl et al. [17] to characterize the antioxidant activity of *Mentha* sp. used as food flavorings and tea. Samples were Soxhlet extracted in turn with chloroform and methanol to fractionate nonpolar and polar constituents, respectively, and extracts were spotted in a corner of polar silica bonded CN and diol micro plates (5 × 5 cm) using an AS-30 applicator; the plates were developed in the first direction with nonaqueous mobile phases (polar modifier dissolved in *n*-heptane) and aqueous mobile phases (methanol–water) in the second direction at right angles using a DS horizontal chamber. Spraying with 0.2% methanolic DPPH* reagent detected antioxidants, and chromatograms were photographed by a Fuji 8 mpx digital camera in daylight. Fingerprints of antioxidant activity were constructed for use in quality control (QC) of *mentha* sp. antioxidant activity. Separate 2D developed plates were sprayed with Naturstoff reagent 2-(diphenylboryoxy)-ethylamine and polyethylene glycol (PEG) 4000 and photographed under 366 nm UV light to obtain 2D chromatographic fingerprints of detected phenols.

Ciesla et al. [18] used TLC–DPPH* to analyze phenolic acid fractions of selected *Salvia* and *Thyme* species, including foods and food preservatives. Plant extracts obtained with a multistep procedure were applied bandwise with an AS-30 sampler

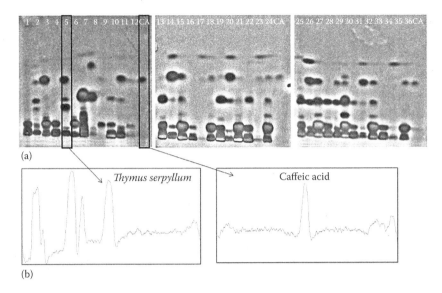

FIGURE 9.2 (a) Videoscans obtained from all analyzed standards after processing with the ImageJ program. (b) Example line program plots obtained after color inversion for one of the samples and the standard compound using the ImageJ built-in option. (Reprinted from Ciesla, L. et al., *J. AOAC Int.*, 96, 1228–1232, 2013. With permission.)

and separated on silica gel 60 F plates with toluene–ethyl acetate–formic acid (60:40:1) mobile phase in a horizontal DS chamber, zones were detected by UV densitometry, plates were sprayed with DPPH*, and the active zones were documented by videoscans with a CAMAG Reprostar CCD camera device and Video Store program, and jpg files from the videoscans were further processed by means of ImageJ software (Figure 9.2). Free phenolic acid fractions and those derived from acidic and basic hydrolysis were analyzed and compared along with caffeic acid standard. *Salvia officinalis* and *triloba* had similar free radical scavenging activity fingerprints obtained for all the analyzed fractions.

Antioxidant activity of bark tissue of *Sonneratia caseolaris* was studied by Simlai et al. [19] with TLC separation of methanol extracts using toluene–ethyl acetate–formic acid (60:40:1) mobile phase and detection with 0.02% methanolic DPPH* spray reagent. Antimicrobial activity was also determined by TLC–bioautography, against *Bacillus coagulans*, *Bacillus subtilis*, and *Proteus vulgaris*. Active fractions recovered by scraping and elution were characterized by GC–MS. The bark was found to be a potential high source of antimicrobial/antioxidant compounds even after pH and thermal treatments, suggesting its use as a natural additive in the food processing industry.

A comparative study on antioxidant potentials of seven edible leafy vegetables obtained from local markets in West Bengal state, India, was reported by Simlai et al. [20]. The constituents of 80% aqueous methanol extracts prepared by intermittent shaking for 2.5 h were separated by TLC on silica gel 60 F plates developed using toluene–ethyl acetate–formic acid (60:40:1) or ethyl acetate–methanol–water

(10:1.35:1) for the separation of less polar and polar substances, respectively. Positions of radical scavenging zones were located by spraying with 0.02% DPPH* in methanol. The study found that *Enhydra fluctuans* and *Corchorus* sp. were the best source of antioxidative compounds for use in the food industry.

Kumar and Pandey [21] revealed the antioxidant activity of *Solanum xanthocarpum* fruit using Soxhlet extraction with hexane, benzene, chloroform, ethyl acetate, acetone, ethanol, and water; manual application of extracts with capillary tubes on silica gel TLC plates; and development with chloroform–ethyl acetate–formic acid (163:63:25). Radical scavenging activity was observed in various extracts by spraying with 0.2% methanolic DPPH* and viewing after 30 min incubation in the dark. Phytoconstituent zones were visualized after spraying with 10% ethanolic sulfuric acid and water-soluble phenolic compounds after spraying with Folin-Ciocalteu reagent (1:1 in water).

Gertrud Morlock's research group has been in the forefront of powerful extended hyphenation of HPTLC with visible mode densitometry, online MS analysis, and EDA, as in the analysis of powdered bilberry, blueberry, chokeberry, acai berry, and cranberry extracts obtained by sonication with acidified methanol [22]. Standard and sample solutions were sprayed with a CAMAG automatic TLC Sampler 4 (ATS 4), and combined two-step development of silica gel 60 F HPTLC plates in a TTC or CAMAG Automated Developing Chamber (ADC 2) was with ethyl acetate–2-butanone–water–formic acid (7:3:0.8:1.2) at controlled 25% humidity for separation of anthocyanins, followed by cutting the plate below the coeluting anthocyaniden fraction and separating these compounds in the upper part by development with ethyl acetate–toluene–formic acid–water (10:3:1.2:0.8). Plate images were photographed with a CAMAG TLC Visualizer documentation system equipped with a 12 bit CCD camera, spectra of zones were recorded with a CAMAG TLC Scanner 4, and they were quantified with the scanner in the visible mode. HPTLC–electrospray ionization (ESI)–mass spectra were recorded directly with a CAMAG TLC–MS interface and Advion single quadrupole mass spectrometer. Detection with 0.5 mM methanolic DPPH* was carried out by dipping in an Immersion Device for 2 s, drying for 90 s in the dark, and heating for 30 s on a CAMAG Plate Heater at 60°C; the yellow radical scavenging zones were monitored for 24 h. 3-Glucosides of delphinidin, cyanidin, malvidin, and peonidin, further cyanidin glycosides, and respective anthocyanidins zones were found in the extracts. Based on results, the method could be used for rapid QC of powdered berry extracts.

The comprehensive hyphenation approach described earlier was further demonstrated by Morlock's group [23] with the addition of UV and fluorescence densitometry (HPTLC–UV/Vis/FLD–EDA–MS) for determination of bioactive components in turmeric and milk thistle. The same instrumentation was used as in the previous study cited, and *Aliivibrio fischeri* (formerly classified as *Vibrio fischeri*) and *B. subtilis* TLC-linked bioassays were performed in addition to TLC–DPPH*. It was shown to be a useful tool for selection of active botanicals and for the profiling of their active ingredients therein. The flexibility of effect-directed detections allowed a comprehensive survey of effective ingredients in samples. Nontargeted effect-driven screening came first, followed by highly targeted characterization of the discovered bioactive compounds. HPTLC–EDA–MS was recommended for bioactivity profiling of food intake, as not only as effective phytochemicals but also unknown bioactive degradation products during food processing or contamination products or residues or metabolites can be detected.

Bag and Chattopadhyay [24] evaluated the synergistic antibacterial and antioxidant efficacy of essential oils of bay leaf, black pepper, coriander seed and leaf, cumin, garlic, ginger, mustard, onion, and turmeric, in combination. Bioactivity-guided fractionation of active essential oils for isolation of bioactive compounds was done using TLC–bioautographic assay, and qualitative and quantitative chemical characterization were done by offline direct analysis in real time (DART)-MS and C18 column HPLC analyses, respectively. TLC was carried out on silica gel 60 GF plates developed with toluene–ethyl acetate (95:5). Antibacterial activity was screened by incubation of a developed plate with *Bacillus cereus* for 24 h followed by spraying with *p*-iodonitrotetrazolium violet solution to give colorless inhibition zones against a purple background on the layer. Antioxidant capacity was screened by spraying with 2.54 mM methanolic DPPH* solution and observing the yellow bands on a purple background after 30 min. PLC was used to isolate bioactive compounds by applying streaks on 20 × 20 cm plates with 1 mm thick layer of silica gel; PLC was used in many studies cited in this review to isolate bioactive compounds identified by TLC–DPPH*. Linalool from coriander seed oil and *p*-coumeric acid from cumin seed oil were found to exhibit synergistic activities, indicating that these substances might be used as an effective and safe natural antimicrobial and antioxidant agents in the food industry.

Jesionek et al. [25] reported the separation and identification of polyphenols (flavonoid aglycones, flavonoid glycosides, and phenolic acids) from extracts of five plant species; these phenols are contained in fruits, vegetables, spices, wines, tea, coffee, and juice and are a very important part of a daily human diet. Silica gel 60 F TLC with a Linomat 5 applicator and horizontal DS chamber was employed, with seven optimized mobile phases and detections using natural product/polyethylene glycol (NP/PEG) reagent and 0.2% methanolic DPPH*. Nine out of ten analyzed polyphenols were proved to be radical scavengers, and identity of target substances was confirmed by HPLC–MS.

Oniszczuk et al. [26] determined an increase of the antioxidant activity of instant corn grits enriched with 5%–20% addition of chamomile by high-temperature extrusion cooking using TLC–DPPH* and a DPPH* spectrometric method. Ethanol extracts showing the greatest activity in a dot-blot test were applied on silica gel 60 F HPTLC plates with an AS-30 applicator and developed with acetonitrile–water–chloroform–formic acid (60:15:10:5) mobile phase in presaturated vertical chambers. Dried plates were immersed for 5 s in 0.2% chloroformic DPPH* solution, and after 30 min were scanned with a flatbed scanner every 5 min for 1 h. JPG scan images were quantified against rutin standard by use of Sorbfil TLC videodensitometer software.

As shown earlier, the spectrometric DPPH* method and the *dot-blot* nonchromatographic plate test are usually applied to the measurement of total antioxidant potential, whereas TLC–DPPH* measures the antioxidant potential of the separated sample components. However, Glod et al. [27] proposed a micro-TLC–DPPH* method to quantify total antioxidant activity of various complex and colored herbs and honeys, particularly meads, based on separation of DPPH* from DPPH-H as well as other colored solutes in the tested matrices. Samples and 1 mM methanolic DPPH* were mixed, shaken, and left to stand for 30 min in the dark. Samples were spotted with a syringe on 5 × 10 cm silica gel 60 or 60 F plates with 0.25 mm layer thickness, and the plates were developed in a horizontal DS chamber for 4 cm with hexane–acetone–ethanol (3:1.9:0.1). The dried plates were then scanned using an HP Scanjet G2410 scanner [8 bit resolution, RGB (red, green, blue) color mode] equipped with Windows-compatible Scion image software. Quantification against gallic acid standards was performed in the green channel. Good correlation was demonstrated with results from the spectrometric method.

An HPTLC method for fingerprinting and for the antioxidant activity evaluation of 27 Romanian red wines was developed by Hosu et al. [28] in order to differentiate them as a function of grape varieties, vintage years, and wineries. Samples of wines were applied as bands with a Linomat 5 onto silica gel 60 F HPTLC aluminum plates, which were developed using ethyl acetate–formic acid–acetic acid–water (20:2:2:4) as the mobile phase in a TTC. Detection was performed under visible and UV light before and after dipping the plates in NP/PEG reagent in an immersion device for 3 s. The antioxidant activity evaluation of wines based on the total measured zone area was performed on separate plates by digitally processing in natural color with ImageJ software and the images taken at 2, 5, 15, and 30 min using a TLC Visualizer after immersion of plates in a 0.04% methanolic solution of DPPH* for 1 s. The fingerprinting method, which was validated for stability, specificity, precision, and robustness, was shown to be useful for QC of the wines as well as their differentiation.

HPTLC, bioautographic assay, and MS were combined to screen and identify antioxidant compounds in leaves of *Hibiscus sabdariffa* L., which are used as a source of beverages, by Wang et al. [29]. Crude extract prepared by 70% aqueous methanol sonication was applied to silica gel 60 plates with a Linomat 5, separated in an ADC 2 with toluene–ethyl acetate–formic acid–methanol (6:6:1.6:1) mobile phase, and antioxidant bands were visualized by dipping in 0.05% methanolic DPPH* reagent applied with an Immersion Device and documented under 254 nm and white light using a Reprostar 3 with Image Analysis software. The antioxidants neochlorogenic acid, chlorogenic acid, cryptochlorogenic acid, rutin, and isoquercitrin were identified by scraping and elution of zones from the HPTLC plate followed by HPLC–MS of the fractions. Principal components analysis (PCA) was carried out to discriminate 10 accessions of the plant.

Sharif et al. [30] also used chemometrics for predicting antioxidant activity and identifying active compounds from the dietary vegetable *Pereskia bleo*. Plant material was extracted by sonication with 100%, 80%, 60%, 40%, 20%, and 0% methanol–water, and samples were applied with a micropipette to silica gel 60 F TLC plates that were developed in a chamber using the mobile phase hexane–diethyl ether–dichloromethane–chloroform (5.6:5:36:3.4) optimized by the PRISMA model. An orthogonal partial least square (OPLS) model was developed consisting of a wavelet converted digital camera-ImageJ software image and DPPH* free radical scavenging activity of 24 different preparations of the leafy cactus as the *x*- and *y*-variables, respectively. The quality of the constructed OPLS model (1 + 1 + 0) with one predictive and one orthogonal component was evaluated by internal and external validity tests. The validated model was then used to identify that the active zones on the TLC plate were mainly glycerol and amine compounds, which were then analyzed by GC–MS after trimethylsilyl derivatization.

Kazemi [31] employed activity guided fractionation of the essential oil of *Achillea tenufolia* by TLC–DPPH* to separate the main antioxidant compound, which was identified as thymol by GC–MS. Extraction of aerial plant parts was by methanol hydrodistillation for 3 h in a clevenger type apparatus, extracts were spotted on silica gel 60 F plates that were developed with *n*-hexane-ethyl acetate (9:1), and antioxidant active zones were detected by spraying with 0.2% methanolic DPPH*. Corresponding zones on an unsprayed plate were scraped off, eluted with chloroform, and analyzed by GC–MS and a DPPH* spectrometric method.

Tian et al. [32] compared the antioxidant activity of crude extracts of differing polarities of the fruits of *Chaenomeles speciosa* (Sweet) Nakai that are eaten after cooking, in jams and jellies, and as flavoring with cooked apples. Extraction was by refluxing with ethanol, and then a sequential extraction was carried out on the extract residue with petroleum ether, ethyl acetate, and *n*-butanol. The different extracts and standards of protocatechuic acid and chlorogenic acids were spotted on polyamide layers that were developed with dichloromethane–ethyl acetate–formic acid (5:4:1) in a glass chamber and then sprayed with 0.04% ethanolic DPPH* solution. It was revealed that the ethyl acetate extract contained the most antioxidant zones, and the two standards applied were the major antioxidant components. The results were confirmed using the spectrometric DPPH* assay and ultraperformance LC (UPLC).

9.4 FUTURE PROSPECTS

It is anticipated that the TLC–DPPH* method will be widely used in the future for characterization of the antioxidant activity of foods and activity-guided isolation [33] of antioxidants because of its advantages over the spectrometry and HPLC methods described earlier. To allow higher sample loading for isolation of greater amounts of active compounds, PLC will be carried out with streaks of sample applied across a thicker layers (e.g., 1 mm [24]) or a thinner analytical in so-called *semipreparative layer chromatography* [34]. Simple to operate and free of charge ImageJ software [15,18,28,30,35] will be increasingly used for documentation of results and quantification, and low temperature TLC [36] will be advantageous for complete analysis of volatile compounds in essential oils being screened for antioxidant potential. SFE [11,37] will be more frequently used as a *green* method for environmentally friendly extraction of plant material to be analyzed for antioxidant potential.

Other trends will be more reports on hyphenation of TLC–DPPH* with MS and other spectrometric techniques in order to obtain the greatest amount of experimental data (a new book in the *Chromatographic Science Series*, edited by Nelu Grinberg, covers all aspects of TLC–MS [38]); preferential use of DPPH* compared to alternative TLC direct bioassays of antioxidant activity employing ABTS* [2,2'-azinobis(3-ethylbenzothiazoline-6-sulfonic acid)], beta-carotene, or Mo(VI) complexed with phosphate [3]; and combination with other available TLC–bioautography assays for estimation of enzyme (e.g., acetylcholinesterase) inhibition [3]; antibacterial activity (e.g., against *B. subtilis*, *B. coagulans*, and *P. vulgaris*) [19], *Escherichia coli* [14], and *Aliivibrio fischeri* using the CAMAG Bioluminizer [22]); and antifungal activity (e.g., against *Cryptococcus neoformans*, *Candida albicans*, *Aspergillus fumigatus*, *Microsporum canis*, and *Sporothrix schenckii* [39]) in order to biologically characterize samples more fully (antibacterial and/or antifungal activity are alternatively termed antimicrobial activity). The TLC–DPPH* method for food analysis has not been used with development techniques such as overpressured layer chromatography (OPLC) and automated multiple development (AMD), and these should be considered to improve resolution in certain analyses of antioxidant solutes.

As its utility becomes better recognized, chemometrics (e.g., PCA [29]; similarity analysis, hierarchical clustering, k-means clustering; neural network, and support vector machine [40]; independent component analysis [ICA], multiway decomposition method [PARAFAC], and partial least squares [PLS] [41]; orthogonal partial least square-discriminant analysis [OPLS-DA] [42]; linear discriminant analysis [LDA] [43]; soft independent modeling of class analogy [SIMCA] [44]; and artificial neural networks [ANN] and k-nearest neighbor [k-NN] [45]) will be more used to discriminate TLC antioxidant fingerprints of different plants used as foods and food ingredients. To help in this recognition of the power of chemometrics in fingerprinting as well as other aspects of TLC, a book on chemometrics in the Chromatographic Science Series that is now in preparation will provide updated comprehensive coverage of the topic [46].

Bernd Spangenberg wrote in an editorial in the *Journal of Planar Chromatography-Modern TLC* (26[5], 385, 2013) that it is often forgotten that TLC is the only separation method that detects in a dry stationary phase, making the use of biological systems possible as detectors. He predicted that the challenges of modern TLC will be more in the field of activity analysis than in classical detection in order to identify classes of analytes defined by a common activity, perform screening for substances not available as a reference standard, and screening for new compounds that show a particular activity. TLC–DPPH* should be an increasingly important food characterization method of this type in the future.

ACKNOWLEDGMENTS

I thank Karen F. Haduck of the Lafayette College Interlibrary Loan Department for her invaluable work in obtaining copies of the publications cited in this chapter, without which it could not have been written.

REFERENCES

1. Kedare, S.B. and Singh, R.P., Genesis and development of DPPH method of antioxidant assay, *J. Food Sci. Technol.*, 48, 412–422, 2011.
2. Plank, D.W., Szpylka, J., Sapirstein, H. et al., Determination of antioxidant activity in foods and beverages by reaction with 2,2′-diphenyl-1-picrylhydrazyl (DPPH): Collaborative study first action 2012.04, *J. AOAC Int.*, 95, 1562–1569, 2012.
3. Ciesla, L., Thin layer chromatography with biodetection in the search for new potential drugs to treat neurodegenerative drugs—state of the art and future prospects, *Med. Chem.*, 8, 102–111, 2012.

4. Pyrzynska, K. and Pekal, A., Application of free radical diphenylpicrylhydrazyl (DPPH) to estimate the antioxidant capacity of food samples, *Anal. Meth.*, 5, 4288–4295, 2013.

5. Bandoniene, D. and Murkovic, M., On-line HPLC-DPPH screening method for evaluation of radical scavenging phenols extracted from apples (*Malus domestica* L.), *J. Agric. Food Chem.*, 50, 2482–2487, 2002.

6. Yrjonen, T., Li, P., Summanen, J., Hopia, A., and Vuorela, H., Free radical scavenging activity of phenolics by reversed phase TLC, *JAOCS*, 80, 9–14, 2003.

7. Lapornik, B., Wondra, A.G., and Prosek, A., Comparison of TLC and spectrophotometric methods for evaluation of the antioxidant activity of grape and berry anthocyanins, *J. Planar Chromatogr.-Mod. TLC*, 17, 207–212, 2004.

8. Jaime, L., Mendiola, J.A., Herrero, M., Soler-Rivas, C., Santoyo, S., Senorans, F.J., Cifuentes, A., and Ibanez, E., Separation and characterization of antioxidants from *Spirulina platensis* microalga combining pressurized liquid extraction, TLC, and HPLC-DAD, *J. Sep. Sci.*, 28, 2111–2119, 2005.

9. Emami, A., Javadi, B., and Hassanzadeh, M.K., Antioxidant activity of the essential oils of different parts of *Juniperus communis* subsp. *hemisphaerica* and *Juniperus oblonga*, *Pharmaceut. Biol.*, 45, 769–776, 2007.

10. Maksimovic, Z., Stojanovic, D., Sostaric, I., Dajic, Z., and Ristic, M., Composition and radical scavenging activity of *Thymus glabrescens* Willd. (Lamiaceae) essential oil, *J. Sci. Food. Agric.*, 88, 2036–2041, 2008.

11. Yeboah, E.M.O. and Majinda, R.R.T., Radical scavenging activity and total phenolic content of extracts of the root bark of *Osyris lanceolata*, *Nat. Prod. Commun.*, 4, 89–94. 2009.

12. Gu, L.H., Wu, T., and Wang, Z.T., TLC bioautography guided isolation of antioxidants from fruit of *Perilla frutescens* var. *acuta*, *LWT-Food. Sci. Technol.*, 42, 131–136, 2009.

13. Hosu, A., Cimpoiu, C., Sandru, M., and Seseman, L. Determination of the antioxidant activity of juices by thin layer chromatography, *J. Planar Chromatogr.-Mod. TLC*, 23, 14–17, 2010.

14. Rossi, D., Guerrini, A., Maietti, S. et al., Chemical fingertprinting and bioactivity of Amazonian Ecuador *Croton lechleri* Mull. Arg. (Euphorbiaceae) stem bark essential oil: A new functional food ingredient?, *Food Chem.*, 126, 837–848, 2011.

15. Olech, M., Komsta, L., Nowak, R., Ciesla, L., and Waksmundzka-Hajnos, M., Investigation of antiradical activity of plant material by thin layer chromatography with image processing, *Food Chem.*, 132, 549–553, 2012.

16. Kusznierewicz, B., Piekarska, A., Mrugalska, B., Konieczka, P., Namiesnik, J., and Bartoszek, A., Phenolic composition and antioxidant properties of Polish blue berried honeysuckle genotypes by HPLC-DAD-MS, HPLC postcolumn derivatization with ABTS of FC, and TLC with DPPH visualization, *J. Agric. Food Chem.*, 60, 1755–1763, 2012.

17. Hawryl, M.A., Niemiec, M.A., and Waksmundzka-Hajnos, M., Micro-two dimensional TLC in search of selected *Mentha* sp. extracts for their composition and antioxidant activity, *J. Planar Chromatogr.-Mod. TLC*, 26, 141–146, 2013.

18. Ciesla, L., Staszek, D., Kowalska, T., and Waksmundzka-Hajnos, M., The use of TLC-DPPH* test with image processing to study direct antioxidant activity of phenolic acid fractions of selected *Lamiaceae* family species, *J. AOAC Int.*, 96, 1228–1232, 2013.

19. Simlai, A., Rai, A., Mishra, S., Mukherjee, K., and Roy, A., Antimicrobial and antioxidant activities in the bark extracts of *Sonneratia caseolaris*, a mangrove plant, *EXCLI J.*, 13, 997–1010, 2014.

20. Simlai, A., Chatterjee, K., and Roy, A., A comparative study on antioxidant potentials of some leafy vegetables consumed widely in India, *J. Food Biochem.*, 38, 365–373, 2014.

21. Kumar, S. and Pandey, A.K., Medicinal attributes of *Solanum xanthocaprum* fruit consumed by several tribal communities as food: An in vitro antioxidant, anticancer, and anti HIV perspective, *BMC Complement. Altern. Med.*, 14, Article Number 112, doi: 10.1186/1472-6882-14-112, 2014.
22. Cretu, G.C. and Morlock, G.E., Analysis of anthocyanins in powdered berry extracts by planar chromatography linked with mass spectrometry, *Food Chem.*, 146, 104–112, 2014.
23. Taha, M.N., Krawinkel, M.B., and Morlock, G.E., High performance thin layer chromatography linked with (bio) assays and mass spectrometry—a suited method for discovery and quantification of bioactive components? Exemplarily shown for turmeric and milk thistle extracts, *J. Chromatogr. A*, 1394, 137–147, 2015.
24. Bag, A. and Chattopadhyay, R.R., Evaluation of synergistic antibacterial and antioxidant efficacy of essential oils of spices and herbs in combination, *PLoS One*, 10(7), Article Number e0131321, doi: 10.1371/journal.pone.0131321.
25. Jesionek, W., Majer-Dziedzic, B., and Choma, I.M., Separation, identification, and investigation of antioxidant ability of plant extract components using TLC, LC-MS, and TLC-DPPH*, *J. Liq. Chromatogr. Relat. Technol.*, 38, 1147–1153, 2015.
26. Oniszczuk, A., Wojtunik, K., Oniszczuk, T., Wojtowicz, A., Moscicki, L., and Waksmundzka-Hajnos, M., Radical scavenging activity of instant grits with addition of chamomile flowers determined by TLC-DPPH* test and by spectrophotometric method, *J. Liq. Chromatogr. Relat Technol.*, 38, 1142–1146, 2015.
27. Glod, B.K., Wantusiak, P.M., Piszcz, P., Lewczuk, E., and Zarzycki, P.K., Application of micro-TLC to the total antioxidant potential (TAP) measurement, *Food Chem.*, 173, 749–754, 2015.
28. Hosu, A., Dancui, V., and Cimpoiu, C., Validated HPTLC fingerprinting and antioxidant activity evaluation of twenty seven Romanian red wines, *J. Food Compos. Anal.*, 41, 174–180, 2015.
29. Wang, J., Cao, X., Qi, Y., Ferchaud, V., Chin, K.L., and Tang, F., High performance thin layer chromatographic method for screening antioxidant compounds and discrimination of *Hibiscus sabdariffa* L. by principal component analysis, *J. Planar Chromatogr.-Mod. TLC*, 28, 274–279, 2015.
30. Sharif, K.M., Rahman, M.M., Azmir, J., Khatib, A., Sabina, E., Shamsudin, S.H., and Zaidul, I.S.M., Multivariate analysis of PRISMA optimized TLC image for predicting antioxidant activity and identification of contributing compounds from *Pereskia bleo*, *Biomed. Chromatogr.*, 29, 1826–1833, 2015.
31. Kazemi, M., Gas chromatography-mass spectrometry analyses for detection and identification of antioxidant constituents of *Achillea tenuifolia* essential oil, *Int. J. Food Prop.*, 18, 1936–1941, 2015.
32. Tian, B.-M., Xie, X.-M., Shen, P.-P., Wu, J., and Wang, J., Comparison of the antioxidant activities and the chemical compositions of the antioxidants of different polarity extracts from the fruits of *Chaenomeles speciosa* (Sweet) Nakai, *J. Planar Chromatogr.-Mod. TLC*, 28, 443–447, 2015.
33. Cheng, M.-J., Wu, M.-D., Chen, I.-S., and Yuan, G.-F., New sesquiterpene isolated from the extracts of the fungus *Monascus pilosus*-fermented rice, *Nat. Prod. Res.*, 24, 750–758, 2010.
34. Jesionek, W., Moricz, A.M., Alberti, A., Ott, P.G., Kocsis, B., Horvath, G., and Choma, I.M., TLC-direct bioautography as a bioassay guided method for investigation of compounds in *Hypericum perforatum* L., *J. AOAC Int.*, 98, 1013–1020, 2015.
35. Kowalska, I., Jedrejek, D., Ciesla, L., Pecio, L., Masullo, M., Piacente, S., Oleszek, W., and Stochmal, A., Isolation, chemical and free radical scavenging characterization of phenolics from *Trifolium scabrum* L. aerial parts, *J. Agric. Food Chem.*, 61, 4417–4423, 2013.

36. Sajewicz, M., Wojtal, L., Staszek, D., Hajnos, M., Waksmundzka-Hajnos, M., and Kowalska, T., Low temperature planar chromatography-densitometry and gas chromatography of essential oils from different sage (*Salvia*) species, *J. Liq. Chromatogr. Relat. Technol.*, 33, 936–947, 2011.
37. Dama, A., Taraj, K., Ciko, L., and Andoni, A., Extraction of essential oils from *Salvia officinalis* L. leaves with different extraction methods, *Int. J. Ecosyst. Ecol. Sci.*, 5, 421–424, 2015.
38. Kowalska, T., Sajewicz, M., and Sherma, J. (Eds.), *Planar Chromatography-Mass Spectrometry*, CRC Press, Boca Raton, FL, 2015.
39. Suleimana, M.M., McGaw, L.J., Naidoo, V., and Eloff, J.N., Detection of antimicrobial compounds by bioautography of leaves of selected South African tree species, *Af. J. Tradit. Altern. Med.*, 7, 64–78, 2010.
40. Tang, T.X., Guo, W.Y., Zhang, S.M., Xu, X.J., Wang, D.M., Zhao, Z.M., Zhu, L.P., and Yang, D.P., Thin layer chromatographic identification of Chinese propolis using chemometric fingerprinting, *Phytochem. Anal.*, 25, 266–272, 2014.
41. Komsta, L., Chemometrics in fingerprinting by means of thin layer chromatography, *Chromatogr. Res. Int.*, 2012(2012), 5. http://dx.doi.org/10.1155/2012/893246.
42. Jandric, Z., Haughey, S.A., Frew, R.D., McComb, K., Galvin-King, P., Elliott, C.T., and Cannavan, A., Discrimination of honey of different floral origins by a combination of various chemical parameters, *Food Chem.*, 189, 52–59, 2015.
43. Geana, E.I., Popescu, R., Costinel, D., Dinca, O.R., Ionete, R.E., Stefanescu, I., Artem, V., and Bala, C., Classification of red wines using suitable markers coupled with multivariate statistical analysis, *Food Chem.*, 192, 1015–1024, 2016.
44. Zhao, Y., Chang, Y.S., and Chen, P., Differentiation of *Aurantii Fructus Immaturus* from *Poniciri Trifoliatae Fructus Immaturus* using flow injection mass spectrometric (FIMS) metabolic fingerprinting method combined with chemometrics, *J. Pharm. Biomed. Anal.*, 107, 251–257, 2015.
45. Tian, R.-T., Xie, P.-S., and Liu, He-P., Evaluation of Chinese herbal medicine: Chaihu (Bupleuri Radix) by both high performance liquid chromatographic and high performance thin layer chromatographic fingerprint and chemometric analysis, *J. Chromatogr. A*, 1216, 2150–2155, 2009.
46. Komsta, L., Vander, H.Y., and Sherma, J. (Eds.), *Chemometrics in Chromatography*, CRC Press, Boca Raton, FL, in preparation.

Index

Note: Page numbers followed by f and t refer to figures and tables respectively.